Volume 1

FOUNDATIONS OF COGNITIVE PROCESSES

FOUNDATIONS OF COGNITIVE PROCESSES

ROBERT W. THATCHER AND E. ROY JOHN

Routledge
Taylor & Francis Group

LONDON AND NEW YORK

First published in 1977 by Lawrence Erlbaum Associates, Inc.

This edition first published in 2022
by Routledge
2 Park Square, Milton Park, Abingdon, Oxon OX14 4RN

and by Routledge
605 Third Avenue, New York, NY 10158

Routledge is an imprint of the Taylor & Francis Group, an informa business

© 1977 Lawrence Erlbaum Associates, Inc.

British Library Cataloguing in Publication Data
A catalogue record for this book is available from the British Library

ISBN: 978-0-367-75066-4 (Set)
ISBN: 978-1-003-18026-5 (Set) (ebk)
ISBN: 978-0-367-75378-8 (Volume 1) (hbk)
ISBN: 978-0-367-75397-9 (Volume 1) (pbk)
ISBN: 978-1-003-16231-5 (Volume 1) (ebk)

Publisher's Note
The publisher has gone to great lengths to ensure the quality of this reprint but points out that some imperfections in the original copies may be apparent.

Disclaimer
The publisher has made every effort to trace copyright holders and would welcome correspondence from those they have been unable to trace.

FUNCTIONAL NEUROSCIENCE VOLUME 1

FOUNDATIONS OF COGNITIVE PROCESSES

ROBERT W. THATCHER
E. ROY JOHN

Brain Research Laboratories
Department of Psychiatry
New York University Medical Center

 LAWRENCE ERLBAUM ASSOCIATES, PUBLISHERS
1977   Hillsdale, New Jersey

DISTRIBUTED BY THE HALSTED PRESS DIVISION OF

JOHN WILEY & SONS

New York   Toronto   London   Sydney

Lawrence Erlbaum Associates, Inc., Publishers
62 Maria Drive
Hillsdale, New Jersey 07642

Distributed solely by Halsted Press Division
John Wiley & Sons, Inc., New York

Library of Congress Cataloging in Publication Data

John, Erwin Roy.
 Foundation of cognitive processes.

 (Their Functional neuroscience; v. 1)
 Bibliography: p.
 Includes indexes.
 1. Higher nervous activity. 2. Cognition.
3. Memory. 4. Neurophysiology. I. Thatcher, Robert, joint author. II. Title.
QP395.J63 vol. 1 612'.82s [612'.82] 76-26461
ISBN 0-470-98926-2

Printed in the United States of America

Contents

Preface

Each of the numerous fields of neuroscience is currently producing an explosion of new facts and techniques. It is easy for the student or worker in these areas to become so immersed in the particulars of a phenomenon that the implications of findings from several areas that bear upon a common function are overlooked. For this reason, we undertook an attempt to integrate the diverse phenomena bearing upon each of a variety of important functions of the brain. Many of the functions upon which we have focused are so-called higher functions, relating to the attentive, affective, evaluative, and cognitive behavior of the organism, rather than vegetative or reflex actions. In part this focus may derive from our orientation as neuropsychologists; in part it comes from the rate of progress in these fields.

Because there still remains such a strong tendency on the part of workers and students to compartmentalize facts according to the disciplines from which they were provided rather than the functions on which they bear, we explicitly attempted to achieve an interdisciplinary integration. It is clear that the neuroscience curricula of today's medical and graduate schools need, perhaps more than at any other time, a strong interdisciplinary and integrative approach. This need is apparent not only because of the multifaceted character of the field of neuroscience, but also because we have reached the point where knowledge recently gained in basic neuroscience can be applied to the understanding and solution of important clinical problems. For these reasons, a major purpose of this book is to provide the student not only with a sound foundation in functional neuroscience, but also to equip him with a detailed understanding of how these facts and methods can be applied to clinical problems.

Neuroscience has evolved to the point where it has important contributions to make in both clinical medicine and remediative education. These areas range from the automatic and computerized identification of patients at risk for neuropathology to the recognition of brain dysfunction related to cognitive impairment in the elderly and the early identification of children who will have organically caused learning disabilities. The more detailed and precise electro-physiological profiles of brain dysfunction become in these various conditions, the more opportunities can be found for early intervention and the more specific the attempts at remediation. Not only must students be trained to meet the pressing societal needs for help in these practical problem areas, but well-trained neuroscientists also will have almost unlimited career opportunities as a consequence of these needs. Neuroscience is entering an era in which it will become an applied science. This textbook derives from our conviction that the development of neuroscience has reached this stage of metamorphosis.

We initially had no intention of writing this book. Dr. Alfred Freedman, chairman of the Department of Psychiatry at New York Medical College, invited us to contribute a few chapters on basic and clinical neurophysiology to *The Comprehensive Textbook of Psychiatry* (A. Freedman, H. Kaplan, & B. Sadock, Eds.). At that time both of us were actively involved in applying electrophysiological methods to the diagnosis and treatment of clinical disorders. Since our prior training was in neuropsychology and neurophysiology, we began to write our contributions to that textbook with the goal of functionally integrating various aspects of neurophysiology so as to provide a groundwork for the understanding of higher order cognitive functions and some of their associated clinical disorders. Robert W. Thatcher wrote the introductory chapters (1–5) on the genesis of the EEG, mechanisms of synchronization, mechanisms of attention and arousal, and the neurophysiology of emotion. Thatcher also wrote Chapters 6 and 7 on information representation and time representation. E. Roy John wrote the chapters on the mechanisms of memory (8–10), statistical theories of learning (11), and neurophysiological processes related to consciousness and subjective experience (12–13). While we worked separately on different chapters, our combined enthusiasm and shared viewpoints soon gave birth to much more material than needed for Dr. Freedman's text. We decided at that point to combine our efforts toward a comprehensive neuroscience text. Actually, it is not much of an exaggeration to say that these volumes literally wrote themselves. The enormous experimental output of neuroscientists from different countries of the world over the last 50 years created the material substrate. Once the integrative viewpoint and functional orientation were adopted, the correlative materials from the different disciplines came readily into focus.

The book was written initially as one volume. However, as this volume grew the hope of comprehensive integration of basic and applied neuroscience within that volume had to be abandoned. There was simply too much ground to be covered, too many areas of basic and clinical research to be reviewed. Thus, the basic groundwork relating fundamental neural processes to the recently developed clinical domain was provided by Volume 1; Volume 2 is primarily concerned with clinical problems. Each volume should be seen as interrelated with and supportive of the other.

The attempt to identify basic phenomena from diverse disciplines which have implications for a set of functions and to construct an integrative description of each of those functions across the various disciplines was not only difficult but hazardous. There exists a great risk of being overly simplistic, as well as of committing inaccuracies. In spite of our attempts to avoid such shortcomings as much as possible, there are undoubtedly numerous instances where we have failed. For specialists in any one discipline, we hope that those defects are outweighed by the overall perspective.

Even though these volumes are extensive, many areas of neuroscience were either not covered at all or touched on lightly. For instance, this work is notably deficient in its treatment of sensation and perception. However, not only are adequate analyses of these processes available elsewhere, but inclusion of those topics in this book would make it substantially larger, and the cost would be prohibitive. Our particular concern has been to present treatments of topics not available elsewhere.

Our efforts would have been wasted without the help of many people. Although the authors assume full responsibility for the viewpoints and statements in this book, special acknowledgments are due to those individuals who have offered advice and suggestions. In particular we would like to thank Dr. Bernard Karmel for his reading of the first drafts, Ms. M. Lobell for her exceptional secretarial and administrative assistance, and E. Schwartz, A. Ramos, P. Easton, D. Brown, and S. Ahn for their advice and help with many advanced computer methods relied upon, especially in Volume 2. We owe much, of course, to our teachers, students, and other colleagues who have criticized and shaped our perspectives through the years and who have provided us with the inspiration necessary to see this project through. Special thanks are also due to Mr. Frank Bartlett for his help with much of the data analysis and, lastly, to our wives Becky and Miriam whose encouragement and support were vital and unfaltering throughout.

ROBERT W. THATCHER
E. ROY JOHN

Brain Research Laboratories
Department of Psychiatry
New York University Medical Center

The bulk of the original research reported here was conducted while both authors were affiliated with the Brain Research Laboratories of the Department of Psychiatry at New York Medical College.

Introduction

The task of understanding the mechanisms of brain function, while no less compelling and challenging than in the past, has recently reached a point of extreme popular interest. Nearly everyone, young and old, is interested in understanding the nature of feelings, basic drives, memories, thoughts, and consciousness. At the same time, neuroscience has reached the level of development where it seems capable of solving some of the most intriguing problems with which men have been fascinated throughout history, the problems of how mental experiences or cognitive functions arise from physical processes.

Although it would be presumptuous to predict a precise date, it is not unreasonable to believe that new fundamental insights into brain function are forthcoming. These new insights and discoveries will come about through a massive but currently loosely coordinated international effort. The multidisciplinary field of neuroscience, including neuroanatomy, neurochemistry, developmental neurobiology, neuropsychology, and molecular biology, to name a few, may naturally converge at some point in the future to provide these new and fundamental levels of understanding. Today scientists debate about the amount of time that will pass before the many microphenomena, which are still in the process of being discovered, can be combined or synthesized into a comprehensive linkage that will elucidate the nature of macrosystem behavior.

It is our belief, however, that unless extremely intensive integrative efforts are begun, the tendency of neuroscience to fragment into methodologically sound but phenomenologically unprogressive subdisciplines will increase and retard the development of vital overall perspectives. Given this view, it was our intention to integrate knowledge from various subdisciplines within each chapter by focusing on functions rather than fragmented phenomena. We chose to emphasize electrophysiological techniques, partly because of our training, but mostly because only with these methods can the moment-to-moment activity of large populations of neurons be monitored. At each moment, interaction between sensory input and coded representations of memories and goals occurs in a complex fashion throughout the nervous system. The monitoring electrode inside or outside the central nervous system detects synaptic activity involving biochemical and neuroanatomic interactions. In this sense one can consider that electrophysiology actually monitors the movements of small molecules or ions such as Na^+ and K^+. The representation of the immediate present involves rapid movements of ionic species at synapses and axons. These spatiotemporal configurations of ionic interchange then become consolidated so as to be represented by more stable macromolecular systems that somehow conserve the initial spatiotemporal representation. Thus, the electrophysiologist who studies behaving organisms has the capability of observing synaptic biochemical and anatomic interactions involved in the representation of the present as well as the multifaceted interrelations between past and future. This perspective, unlike most others in the field of neuroscience, lends itself to the development of techniques for the diagnosis and evaluation of treatments for the cure of functional and organic neuropathology. These techniques are discussed in Volume 2. With the subject of the second volume clearly in mind, our primary focus in this first volume is on basic neurophysiological processes that underlie higher cognitive functions, particularly those that can be studied readily by electrophysiologists.

This volume can be divided into three parts. Part 1, which includes Chapters 1–5, is concerned with the integration of neuroanatomic, neurochemical, and neurophysiological processes. The major focus is on the genesis of electroencephalogram (EEG) evoked potentials and alpha rhythms, on the mechanisms of coherence and synchronization, on processes of attention and arousal, and, in Chapter 5, on processes of emotion and basic drive states. Part 2, consisting of Chapters 6–11, is concerned with mechanisms of learning and memory and time representation. The starting point is on how representational systems are built (Chapter 6) and reflected electrophysiologically. A statistical model of neural interaction is advocated in Chapters 8–11 as a unifying concept for the representation of memory and global functions in general. In these chapters the chemistry of memory consolidation is also considered. Part 3, consisting of Chapters 12 and 13, is concerned with extending the statistical model to the subject of consciousness and daily subjective experience. These chapters, which rely on understanding the earlier basic chapters, are an attempt to integrate disparate facts into a comprehensive view of the self, of sensory awareness, and of the construction of consciousness.

Throughout this volume certain basic principles of cellular interaction are emphasized. One (within Chapters 2–6) is the dynamics of inhibitory and excitatory synaptic interactions which are constrained by phylogenetic experience (that is, neuroanatomy) and environmental input. Another (Chapters 3–10) is the notion of coherence or synchronization in which disparate cell groups are activated simultaneously. Not only do such processes underlie EEG and evoked potential rhythmic phenomena, but they also constitute the basis for representation of information in which large populations of coherent cell groups form functional assemblies. Coherent cellular discharge, in contrast to the random fluctuations of background activity, constitutes a "signal" embedded in "noise." The creation of that signal and its many transformations are the subject of Chapters 6–13. Another basic principle is the notion of reciprocal interactions. This principle operates throughout the volume. In Chapter 5 reciprocal interactions take the form of a complex and dynamic balance of opposites. Thus, the physical substrate for feeling states and basic drives is believed to involve reciprocal but fluid and finely tuned interactions. Finally, the notion of "signal-to-noise" ratio is developed to explain how particular memories are selected for retrieval (Chapter 10) and how different anatomic systems, while participating to varying degrees in the mediation of a given function, are nonetheless equipotential in their qualitative contributions.

1
Basic Neurophysiology

I. FUNDAMENTALS OF NEUROPHYSIOLOGY

This chapter is intended to provide a brief outline of the fundamentals of neurophysiology. More detailed information on this subject can be found in the following references: Brazier (1958); Ochs (1965); Purpura (1959); Quarton, Melnechuk, and Schmitt (1967); and Ruch, Patton, Woodbury, and Towe (1965).

A. Basic Structure

The brain is constituted of three basic compartments: glial cells, the extracellular space, and nerve cells or neurons. Glial cells, consisting primarily of astroglia and oligodendroglia, form a close and interactive relationship with neurons. They provide metabolic and structural support for the nerve cells. They also appear to serve a reciprocal, "symbiotic" function by exchanging proteins with neurons (Hydén, 1967) and by regulating ionic concentrations in both the intracellular and extracellular space (Karahashi & Goldring, 1966; Nicholls & Kuffler, 1964). The extracellular space and glial cells together provide the structural and functional matrix in which the neurons are embedded.

The extracellular space (sometimes referred to as the intercellular space) is the smallest (15 to 20%) of the compartments. However, through this space flow essential metabolic substances and ionic materials related to the polarization and depolarization of neuronal membranes and related as well to the relatively powerful currents generated by the activity of synchronized masses of neurons. The term "extracellular space" is misleading. Actually, this space is occupied by an extracellular structure. Electronmicrographs reveal an extracellular "fuzz" made of branching mucopolysaccharides and glycosaccharides extending from a

1

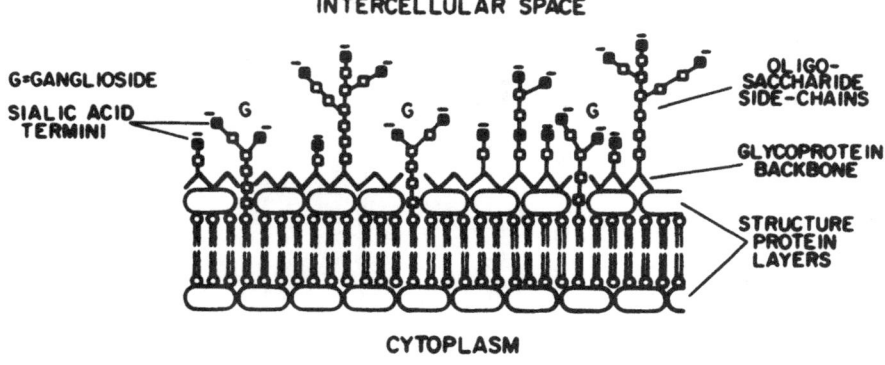

FIG. 1.1 Model of the neuronal membrane illustrating the glycoprotein and polysaccharides that form the extracellular "fuzz." (From Lehninger, 1968.)

protein backbone on the outer surface of neuronal membranes (Fig. 1.1). The complexly branching saccharides contain positively and negatively charged sites that bind extracellular ions. Weak electrical currents, such as those used in measurements of tissue impedance, pass through the extracellular space with very little, if any, penetration of the membranes of neurons or glia (Cole, 1940). Extracellular fluids have a specific resistance of approximately 4 Ω cm^{-2} as compared with resistances of neurons or neuroglia which are approximately 5000 Ω cm^{-2} (Coombs, Curtis, & Eccles, 1959; Nicholls & Kuffler, 1964).

Impedance measurements which reflect conductance within the extracellular channels demonstrate changes during learning, suggesting that structural alterations of the extracellular matrix underlie information storage and retrieval (Adey, Kado, & Didio, 1962; Adey, Kado, Didio, & Schindler, 1963; Adey, Kado, McIlwain & Walter, 1966b). The greatest concentration of extracellular mucopolysaccharides is around synaptic junctions, the axon hillock, and axonal nodes, three regions which are fundamental in the generation of electrical impulses in neurons (Meyer, 1969). Conformational changes in the extracellular matrix may modify neural excitability by affecting ionic binding affinities at these critical sites. As will be emphasized in later sections, the electroencephalogram (EEG) involves complex current flows within small channels that branch profusely throughout this extracellular matrix.

The third compartment is comprised of neurons. The neuron consists of three functional portions: (a) a cell body or *soma* which contains genetic material and protein synthesizing mechanisms. The soma also contains specialized receptive membrane structures (at *synapses*) that allow the neuron to be influenced directly by other neurons. Extending from the soma are (b) *dendrites* which can be considered as receptive surfaces and (c) an *axon* which transmits impulses

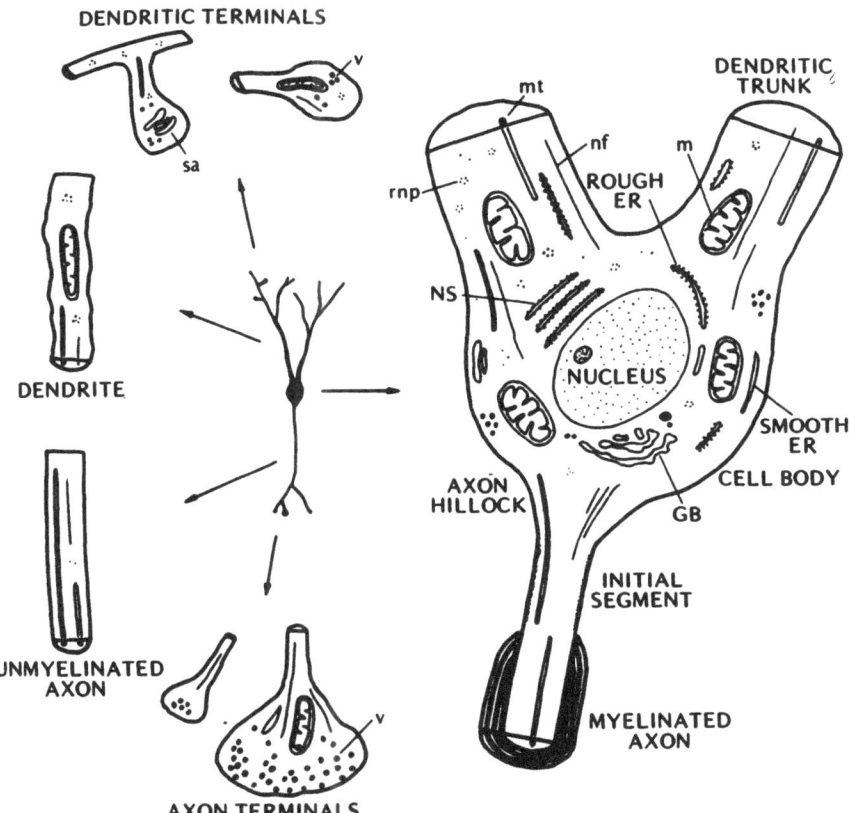

FIG. 1.2 An illustration of a neuron showing the cell body, axon, and dendrites as well as the interneuronal structure. ER, endoplasmic reticulum; GB, Golgi body; NS, Nissl substance; mt, microtubule; nf, neurofilament; rnp, ribonucleic particles; sa, spine apparatus; v, vesicles; m, mitochondria. (From Shepherd, 1974.)

(Fig. 1.2). An axon is usually several times longer than its diameter and in the peripheral nervous system can be as great as one meter in length.

Neurons are electrically excitable. Slow electrical potentials generated in the receptive regions (called synaptic potentials) give rise to impulses (called spikes or action potentials) that are transmitted down the axon. The axon arises from the soma at a specialized region called the *axon hillock.*

In an electronmicroscope, the soma can be seen to be limited by a double membrane structure which encloses a number of subcellular constituents: endoplasmic reticulum and ribosomes, nucleus and nucleolus, mitochondria, golgi apparati, and neurotubules. The neurotubules form a complex network serving the function of transporting cellular constituents to dendrites and axon termi-

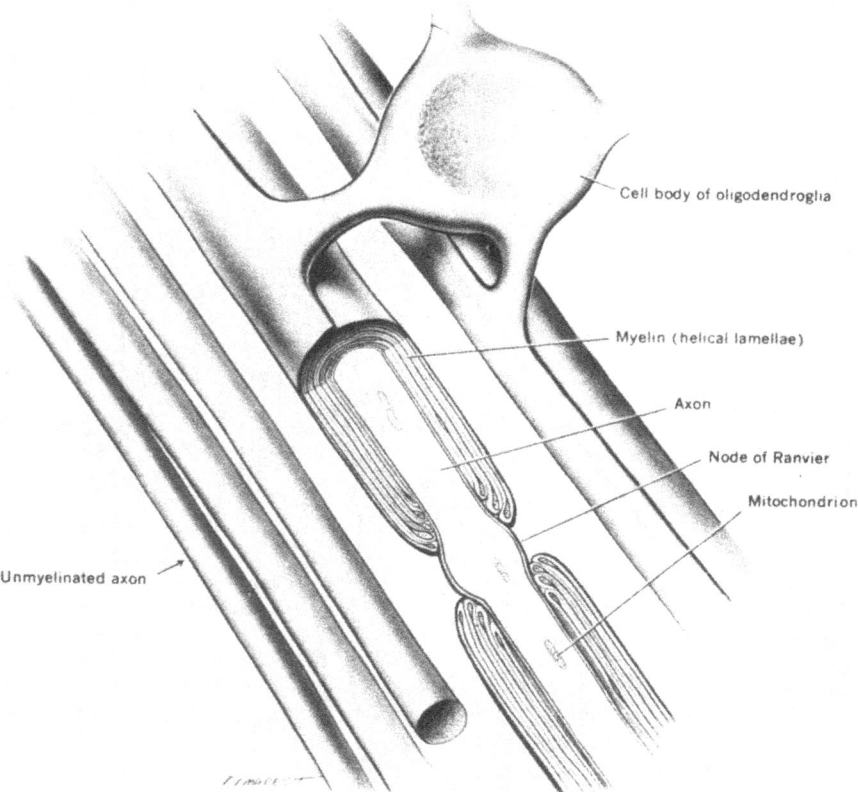

FIG. 1.3 The formation of myelin by oligodendroglia. (From *The Human Nervous System* by C. R. Noback & R. J. Demarest. Copyright © 1975 by McGraw-Hill Book Company. Used with permission of McGraw-Hill Book Company.)

nals. Recent evidence shows that cell bodies strategically located in the brain stem (nucleus coeruleus, nucleus raphe, nucleus substantia nigra) give rise to axons that travel great distances to synapse with the limbic system, the cortex, and the extrapyramidal motor system. Through these axons flow chemical transmitters (norepinephrine, serotonin, dopamine) which play important roles in the regulation of emotion, arousal, and motor–sensory coordination. These transmitters are probably transported through the neurotubules. The function of these chemicals will be discussed in Section III.

The importance of neurotubules has been further emphasized by the discovery that aging is accompanied by a flattening of the dendritic neurotubules (Wisniewski, Coblentz, & Terry, 1972). Malfunction of the transport mechanisms appears to be responsible, in part, for diminished function of brain cells in the very old.

In addition to the double membrane structure seen in the soma, long axons are enclosed by an insulating cover (myelin sheath) which functions to enable rapid and constant transmission velocities. The myelin sheath is in reality two membranes formed by specialized glial cells (Schwann cells) that wrap around the axon many times to form a multilayered structure (Fig. 1.3). The myelin is interrupted at regular intervals of approximately 0.5 to 1.5 mm along the fiber, by *Nodes of Ranvier*. The distance between the nodes, the internodal length, vary with axon diameter, age, and species (Hiscoe, 1947; Tasaki, 1953). The myelin sheath is a prominent feature of the long-axoned neurons; that is, those which form the peripheral sensory and motor systems and the long association bundles within the central nervous system (CNS). In the shortest axoned cells only one layer of myelin is present. As a consequence, a myelin sheath is not evident in the electronmicroscope. Over 97% of the cells comprising the CNS (called interneurons or association neurons) have very short axons without myelin sheaths.

B. Action Potential

In the resting state, the inside of the axon is approximately 90 mV more negative than the outside. This potential difference, sometimes referred to as the *transmembrane potential* or simply the *resting membrane potential,* is caused by an imbalance in the concentration of ions (particularly sodium and potassium) across the membrane. Figure 1.4 illustrates the various ionic constituents that give rise to the potential difference. The concentration gradients are created by an active metabolic pump mechanism. When a metabolic poison such as Ouabain is administered, the gradients disappear.

There are two forces that act to alter the distribution of the ions across the membrane. One is caused by the individual concentration gradients for each ion, which tend to drive sodium into the cell and potassium and chloride out of the cell. The second is caused by the electromotive force produced by the imbalance of total ions across the membrane, which tends to drive potassium and sodium into the cell and chloride out of the cell. A succinct description of these interactive forces is given by the Nernst equation which expresses the equilibrium potential, that is, the condition where the potential difference across the membrane for a given ion is zero. The Nernst equation is

$$V = RT/FZ \log_e (C)_o/(C)_i$$

where V is the equilibrium potential for the ion, R is the gas constant, T is the absolute temperature, Z is the ionic valence, F is the Faraday (which equals 96,500 coulombs), and the quantities in parentheses are the concentration of the ion inside the membrane, $(C)_i$, and outside the membrane, $(C)_o$.

Due to concentration differences, the tendency for Cl^- ions and K^+ ions to diffuse out of the cell is almost counterbalanced by the resting membrane potential. This balanced situation is not present in the case of Na^+. Both the

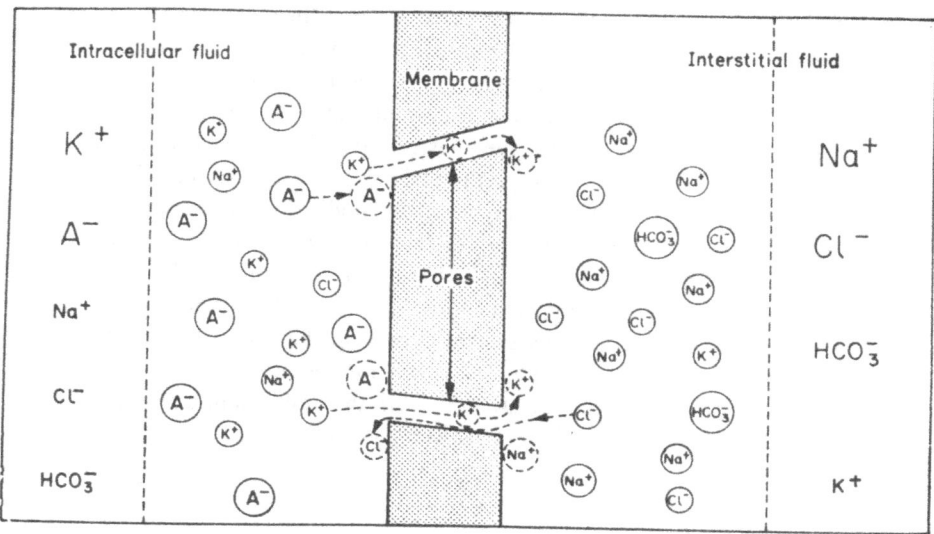

FIG. 1.4 Development of transmembrane voltage by an ion concentration gradient. Diagram of an intracellular fluid-membrane-interstitial fluid system. Membrane shown has some, but not all, properties of a real cell membrane. Hypothetical membrane is pierced by pores of such size that K^+ and Cl^- ions can move through them easily, Na^+ ions with difficulty, and A^- not at all. Sizes of symbols in left- and right-hand columns indicate relative concentrations of ions in fluids bathing the membrane. Dashed arrows and circles show paths taken by K^+, A^-, Na^+, and Cl^- ions as a K^+ or Cl^- travels through a pore. Penetration of the pore by a K^+ or Cl^- follows a collision between the K^+ or Cl^- and water molecules (not shown), giving the K^+ or Cl^- the necessary kinetic energy and proper direction. An A^- or Na^+ unable to cross the membrane is left behind when a K^+ or Cl^-, respectively, diffuses through a pore. Because K^+ is more concentrated on left than on right, more K^+ diffuses from left to right than from right to left, and conversely for Cl^-. Therefore, right-hand border of membrane becomes positively charged (K^+, Na^+) and left-hand negatively charged (Cl^-, A^-). Fluids away from the membrane are electrically neutral because of attraction between + and − charges. Charges separated by membrane stay near it because of their attraction. (From Regan, 1972; after Ruch *et al.*, 1965.)

concentration gradient and the electrochemical force tend to drive Na^+ into the cell.

The resting equilibrium potentials expressed by the Nernst equation are ~ -97 mV for K^+ and ~ -90 mV for Cl^-. The electromotive force necessary to achieve equilibrium for a resting membrane potential of -90 mV, is -7 mV for K^+ and 0 mV for Cl^-. On the other hand, the equilibrium potential for Na^+ is $\sim +66$ mV, which means an electromotive force of −156 mV must act on this ionic species to maintain a resting membrane potential of −90 mV. This state of extreme disequilibrium with regard to Na^+ led to the formulation of the *sodium pump hypothesis* (Dean, 1941; Krogh, 1946). This theory postulates an active transport mechanism which maintains a Na^+ concentration gradient in the face of the electromotive force. More recently, a sodium–potassium pump has been postulated to account parsimoniously for the maintenance of both the Na^+ and K^+ disequilibria (Davies & Keynes, 1961).

In the resting state, the negative 90 mV membrane potential is maintained by an active process. When a current is injected into an axon so as to depolarize the membrane (i.e., drive the transmembrane potential toward zero), then an active and powerful shift in the Na^+ and K^+ concentrations may occur. This shift happens if a critical threshold level of depolarization is reached. The threshold value is approximately the potential difference for which the inward flux of Na^+ just balances the outward flow of K^+. Depolarization beyond this point results in the initiation of a positive feedback process: Na^+ ions enter the cell at an accelerating rate, the slope of which increases the more depolarized the membrane becomes. Because of the Na^+ influx, the membrane potential very rapidly approaches zero and even exceeds zero so that the inside of the cell becomes more positive than the outside. At this point, a deactivation of sodium influx occurs and, simultaneously, K^+ ions are extruded from the inside of the cell to the outside, a process that rapidly restores the membrane potential to -90 mV. This entire process of Na^+ activation, Na^+ deactivation, K^+ activation, constitutes the *action potential*, and is completed within 2 msec. The repolarization process involving restoration of the K^+ gradient produces a series of lingering oscillations (usually positive, negative, then positive) called *afterpotentials*. Afterpotentials have time courses similar to *synaptic potentials* and can make significant contributions to EEG waves, particularly those recorded from fiber tracts (Laufer & Verzeano, 1967; Verzeano, 1972). This will be discussed in a later section (see Chapter 2, Section IIF).

The action potential is referred to in many ways, including the *spike,* neuronal discharge, the firing of a nerve cell, or *unit activity.* The action potential is usually initiated at the point where the axon leaves the soma, the so-called *axon hillock.*

The action potential propagates the length of the axon in a nondecremental manner, analogous to the combustion of gun powder in a fuse. The depolarization produced by the action potential spreads electronically from the axon hillock to adjacent regions of the membrane. Depending on the space constant of the membrane (the space constant is related to the resistive and capacitative properties of the membrane and determines the distance depolarization will spread from a point source) adjacent regions will become sufficiently depolarized to reach threshold and thus sustain the regenerative spike process; this adjacent spike in turn depolarizes the adjacent membrane which results in the continued propagation of the spike, and so on. This regenerative process is a communication engineer's dream, since spike amplitude does not decrement with distance and the spike requires little expenditure of energy.

As mentioned earlier, small nodes or gaps in myelin occur at regular intervals along the length of myelinated axons. Because of the considerable resistance to ion flow offered by myelin, the nodes are regions where current density is very high. As a consequence, in such axons action potentials are generated only at nodes. If one node is active and an adjacent node is inactive, a local current flows between the nodes and depolarizes the inactive node. For this reason, in myelinated axons the propagating spike jumps serially from node to node. This form of propagation is called *saltatory conduction.*

C. Synaptic Transmission

Action potentials are initiated near the axon hillock and then propagate the length of the axon where they invade the terminal region of the axon called *terminal arborization.* Here, the axon ramifies into a spray of fine fibers, each ending in a *terminal bouton,* in close proximity to the membrane of another neuron. This junction is called a *synapse.* Although in the peripheral nervous system one neuron usually synapses with one or very few neurons, in the central nervous system one neuron usually sends synaptic outflow to many other neurons and receives synaptic inputs from a large number of neurons. Relationships are one-to-many and many-to-one rather than one-to-one. Synapses usually occur between axons and dendrites. Some synapses are axosomatic, and recently axoaxonal synapses have been reported (Shepherd, 1974). At the terminal bouton, the wave of depolarization caused by the action potential releases bound Ca^{2+} which interacts with small vesicles or sacks, causing them to release chemicals into the synaptic gap between the bouton and the postsynaptic membrane of another neuron (see Fig. 1.5). When the synaptic chemical engages a receptor site on the postsynaptic side a slow electrical process, called a synaptic potential, is generated (5 to 20 mV at peak amplitude and 20 to 150 msec in duration).

The released chemicals are called neurohumors or, more commonly, synaptic transmitters. A variety of synaptic transmitters have been identified. Transmitters vary from anatomic region to region, and even within regions. Acetylcholine, epinephrine, gamma-aminobutyric acid (GABA), norepinephrine, dopamine, serotonin, and glycine are all believed to be transmitters. Whatever the transmitter, it interacts with receptive structures on the postsynaptic membrane, depolarizing the membrane if the transmitter is excitatory (e.g., epinephrine or acetylcholine) or hyperpolarizing the membrane if the transmitter is inhibitory (gamma-aminobutyric acid or glycine). In some cases, the process of *exocytosis* is believed to be the mechanism that opens synaptic vesicles and empties their content (see Fig. 1.5).

One proposed mechanism for the postsynaptic action of transmitters is shown in the lower right of Fig. 1.5 in which the transmitter T interacts with a receptor surface that activates adenylcyclase to produce 3′,5′ cyclic AMP. The cortex has

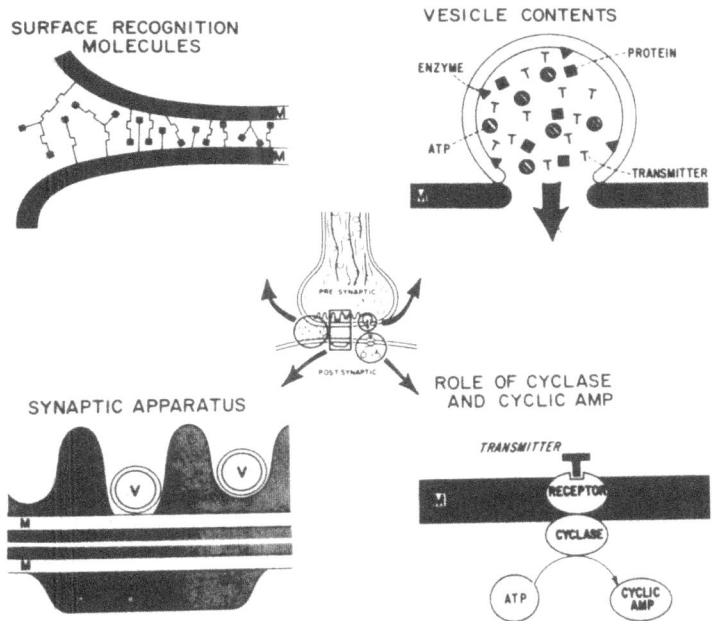

FIG. 1.5 Models of presynaptic and postsynaptic membrane interfaces showing the synaptic apparatus and various biochemicals involved in synaptic transmission. (From Schmitt & Samson, 1969.)

the highest concentration of adenylcyclase of all body tissues. Recent evidence has shown that cerebellar synaptic transmission is dependent on cyclic AMP (Bloom, Nicholson, Ungerlei, & Ledley, 1973; Cooper, Bloom, & Roth, 1974; Rall & Gilman, 1970). As can be seen in Fig. 1.5, the membrane is made of a serrated surface called a *presynaptic complex* in which transmitter molecules are believed to lodge. The extracellular mucopolysaccharide matrix, mentioned earlier, is also depicted in Fig. 1.5 (upper left).

Generally, as mentioned earlier, there are multiple branchings of the end region of the axon so that a given axon distributes its output over a large surface area. All boutons in such a terminal arborization contain the same chemical transmitter. Since the multiple boutons are activated in concert the terminal arborization of the axon is called a *functional synaptic unit* (Elul, 1972) in analogy with the functional motor unit of skeletal muscle endings which act synchronously. For this reason, as will be discussed later, functional synaptic units are primary contributors to EEG waves.

There are two types of synapses. One is excitatory (depolarizing) and acts to increase the probability of spike discharge by driving the membrane toward threshold. The other is *inhibitory* (hyperpolarizing) and acts to decrease the probability of spike discharge by driving the membrane potential away from threshold. In large neurons, inhibitory synaptic endings are located largely on cell bodies and the trunk of dendrites. Some excitatory synapses are located on the soma but most occur on distal dendritic regions. Somewhat different synaptic structures help identify inhibitory and excitatory synapses in electron-micrographs (Akert, Pfenninger, Sandri, & Moore, 1972; Shepherd, 1974). Excitatory and inhibitory synaptic potentials also exhibit different dynamic characteristics and are not simple mirror images of each other. Excitatory synaptic potentials are intense and of short duration (seldom exceeding 50 msec) while inhibitory synaptic potentials are typically less intense (Schlag & Villa-blanca, 1968) and of longer duration (up to 600 msec in cortical and hippo-campal neurons; Purpura & Shofer, 1964; Purpura, 1972). *Excitatory post-synaptic potentials* and *inhibitory postsynaptic potentials* usually are abbreviated EPSPs and *IPSPs*, respectively.

D. Neuronal Integrative Function

When a neuron is sufficiently excited by summation of excitatory synaptic processes (EPSPs) at different places along its membrane, a critical level of depolarization ("threshold") may be reached in the region of the axon hillock, where the consequent electrochemical self-propagating spike is initiated.

The nature of synaptic slow potential and spike initiation processes allows the neuron to perform an "integrating" function. The integration arises from the fact that more than one synaptic process is required to initiate a spike. Typically, thousands of synaptic potentials arising along long stretches of the dendritic or somatic membrane summate algebraically (interaction of depolarizing and hyperpolarizing influences). Varying intensities of output (in terms of the frequency of spike discharge) are produced, depending upon the rate and magnitude of the resultant depolarizations.

Two forms of slow potential integration occur in neurons. One is *spatial summation,* in which a group of spatially separated synapses are activated simultaneously. These spatially disparate processes summate and may, if there are a sufficient number, reach threshold for spike initiation. The second type of integration is called *temporal summation.* This process can involve a single synapse activated repetitively. If a sufficient number of synaptic potentials from a small group of synapses occur within a sufficiently short period of time, then threshold may be reached.

In the previous section, we described the constituents of which the nervous system is composed. Although important roles are played by the glia cells and the extracellular structure, the most essential role in information processing is probably played by the neurons. Neurons integrate the many diverse influences impinging upon them and in turn transmit the result of such integration as influences affecting many other neurons. Obviously, the quality of the afferent influences acting upon a neuron depends in part upon their anatomic origins.

II. FUNCTIONAL NEUROANATOMY

The goals of this section are to familiarize the reader with certain anatomic terms which will be a prerequisite for the understanding of later chapters, and to provide a description of the anatomic organization and interrelations of the major functional neural systems. No attempt will be made to discuss neuroanatomic connections and pathways in great detail. Several excellent books on this subject are available (Krieg, 1953; Mettler, 1948; Ranson, 1936; Truex & Carpenter, 1964). Our purpose here is only to provide a minimal anatomic organization within which we can provide further structural details as they become relevant. Much of this anatomical perspective is discussed by Luria (1966).

A. Definitions

Before proceeding it is necessary to define several important anatomic terms. It is common to discuss the relation of parts of the body in terms of three imaginary planes (sagittal, horizontal, and coronal; see Fig. 1.6). The *sagittal* plane divides the body into left and right. The *horizontal* plane divides the body into upper and lower parts. The *coronal* (or frontal; Fig. 1.6) plane divides the body into front (anterior) and back (posterior) parts. The coronal plane is at right angles to both the sagittal and the horizontal planes. There are several terms that orient one to specific points of reference. For instance, *anterior* refers to the front of the body, *posterior* refers to the back, *superior* refers to higher structures, *inferior* refers to lower ones. *Rostral* refers to higher structures (i.e., toward the head) and *caudal* refers to lower structures (toward the tail). Structures near the front of the body are *ventral,* structures located near the back are *dorsal.* It is helpful to discriminate ventral from dorsal structures by imagining the head tilted back. Structures located near the upper neck and under the chin are *ventral* while structures located toward the top of the head are *dorsal.*

In the pages to follow the word *afferent* will refer to incoming activity, usually flowing in a peripheral to central direction. Outflow, usually in the central to peripheral direction is called *efferent.* In general, the conduction of activity from caudal to rostral brain regions is often referred to as *centripetal* while conduction from rostral to caudal is called *centrifugal.*

B. The Organization of Sensory Systems

In this section, the visual system will be discussed in detail most, since the anatomic organization of the other sensory systems follows a similar plan. The visual, somatosensory, and auditory systems all involve a set of specialized peripheral receptors capable of transducing specific environmental energies into

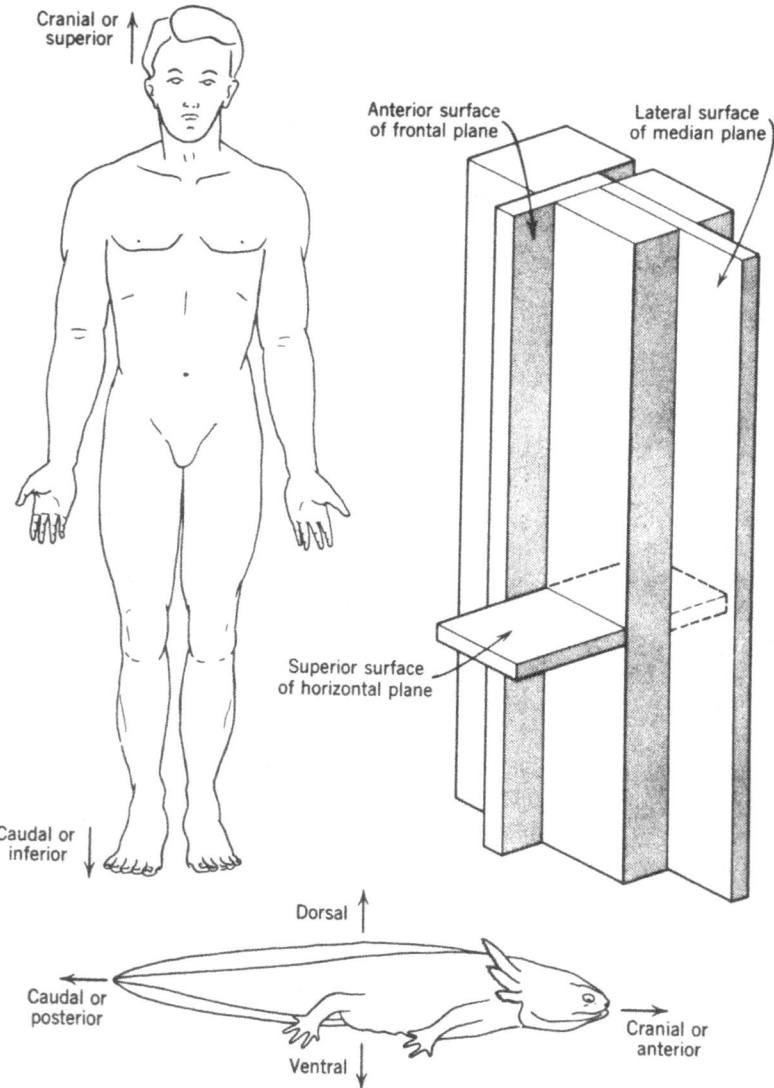

FIG. 1.6 The anatomic position and corresponding terms of direction; a block diagram showing the primary planes and surfaces; and an amphibian illustrating comparative anatomic nomenclature. (From Gardner, 1963.)

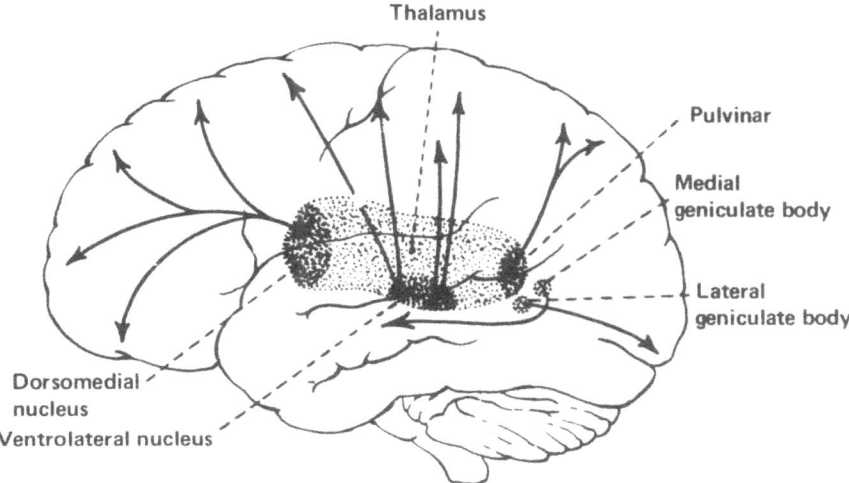

FIG. 1.7 Diagram of principal thalamocortical projections. (Fig. 1-17 Reproduced with permission From Chusid, J. G.: *Correlative Neuroanatomy and Functional Neurology,* 15th ed., Lange, 1973.)

nerve impulses which are conducted in what are called *primary sensory* pathways to *relay stations* located in a structure called the *thalamus* (see Figs. 1.7 and 1.8). The thalamus, which in humans is about the size of a closed fist, occupies the very middle portion of the head. Nearly all sensory input interacts either directly or indirectly with neurons in the thalamus. The thalamus has been likened to a "Grand Central Station," carrying out the switching or "gating" of peripheral sensory drives to specific cortical regions as well as integrating sensory–motor outflow activities. In Chapter 3 it will be seen that the thalamus plays a fundamental role in the production of alpha rhythms and other EEG phenomena.

Figure 1.9 diagrammatically illustrates the organization of the thalamus. The thalamic relay nuclei for the primary sensory systems are located primarily in the lateral posterior regions. The relay nucleus for vision is the *lateral geniculate,* for audition the *medial geniculate,* and for somathesis the *ventroposterior* and ventromedial nuclei. These nuclei have been considered relay nuclei because they typically involve only one synapse in the transmission of information to specific cortical regions. In the relay nuclei, modality specificity is rigidly maintained through the action of two factors: (1) a close mapping of primary afferent terminals onto a small number of relay neurons which accomplish transmission to the cortex, and (2) a complete absence of interaction with other relay nuclei. The lateral geniculate projects to the occipital pole or *area 17* of the neocortex,

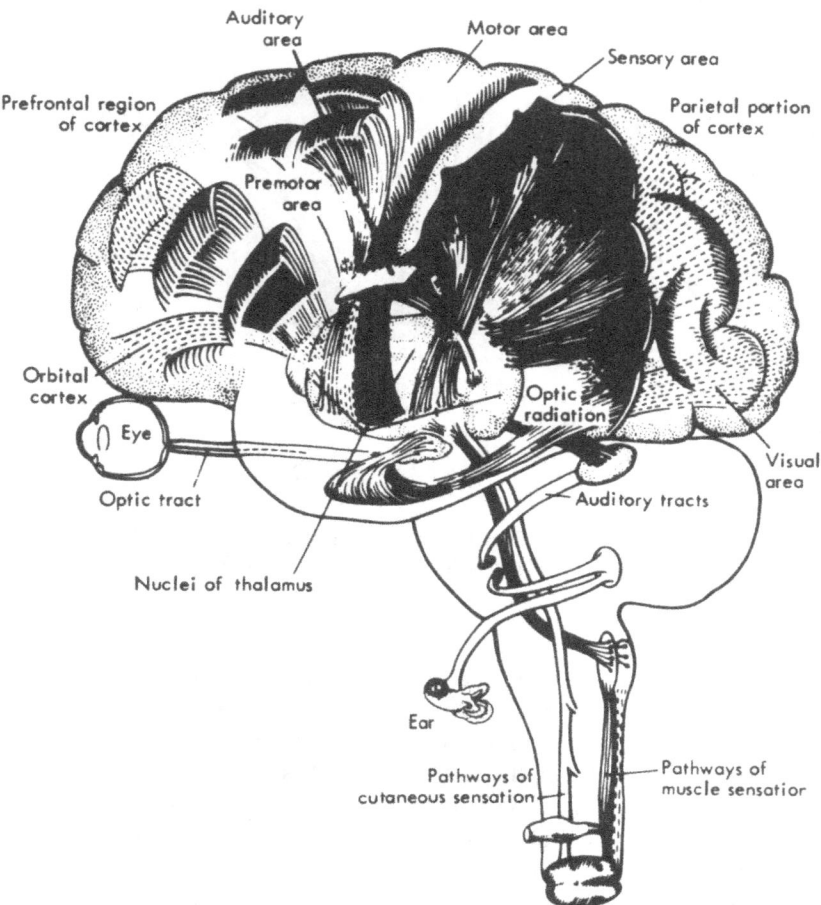

FIG. 1.8 Cortical sensory and motor projection systems illustrating central thalamic projections. (From Luria, 1973.)

the medial geniculate projects to the temporal cortex (area 41) and the ventroposterior nucleus projects to the postcentral gyrus (areas 1, 2, and 3). Some of the cortical projections of the primary sensory system are seen in Fig. 1.10.

In addition to inputs to relay nuclei with modality-specific thalamocortical projections, there are collateral projections from peripheral afferent fibers to other brain stem nuclei. In the case of vision, collaterals from *optic tract* fibers, arising just anterior to the lateral geniculate, course caudally to terminate in nuclei in the *midbrain* (see Fig. 1.8), called the *pretectum* and the *superior colliculus*. These nuclei also receive fibers from the lateral geniculate. This midbrain component of the visual system is involved in the control of eye

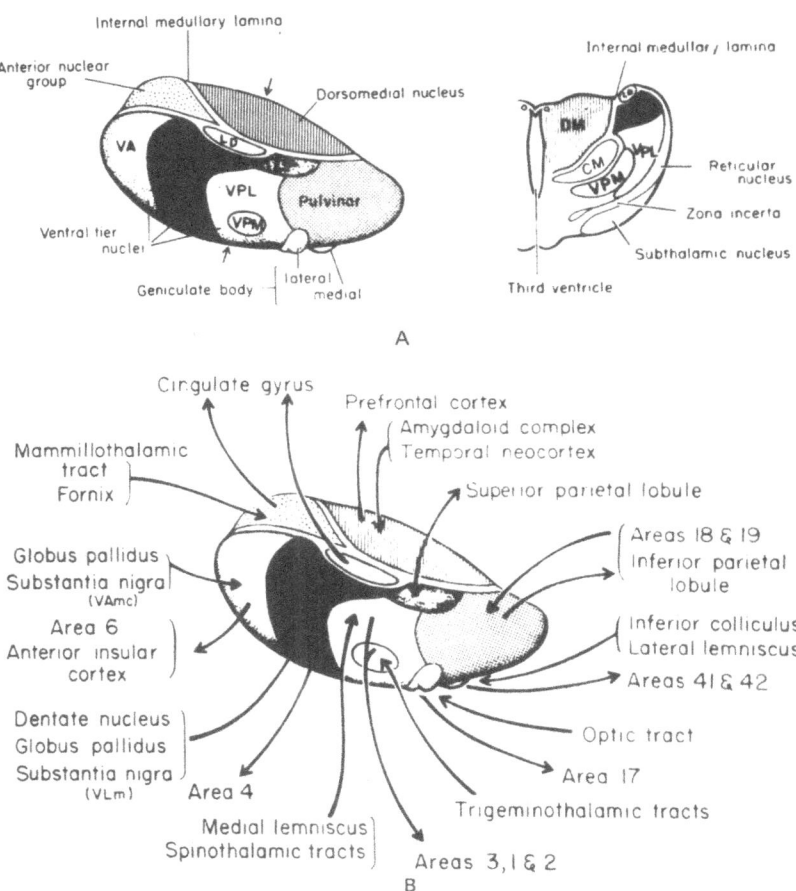

FIG. 1.9 Diagram of major thalamic nuclei and principal efferent and afferent fiber connections. (A) Identification of thalamic nuclei (left) and transverse section (right); (B) major efferent and afferent fiber connections (see Fig. 1.14 for cortical projections). (From Truex & Carpenter, 1973. © 1973 The Williams & Wilkins Co., Baltimore.)

movement and the pupillary reflex. In the case of auditory sensation, nuclei called *cochlear nuclei* receive the primary auditory afferents and transmit their output either directly or by collaterals to brain stem nuclei such as the *inferior colliculus,* as well as to the medial geniculate.

The system of connections thus described, that is, the peripheral afferent fibers, the thalamic relay nuclei and their cortical projections, and the sensory–motor brain stem connections, constitutes what is called the *primary sensory system.* A *secondary sensory* system involving associative or sensory interactive processes parallels the primary sensory system. Neurons comprising the second-

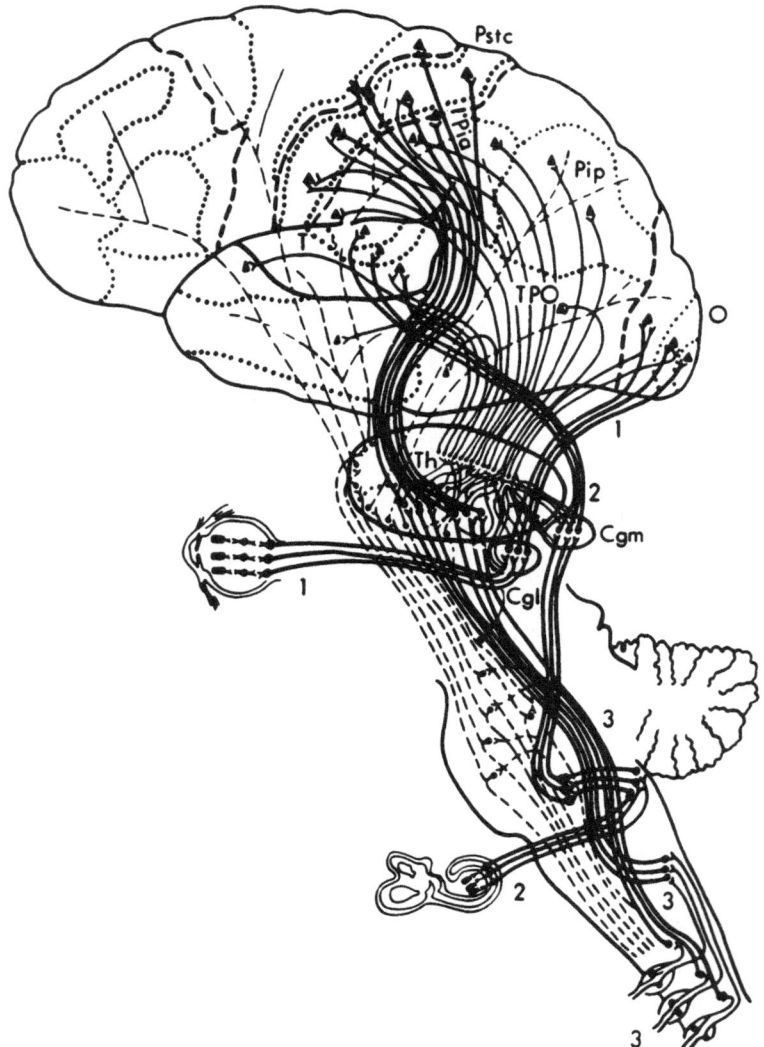

FIG. 1.10 Scheme of the cortical–subcortical relationships of the primary, secondary, and tertiary sensory systems. The thick lines represent systems of sensory analyzers with their relays in the subcortical divisions. (1) Visual analyzer organ; (2) auditory analyzer; (3) somatosensory analyzer; (T) Temporal region; (O) occipital region; (TPO) temporoparietal-occipital region; (Th) thalamus; (Cgm) medial geniculate body; (Cgl) lateral geniculate body; (Pip) area 30; (Pia) area HO; (Pstc) postcentral gyrus. (From *Higher cortical functions in man,* by A. R. Luria. Authorized translation from the Russian by B. Haugh, © 1966 Consultants Bureau, Inc., and Basic Book, Inc., Publishers, New York.)

ary sensory system are commonly multimodal in nature, that is, they have connections with more than one modality of sensation. Anatomic divisions between secondary (associative) and primary sensory systems can be seen in both the thalamus and cortex.

Convergence of different sensory inputs onto single neurons is a logical prerequisite for cross-sensory integration. Thus, the secondary sensory system is involved in much more integrative or associative activity than the primary. As is discussed in detail in Chapter 6, the primary sensory systems seem to be involved in extracting the initial and elementary aspects of sensation while the secondary systems play an important role in synthesizing these elementary sensory forms into more complete wholes.

As a result of numerous afferent connections and feedbacks with the thalamic relay stations, the primary cortical fields are important for very fine discriminations between stimuli within a modality (see Chapter 4). By efferent outflow causing adaptations of the sense organs, the cortex plays a role in arranging the receptor surfaces so they are optimally "tuned" and oriented toward stimuli that are to be analyzed. In man the primary sensory cortex is important for minute differentiation between stimuli. Lesions of primary projection area 17, for example, result in *scotomas,* or "blind spots" in the receptive field. Electrophysiological analyses demonstrate that neurons in area 17 respond optimally to very simple stimuli such as lines or slits with a particular orientation. Similar elementary discriminative analyses are performed in the primary sensory systems of the somasthetic and auditory modalities.

The nuclei of the thalamus can be subdivided into two groups: *specific relay* nuclei and *association* nuclei. The association nuclei are principally located in the midline regions (as well as dorsal, posterior, and anterior locations) and project to the secondary cortical areas (see Fig. 1.11). An example of an association nucleus is the *pulvinar,* located in the posterior region of the thalamus (see Fig. 1.9). This nucleus is part of the secondary visual system since it projects to and receives fibers from the secondary visual cortex (areas 18 and 19; see Fig. 1.10). However, the pulvinar also receives inputs from specific relay nuclei (such as the lateral geniculate, in addition to those from secondary cortical regions).

The secondary sensory system also is engaged in collateral interactions with the *midline reticular formation.* The reticular formation is a complex network of interconnected neurons located in the brain stem. As discussed in detail in Chapter 4, the greatest degree of interactive convergence and divergence of neural connections occurs in the midline systems of the thalamus and reticular formation. The midline systems appear to be at the hub of the integration "center" of the brain. In this Volume particular emphasis is placed on the function of the midline integrative centers of the thalamus and reticular formation. It is these centers that are involved in the control of arousal and attention

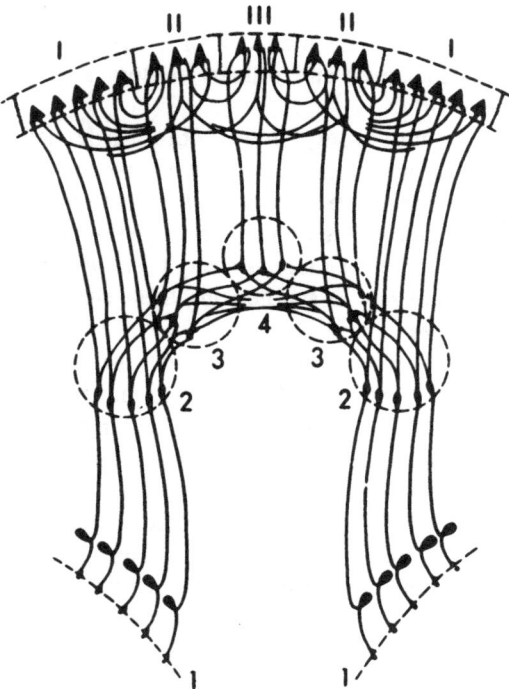

FIG. 1.11 Scheme of relays in primary, secondary, and tertiary sensory systems. (1) Peripheral receptor surfaces; (2–4) relays in the subcortical divisions of the nervous system; (I) primary cortical zones, (II) secondary cortical zones, (III) tertiary sensory regions showing the highest degree of convergence and sensory overlap. (From Luria, 1966.)

(Chapter 4), the production of alpha rhythms and related EEG phenomena (Chapter 3), the initiation of the orienting reflex and the creation of representational systems (Chapter 6), recall from memory (Chapter 9), the integration of eletrically pulsed information (Chapter 11), and it is these midline systems that make fundamental contributions to the maintenance of consciousness and sensory awareness (Chapters 12 and 13).

The secondary sensory, or association, cortex is located adjacent to the primary cortical fields and exhibits architectonic and physiological characteristics distinctly different from the primary fields. For instance, in man lesions of visual areas 18 and 19 leave intact the perception of elementary visual form but severely impair the perception of complex groups of stimuli or the relations between stimuli (Luria, 1966; see Chapter 6).

As discussed in Chapter 6, the cytoarchitecture of the secondary cortical areas is distinctly different from the primary areas in that the secondary areas possess larger numbers of small granule cells sometimes referred to as associative inter-

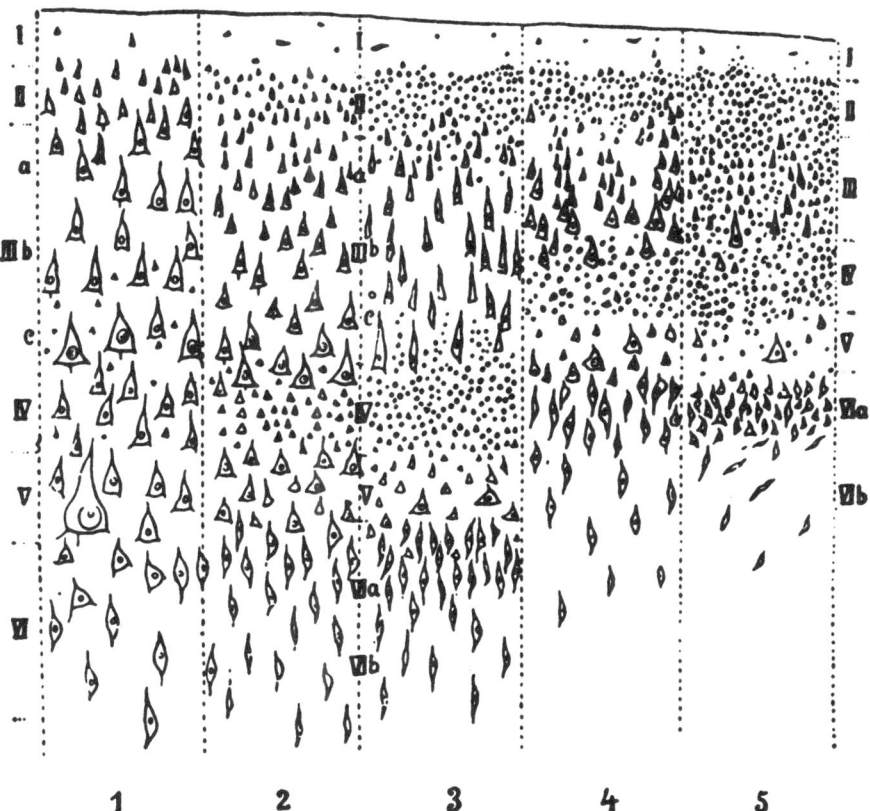

FIG. 1.12 Cortical cell layers and the five structural types of cerebral cortex. (1) Agranular (thick, without small granule cells found in premotor cortex, area 6); (2) frontal (contains granule cells and large cells); (3) parietal (thick cortex with increased number of granule cells); (4) polar (thin cortical structures found near the occipital and frontal poles); (5) granulous (reverse of type (1), that is, thin and composed mainly of densely packed granule cells found in areas 17 and 3). Nomenclature for layers at right of figure: I, molecular layer; II, external granular layer; III, external pyramidal layer; IV, internal granular layer; V, internal pyramidal layer, and VI, the multiform layer. (From Truex & Carpender, 1973. © The Williams & Wilkins Co., Baltimore, after Economo, 1929.)

neurons. Subcortical afferents, which pass through the granule cell layer, terminate on large neurons in a lower layer of this cortex than in the primary cortex (see Fig. 1.12). This structural substrate plays an important role in analyzing and synthesizing the individual stimuli differentiated by the primary fields and in accomplishing functional integration between the primary fields of different modalities. Thus, the secondary fields are primarily involved with relatively more complex integrative functions. Consistent with this greater complexity of sec-

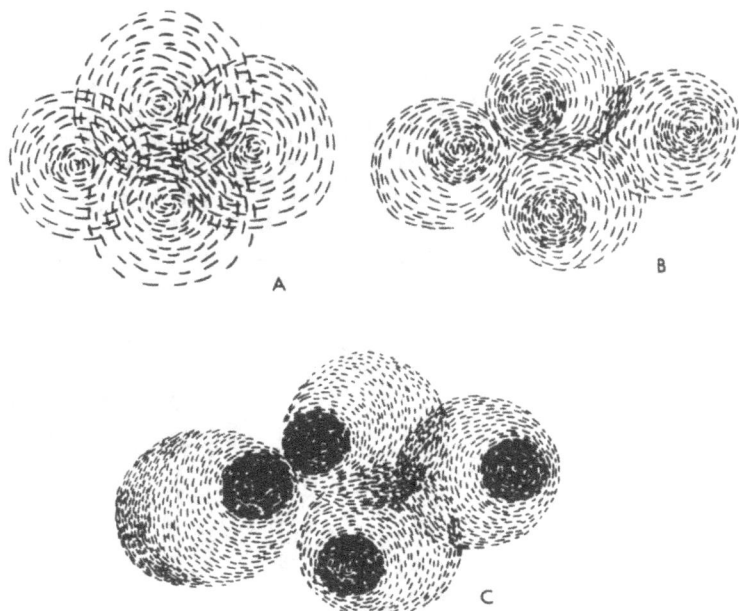

FIG. 1.13 Phylogenetic differentiation of primary, secondary, and tertiary regions of cortex in mammals. (A) Insectivores; (B) carnivores; (C) primates and man. Central darkly shaded areas represent primary systems, lighter peripheral shading represents secondary systems, and areas of overlap constitute the tertiary systems. Note that phylogenesis involves an interlocking development of the three sensory systems. (From Luria, 1966.)

ondary sensory function is the fact that secondary sensory cortical regions receive input from a more complex system of relays in the subcortical (mainly thalamic) internuncial neurons than do the primary fields. This increase in complexity is schematically illustrated in Fig. 1.11. Afferents reach the secondary cortex through a larger number of relays in the associative nuclei of the thalamus than do the afferent impulses traveling to the primary areas from the thalamic relay nuclei. Thus, the receptor impulses which eventually reach the secondary regions undergo considerable preliminary processing and integration while still at the subcortical level.

A *tertiary sensory* field arises from a further elaboration of cytoarchitectural connections involving additional integration and overlap between the secondary sensory areas of disparate sensory modalities (see Fig. 1.11). These cortical formations involving the *inferior parietal, midtemporal,* and junctional *parietal-temporal–occipital* zones are associated with the most complex integration of simultaneous visual, auditory, and somatesthetic activity. In man, lesions or stimulation of the tertiary zones are not accompanied by any marked loss or modification of the function of the primary sensory systems. Instead, functions related to the highest levels of multisensory integration, such as those involved in the combined action of several sense modalities, are severely effected (Luria, 1966, 1973). A schematic overview of differentiation and overlap of primary, secondary, and tertiary cortical fields across phylogenetic development is shown in Fig. 1.13.

C. The Organization of the Motor System

The primary, secondary, and tertiary areas of the cortical sensory systems are located posterior to the *central sulcus* (see Fig. 1.14, PB). The frontal cortex and motor cortex which are cytoarchitecturally as well as functionally distinct from the sensory cortex are located anterior to the central sulcus. In principle the anterior motor areas are functionally organized, much like the posterior cortex, into primary, secondary, and tertiary fields. Throughout phylogenesis the precentral region, bounded anteriorly by the postcentral gyrus, became differentiated into a primary, large-celled motor region (area 4, see Fig. 1.14) and a secondary premotor region (areas 6 and 8). Stimulation of the surface of area 4

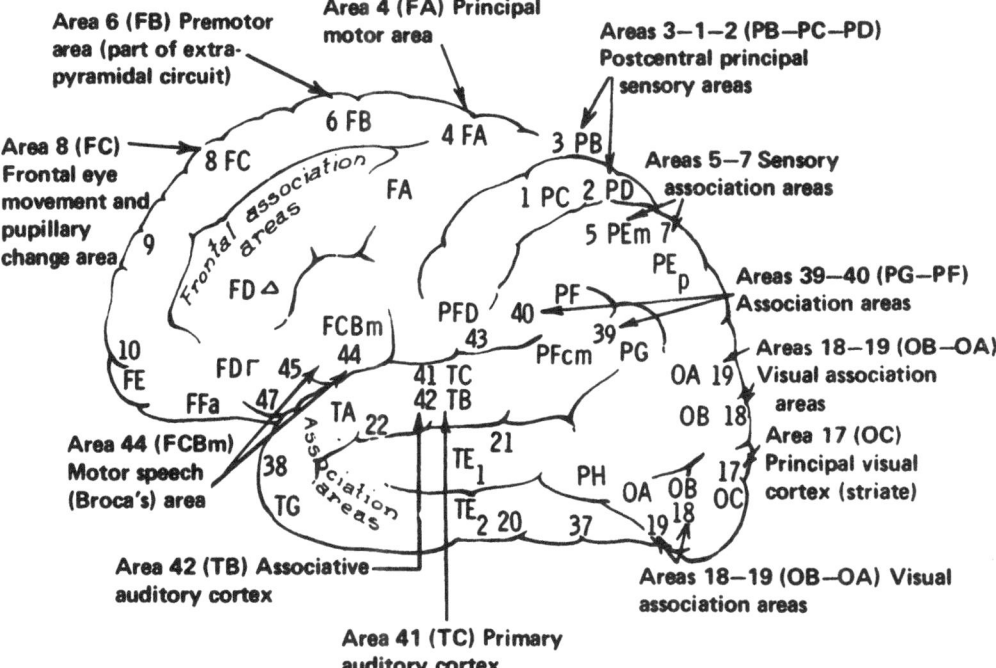

FIG. 1.14 The lateral cortical areas shown according to Brodmann (numbers) and Economo (letters), with functional localizations. (Fig. 1.10 Reproduced From Chusid, J. G. Correlative *Neuroanatomy and Functional Neurology*, 15th ed. Lange, 1973.)

can give rise to motor movements in select and differentiated muscle groups. In contrast, the degree of specificity of motor movement is less and total area of activation greater upon stimulation of areas 6 and 8. A striking cytoarchitectural peculiarity of area 4, in contrast to the sensory cortex, is the absence of a clearly distinguishable layer IV and the presence of very large *Betz* cells in layer V (see Fig. 1.12). Betz cells send axons into the *pyramidal tract* which convey impulses mediating voluntary movements to the brain stem and spinal cord. Area 4 receives afferents from the ventrolateral nucleus of the thalamus, which is primarily a gating and relay station for impulses arising in the *cerebellum* and *basal ganglia.* Area 4 of the motor cortex, like area 17 of the sensory cortex, is characterized by short and direct subcortical cortical connections. This is in contrast to the secondary areas of sensory and motor cortex which involve a large number of synaptic relay steps and indirect connections between cortical and subcortical regions.

The secondary premotor area 6 does not contain Betz cells, has a marked development of large *pyramidal* cells in layer III, and contributes fewer fibers to the pyramidal tract than does area 4. Physiological and clinical data indicate that areas 6 and 8 participate in the performance of complex and coordinated movements, particularly those taking place over a long period of time and involving diverse muscle groups (Luria, 1966). Areas 6 and 8 contribute large numbers of fibers to the *extrapyramidal tract* which, in contrast to the direct pyramidal tract, passes through a complex series of relays in the thalamus, basal ganglia, and brain stem, before reaching the spinal cord.

The secondary premotor areas 6 and 8 and the primary motor system form a single integrative complex possessing bilateral afferent and efferent interactions at both the cortical and subcortical levels. Massive cortical association fiber systems intercommunicate between sensory and motor cortex. This associative interaction of sensory and motor systems on the cortical level is paralleled by similar interactions within the thalamus. However, anatomic evidence shows clearly that the source of input to the motor cortex is different from the afferent source for the sensory cortex. This is illustrated most clearly in Fig. 1.15. The afferent impulses to the motor regions are relayed by a different group of thalamic nuclei than for the posterior cortex. In particular, movement relevant inputs from the dentate nucleus of the cerebellum, the basal ganglia, brain stem, and midbrain nuclei are relayed via the ventrolateral and ventro-anterior nuclei. In general, the anterior or rostral regions of the thalamus project to the motor cortex while the posterior regions project to the posterior or sensory cortex.

It is important to emphasize that nearly all of the connections from thalamus to cortex are reciprocal in nature. Clearly, dynamic feedback is inherent to cortical–subcortical systems and must be vital in performing match–mismatch operations, which are best understood in the case of the motor system. The principle of feedback is fundamental to the organization, planning, and execu-

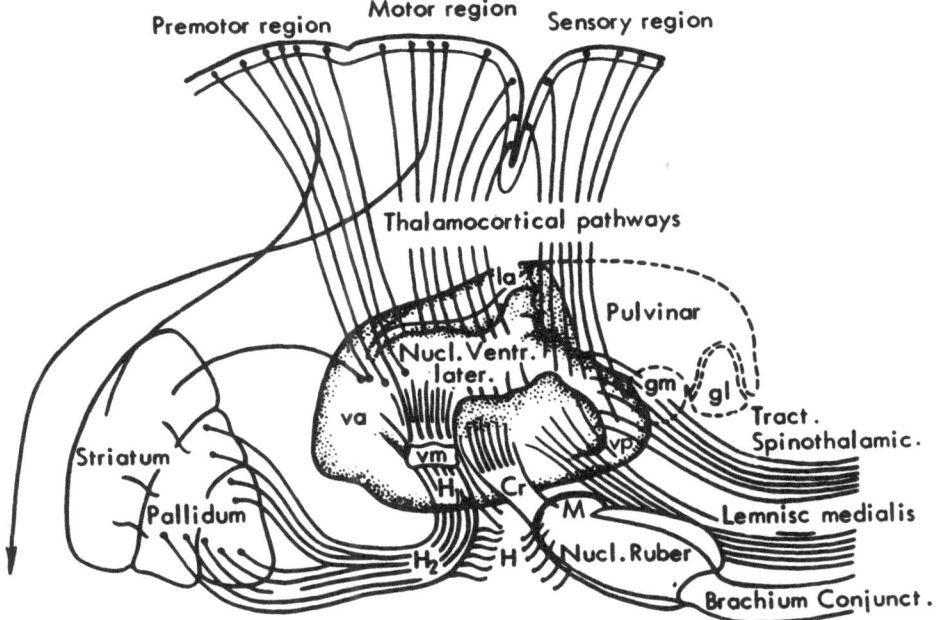

FIG. 1.15 Various afferent systems of the sensory and motor regions of the cortex, emphasizing thalamic relations. (From Luria, 1966.)

tion of voluntary motor movements since it is the progressive success of movements that modifies the actions performed. The physiological role of the motor cortex largely involves "matching" motor programs with analytical and integrative information, derived from the posterior sensory regions, about the progress of motor performance, that is, about the successes and errors of movement. A mismatch between expected motor output and actual movement leads to rapid adjustment of motor behavior to conform to expectations.

A similar principle of match–mismatch appears to operate in the sensory system as well. For instance, neuronal models of the environment, called *representational systems,* are compared and analyzed on a multidimensional basis in sensory systems.

The phylogenetic development of elaborately organized adaptive motor actions paralleled the development of the *tertiary motor* or *frontal cortex.* The perfection of the whole domain of voluntary, goal-directed acts is associated with the development of these tertiary fields (see Luria, 1966). In man, the frontal fields constitute approximately one quarter of the entire surface of the cortex. These tertiary fields involved the expansion anteriorly of the neural elements located in the motor and premotor regions. Just as the tertiary sensory regions are devoted to generalized and integrated forms of sensory perception, so the tertiary motor fields are involved in the highest level of goal-directed acts including complex sequencing, the creation of long- and short-term plans, and the internal manipulation of representational systems which is basic to abstract thinking.

The anatomic interconnections of the frontal cortex are extensive. They involve massive cortical–cortical association fibers, particularly with posterior cortex, considerable thalamic and basal ganglia interconnections, as well as connections with the *limbic* or "emotive" subcortex. The latter set of connections illustrates the crucial position of the frontal cortex in the hierarchical control of behavior, emphasizing the intimate relations of this region with the structures mediating motivational and primal feeling states of man. The extensive connections of the frontal cortex with subcortical and cortical sensory, motor, and limbic systems would seem to constitute an essential anatomic substrate for functional unification of the higher mental processes.

D. The Organization of the Limbic System

The *limbic system* is comprised largely of phylogenetically older brain structures (extending inferiorly from the hypothalamus and superiorly to the mediobasal portions of the neocortex.) A diagrammatic illustration of the mediobasal–subcortical connections is shown in Fig. 1.16. The mediobasal zone is made up of

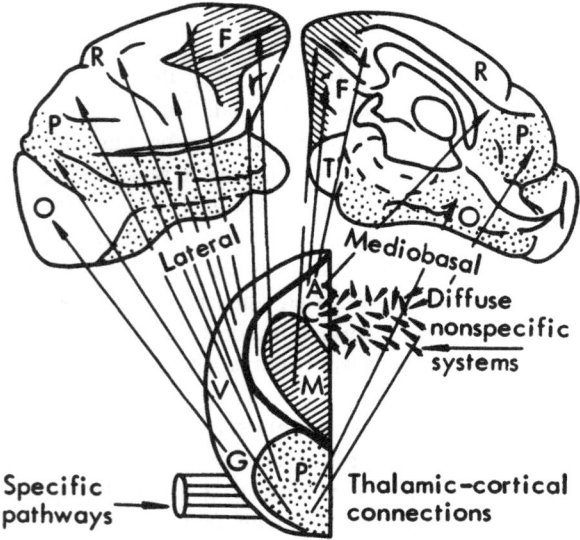

FIG. 1.16 Connections of the lateral and mediobasal regions of cortex. (M) Medial nucleus of the thalamus; (V) ventral lateral nucleus; (P) pulvinar; (G) geniculate bodies; (O) occipital regions; (T) temporal cortex; (R) central cortical region; (F) frontal cortical region. (From Luria, 1966; after Pribram, 1971.)

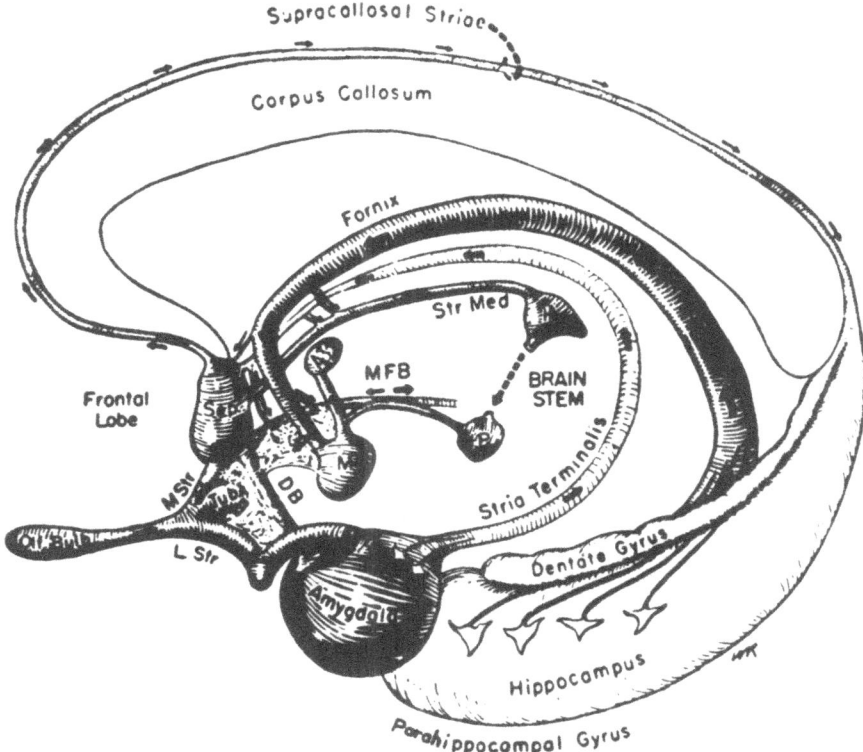

FIG. 1.17 Principal connections of the limbic system. MStr, LStr, medial and lateral olfactory striae; Str Med, stria medullaris; Tub, olfactory tubercles; DB, diagonal band of Broca; Sep, septum; AT, anterior nucleus of the thalamus; M, mammillary body; H, habenula; P, interpeduncular nucleus; MFB, medial forebrain bundle. (From MacLean, 1949.)

two parts: (1) a neocortical zone comprising the insular cortex, the medial temporal cortex, and the basal frontal cortex (Luria, 1966), and (2) the phylogenetically older cortical regions—the archicortex, paleocortex, intermediate hippocampal cortex, and the adjacent entorhinal cortex. Some of the subcortical structures which are part of the limbic system are shown in Fig. 1.17. The subcortical structures include hypothalamic, septal, thalamic, and basal ganglia (amygdaloid complex) contributions.

Recent clinical and physiological investigations have shown that the limbic system is involved in many different activities, the most notable being motivational control, autonomic or "visceral" regulation, the elicitation of feeling states, sensory selection or attention processes, and individual as well as species preservation (see Chapter 5). Phenomena such as positive and negative reinforcement can be elicited by electrical stimulation of widespread regions of the limbic system (Olds, 1956). Marked changes in affective state, involving placidity or tameness in some cases and rage in others, follow lesions of particular areas of the system (Goddard, 1964). Epileptic patients with foci located in the temporal lobes and limbic structures frequently experience a preepileptic warning or "aura," heralded by strong emotional changes. For instance, epileptic auras that precede a seizure may involve feelings of conviction, absolute truth, deja vu, strangeness, or other subtle feelings, such as in the case of Dostoyevsky who experienced "a sensation of existence in the most intense degree" (MacLean, 1970). These reports indicate that the most significant and personal of man's internal states are related to activity in the limbic lobes.

As discussed in Chapter 5, a major principle of operation in the limbic system involves the dynamic balance of opposites. Interrelated septal, amygdaloid, hypothalamic, hippocampal, and cingulate cortex processes form a multipolar system involving dynamically balanced operations. The motivational forces driving animals toward or away from stimuli are generated in this system. The strong reciprocal connections with tertiary frontal cortex give rise to emotionally laden goal-oriented behaviors. The reciprocal interaction between the primitive subcortical nuclei and the frontal and mediobasal cortex allows the regulation and evaluation of internal affective states and the integration of these states with events taking place in the external environment.

While the foregoing material gives far from a complete description of the structure of the central nervous system, it has not been our purpose to do so. Numerous excellent texts, cited at the beginning of this section, can be consulted by the reader who wishes more detail. However, what we have provided is a picture of the salient *organizational* features of the most important functional anatomic systems of the brain. That picture imparts the fundamental knowledge of structure, function, and terminology necessary for understanding the material we now wish to present. Additional anatomic detail will be added as it becomes relevant.

III. NEUROCHEMICAL NEUROANATOMY

The development of fluorescent histochemical techniques (Dahlström & Fuxe, 1965) created an important new discipline called *neurochemical neuroanatomy*. This discipline is based upon the fact that proposed transmitter substances, such as noradrenalin (norepinephrine), dopamine, and serotonin are manufactured in the soma of neurons and then transported down the axon. These transmitter substances, after they selectively bond to certain dyes, fluoresce with characteristic colors when examined under ultraviolet light. Examination of properly prepared slices of brain tissue in the fluorescence microscope permits identification of the transmitter substances in particular anatomic regions. For example, norepinephrine produces a yellow color, serotonin a green color, and so on. Further, the fluorescence technique permits tracing of complex axonal outflow trajectories across large domains of tissue. The major advantages of this technique over those used by classical neuroanatomists are that it does not rely on lesions or axonal degeneration, nor does it depend upon stains which permit visualization of only a small fraction of all the cells, but instead labels, in three dimensions, what appear to be functional systems as defined by a shared transmitter substance.

The transmitters initially examined with this technique were biogenic amines. In general, biogenic amines are made up of catecholamines (i.e., dopamine, epinephrine, and norepinephrine) and serotonin (5-hydroxytryptamine). These chemicals are implicated in the control of mood and emotion. Alterations of selected amines either by increasing or decreasing their concentration or by altering the substrate on which they act can result in depression, elation, sleep, dreaming, tranquilization, psychic energization, mood elevation, and a multitude of other effects and combinations of such effects now well known to the psychiatric profession.

For instance, one of the first tranquilizers to be discovered was reserpine which depletes the effective concentration of biogenic amines within the hypothalamic–limbic substrate. Iproniazid, a mood elevator, acts by inhibiting an enzyme (monoamine oxidase) that destroys serotonin and dopamine. Amphetamines act by releasing norepinephrine at physiologically active sites (Cooper, Bloom, & Roth, 1974; Kety, 1967). Other popular drugs such as imipramine (a psychic energizer) act by preventing reuptake of serotonin. Cocaine, like morphine, inhibits the turnover rate of serotonin in forebrain structures, but not in the midbrain (Costa, Grappetti, & Revuelta, 1971; Costa & Revuelta, 1972; Knapp & Mandell, 1972). Analyses of serotonergic nerve endings in the septal nucleus show that while cocaine and morphine have a similar effect on serotonin, their mechanisms of action are completely different (Knapp & Mandell, 1972). Morphine causes a short-term but immediate decrease in the activity of tryptophan hydroxylase, the rate-limiting enzyme in the biosynthesis of serotonin. At the same time, morphine has no effect on the uptake of trytophan, the serotonin precursor. Conversely, cocaine inhibits the uptake of tryptophan, but has no effect on tryptophan hydroxylase activity. Cocaine and heroin have distinctly different psychological effects. These differences could be due to the fact that the drugs act on different receptive surfaces in different systems of the brain (Jacquet & Lajtha, 1973) or because they affect complex equilibrium processes in which serotonin is not a differentiating member. These drugs also have multiple effects. Another effect of cocaine is the potentiation of adrenalin by inhibiting its reuptake (Kopin, 1967).

There are three different but interrelated biogenic amines, each distributed in somewhat different regions of the brain. These amines are *dopamine, noradrenalin,* and *serotonin.* Catabolic and anabolic mechanisms related to these chemicals

TABLE 1.1
Summary of Amine Pathways[a]

A. Dopaminergic pathways
   1. Substantia nigra to caudate and putamen (corpus striatum)
   2. Midbrain to nucleus accumbens and tuberculum olfactorium

B. Noradrenergic pathways
   1. Reticular formation to neocortex
   2. Reticular formation to hypothalamus and limbic forebrain

C. Serotonergic pathways
   1. Raphe nuclei to hypothalamus and limbic forebrain
   2. Raphe nuclei to neocortex

[a]After McGreer (1971).

have been discussed in many sources (Cooper *et al.*, 1974; Kety, 1967, 1970; Kopin, 1967; Stein, 1968).

The respective amine anatomic pathways are represented in Table 1.1 and Figs. 1.18 and 1.19. The biogenic amines that are implicated in the control of mood (noradrenalin and serotonin) are located primarily in the limbic system, frontal

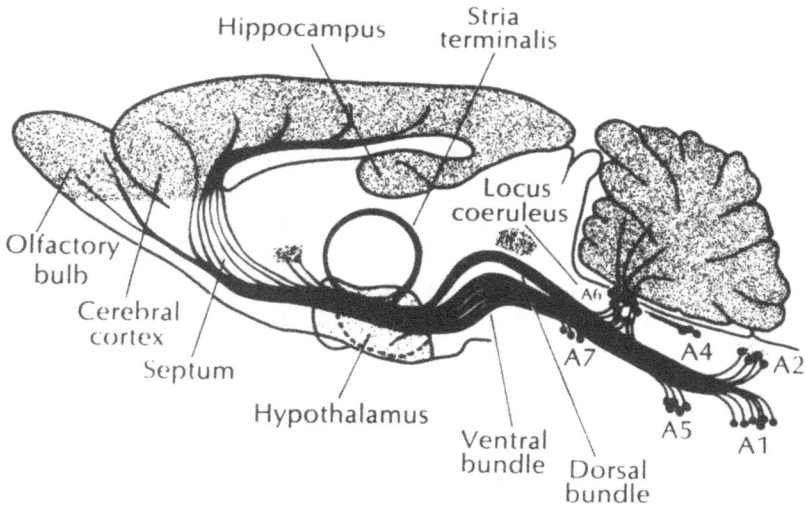

FIG. 1.18 Schematic diagram showing the distribution of the main ascending neuronal pathways containing norephinephrine. The strippled regions show the major nerve terminal areas. A1 and A2, caudal reticular formation fiber groups; A4, dorsal reticular fiber groups, A5, ventral reticular fiber groups; A6, locus coeruleus; A7, rostral reticular fiber groups. For details of anatomic nomenclature see Dahlstrom and Fuxe (1965). (From Cooper *et al.*, 1974; after Ungerstedt, 1971.)

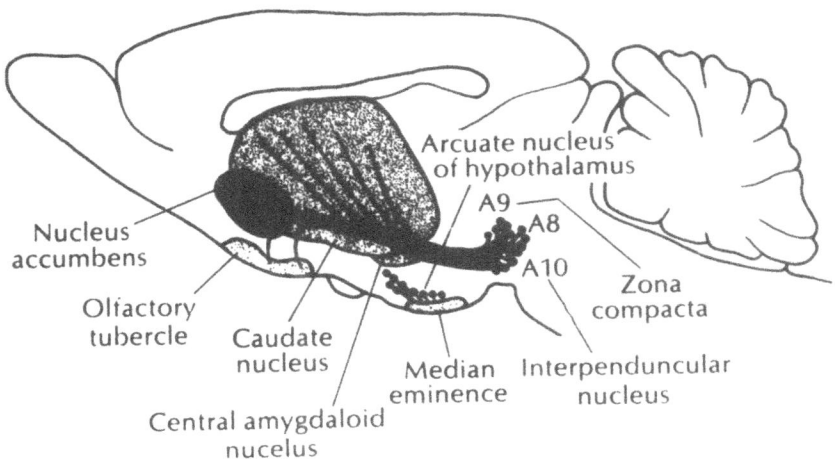

Nucleus accumbens

Olfactory tubercle

Caudate nucleus

Central amygdaloid nucelus

Median eminence

Arcuate nucleus of hypothalamus

A9

A8

A10

Interpenduncular nucleus

Zona compacta

FIG. 1.19 Schematic diagram illustrating the distribution of main ascending pathways containing dopamine. A10, cell groups near nucleus interpenduncularis; A9, substantia nigra; A8, cell groups caudal to A9 and dorsal to the medial lemnisius. Anatomic nomenclature according to Dahlstrom and Fuxe (1965). (From Cooper *et al.*, 1974; after Ungerstedt, 1971.)

and temporal cortex, and the pontine tegmental reticular formation. A recent study has traced noradrenalin manufactured in neural cell bodies located in the tegmental reticular formation (locus coeruleus) to the hippocampus, hypothalamus, thalamus, septum, and frontotemporal cortex (Moore, Jones, & Halaris, 1973).

A second and separate biogenic amine system (dopamine), implicated in the regulation of motor movement is located in the caudate nucleus, putamen, and substantia nigra. Dopamine is manufactured largely in the substantia nigra and transported to the other structures (see Fig. 1.19). It has been found that damage to dopaminergic synaptic endings in the extrapyramidal motor system causes Parkinson's disease.

The third biogenic amine system (serotonin) is located nearly contiguously with the noradrenalin system. This chemical is manufactured in the medullary and pontine reticular formation and transported to the hypothalamus, cortex, and limbic system. A close interplay of serotonergic and noradrenergic systems in the brain stem and diencephalon is responsible for the regulation of mood and also for the control of the sleep–wakefulness cycle.

A fourth chemical transmitter which is not an amine is acetylcholine. This chemical is located throughout the neuroaxis with heavy concentrations in the cortex, corpus striatum, and limbic system. It has been linked specifically with Parkinson's disease, food and water intake, and aggression (McGreer, 1971). In the latter instance, it has been shown that deposition of cholinergic agents in the hypothalamus (Smith, King, & Hoebel, 1970) and septal–amygdala areas of rats (Igic, Stern, & Basagic, 1970) elicits killing behavior.

Thus, we see that neurons are differentiated not only by their micromorphology and anatomic connectivity, but also by the transmitter substances to which they respond and which they produce. Yet, in spite of all this diversity of structure, location, and chemistry, neurons share the properties of (1) integrating afferent influences which alter the resting membrane potential at synapses, and (2) generating propagating action potentials when activation thresholds are exceeded. Informational transactions, whatever their structural and chemical fine structure, share common electrical indices: postsynaptic potentials are relatively slow electrical processes while action potentials are relatively fast. In succeeding chapters we shall rely heavily upon these shared electrical indices for our analysis of how the brain mediates various functions.

2
Functional Electrophysiology

I.   INTRODUCTION

In the preceding chapter we presented an overview of the brain: the compartments into which it is divided, the structure of the neuron, the mechanisms by which a neuron receives and transmits impulses, the anatomic organization of the major functional systems, and the anatomic distribution of various neurochemical transmitters. While there will be many occasions in which the discussion will be conducted on one or another of these levels, our analysis of the mechanisms by which the brain accomplishes the functions that are the focus of this book will rely most heavily on electrophysiological evidence. This is because only electrophysiological techniques permit direct observation of the dynamic transactions between neurons and the coordinated processes which produce organized patterns of activity within and between cell populations. This, in our view, is the essence of information processing, of behavior, and of subjective experience. Thus, before we can proceed with our analysis of such functions it is obligatory that we examine thoroughly what is known of the genesis of the electrical activity which is one of the most unique characteristics of nervous tissue.

That important insights into brain function can be achieved through the analysis of EEG seems more probable today than ever before. Four decades have passed since the initial report on brain waves in humans (Berger, 1929). During this time a variety of theories were generated to explain EEG and high expectations were generated about the valuable new information which it would provide. As these theories were discredited by experimental observations, and as the limitations became apparent of what could be learned by visual analysis of the scalp recorded EEG, those high expectations were replaced by disappointment. Many researchers concluded that little information about brain function could be derived from the study of slow waves, and turned instead to the study of the activity of single nerve cells examined with the aid of very fine microelectrodes. With the development of the capillary microelectrode and the advent of intracellular recording techniques, the pendulum of neurophysiological interest swung even further toward the single cell. The last 15 years have seen the introduction of methods of long-term or chronic implantation of electrodes, and the development of the special purpose average response computer and the dedicated laboratory minicomputer. Most of all, there has come the gradual realization that the probability of understanding complex integrative functions by intensive observation of single components of an enormously complex system was extremely low, notwithstanding the impressive contributions to our understanding of sensory and motor processes which these methods have yielded. As the new tools were applied to the analysis of the EEG and the average sensory evoked potential, important new information about brain function and especially higher integrative functions was obtained, as we shall see. The pendulum has paused. Hopefully, it will not swing back to a preoccupation with slow waves and a neglect of unitary processes, but research in this field will strike a balance, with investigations proceeding on both levels of analysis so we achieve an understanding of how single cells can participate in cooperative processes.

The experimental data which have been obtained in this period provide a broad general outline of the genesis of the EEG. Although our understanding of this complex phenomenon is still developing, nevertheless considerable progress has been achieved. Once again hopes are raised that subtle psychological processes, including thought processes, can be brought under direct scrutiny. It is particularly important for neuroscientists to gain an understanding of the neural basis of the EEG and the evoked potential, since a variety of practical clinical applications for the diagnosis of functional and organic pathology are already possible and undoubtedly will become more prevalent in the future. Such applications are discussed in detail in Volume 2 of this work.

II. THE GENESIS OF EEG

This chapter, then, is a review of knowledge regarding the neural basis of EEG. A broad overview, both historical and functional, is required to acquaint the reader with the various processes which contribute to neural electrical activity. It is important to understand the role of the extracellular space as the medium through which electrical currents flow and interact. It is also important to comprehend, in a quasiquantitative manner, the relative contribution made by each of the different types of generators since emphasis must be placed squarely on one (synaptic potentials) without detracting from the significance of the others (afterpotentials, action potentials, and intrinsic oscillations). A thorough understanding of the genesis of EEG is all the more important since the possibility exists that currents related to EEG voltages, flowing in the extracellular space, can "feedback" onto neurons and alter their excitability. If indeed this does happen, there may be a substantial class of cooperative neural interactions which depend to an important extent upon such nonsynaptic influences. These hypothetical influences may be particularly relevant to integrative functions (see Section V.B, Chapter 12).

A. Unitary Sources of EEG

Early theories of the electrogenesis of EEG postulated that slow surface cortical waves arise from the summation of sequential action potentials generated in the neuropile below (Adrian & Matthews, 1934). It was known at that time that synchronous action potentials occurring in neighboring cells could produce an EEG spike 20 to 100 msec in duration. It was not unreasonable to propose that a sequence of neural firing could produce a summated envelope, that is, a wave of long duration. No mechanism for coordinating such neural discharge was postulated nor was fundamental knowledge, such as the synchronizing role of the thalamus or the occurrence of dendritic synaptic potentials, available at this time. For these reasons, it is remarkable that the envelope of spikes theory was disproven as rapidly as it was.

Two years after Adrian and Matthews' model, R. Gerard (1936) observed the persistence of wave activity in small fragments of brain tissue. If one adopted the spike envelope hypothesis, these fragments were too small to account for the amplitude of the slow waves. Gerard (1936) concluded that the waves were produced from within individual neurons and were, to some extent, independent of spike discharge. However, in the 1930s spikes were thought to be the cardinal activity of brain cells and Gerard's proposal, as well as his later findings of slow waves occurring in the absence of action potentials (Gerard & Libet, 1940), went unheeded. It was not until the advent of the micropipette which allowed intracellular impalement, that the spike envelope hypothesis was definitively discarded.

B. Intraneuronal Slow Waves

Intracellular recordings from neurons have long dominated the field of neurophysiology, beginning in the middle 1940s and early 1950s, and today still constitute a very powerful tool (Purpura, 1959, 1972). Figure 2.1 shows an intracellular record from a cortical neuron during different brain states and provides a direct comparison between intracellular slow waves and the EEG. Figures 2.2 and 2.3 show superimposed intracellular records that correlate with simultaneous waves from the cortical surface EEG.

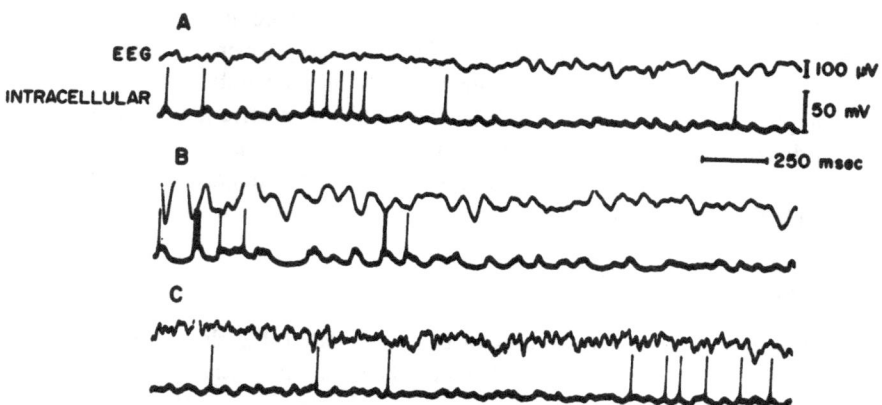

FIG. 2.1 Neuronal wave activity recorded with an intracellular microelectrode. Animal awake (A), sleeping (B), and intensely aroused (C), with EEG patterns characteristic of each of these states. Note corresponding changes in form of the neuronal waves. Recordings are from same cell in posterior suprasylvian cortex, 750-μ depth, EEG taken from anterior suprasylvian cortex on contralateral hemisphere. (From Elul, 1968.)

Renshaw *et al.* (1940) were the first to establish, using micropipettes, the existence of two fundamentally different types of neural potentials: brief, all-or-none action potentials and slow, graded potentials later attributed to the activity of synapses (for a comprehensive review of the early literature, see Purpura, 1959). Through intracellular analyses, it became immediately apparent that spikes are distinguishable from slow waves and that the intracellular slow waves bear a marked resemblance to the gross EEG. Spectral analyses of the intracellular slow wave record and the surface EEG during different brain states reveal a nearly identical frequency content (Elul, 1968) which changes concordantly with brain state as shown in Fig. 2.1. Intracellular analyses from hippocampal neurons during the appearance of theta waves in the EEG reveal waves that resemble the hippocampal EEG to such an extent that Fujita and Sato (1964) termed such activity "Intracellular theta rhythm." The major difference between slow waves (EEG) recorded intracellularly and extracellularly is that the amplitude of the intracellular waves is 100 to 500 times *greater* than those recorded outside the cell. This indicates that the sources of the extracellular slow waves must be intracellular generators.

The intracellular studies established the unitary cellular origin of the EEG. However, questions remained as to exactly what these intracellular potentials were and how they were organized to create the dynamic features of the EEG. A further, more specific question was: "Are there additional sources of EEG, that is, exactly to what extent do action potentials, or afterpotentials or glial cells contribute to the EEG?"

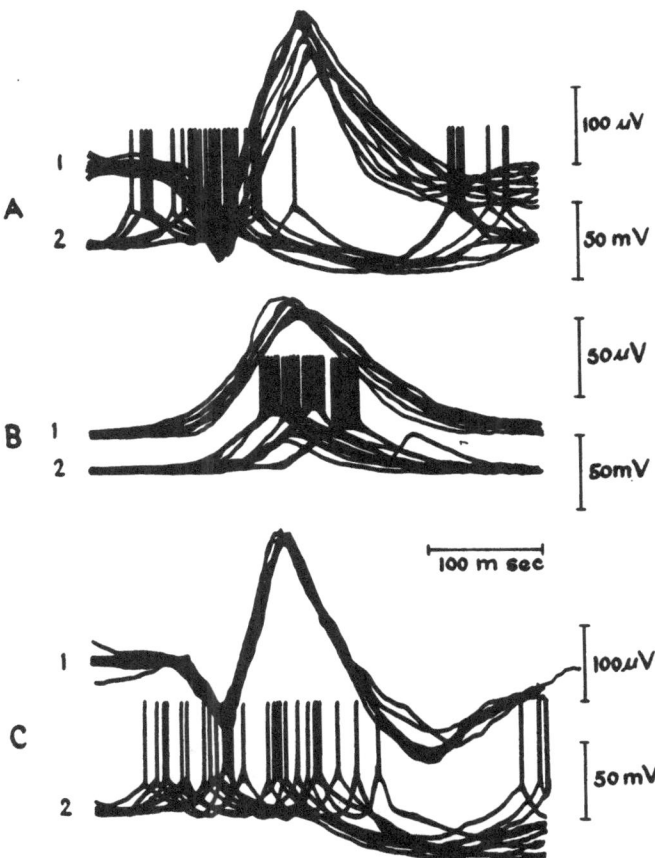

FIG. 2.2 Superimposed line drawings of coherent wave shapes (coherence judged by visual inspection) extracted from the surface record during spontaneous alpha bursts (channel 1), together with intracellular unit record (channel 2) corresponding to the EEG segment selected. Ten to twenty superpositions in each case. (A), (B), and (C) are three different cells. Negativity is up for channel 1 (surface record) and down for channel 2 (intracellular record). (From Morrell, 1967.)

C. Contributions of Glia Cells

Intracellular analyses of glial cells fail to reveal slow wave oscillations in the EEG frequency range (Hild & Tasaki, 1962; Kuffler & Potter, 1964; Grossman, Whiteside, & Hampton, 1969). Also, glial cells do not appear in large numbers in the neocortex of kittens until the third week of postnatal development (Altman, 1967; Brizzee, Vogt, & Kharetchko, 1964), although EEG is present during the

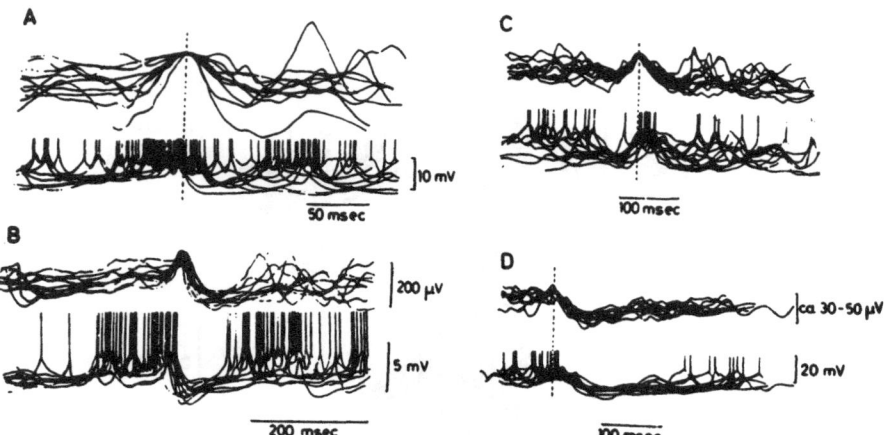

FIG. 2.3 Superimposed line drawings of two types of EEG waves (upper traces) and the corresponding intracellular records from pyramidal cells (lower traces). The EEG waves were collected from two experiments according to their shape. (A) and (C) mainly show surface negative symmetric waves that correspond to pure EPSPs with cellular discharges. (B) and (D), negative–positive waves. The surface negativity coincides with an EPSP and the surface positive wave to the following IPSP. (From Creutzfeldt *et al.*, 1966b.)

first week of life (Purpura *et al.*, 1964). This shows that glial cells are not significant electrocontributors to the EEG. Glial cells more likely contribute to DC steady potentials (and the *contingent negative variation,* or *CNV*; Walter, Cooper, McCallum, & Cohen, 1965) since the time course of glial cell depolarization following uptake of extracellular potassium involves several seconds (Kuffler & Potter, 1964; Kuffler, Nicholls, & Orkard, 1966) and is dependent upon synchronous and prolonged discharge of neurons (Karahashi & Goldring, 1966). It has also been demonstrated that in kittens cortical DC potentials are not well developed until the third week of postnatal life, which is coincident with glial cell development (Purpura *et al.*, 1964).

D. Contributions From Action Potentials

The contribution of spike discharges to the EEG is also minor. This is particularly true when neuron activity is desynchronized. The biophysical properties of brain tissue provide strong capacitative effects (Elul, 1972) so that the fields from rapid transients, such as spikes, diminish sharply with distance from the source. Action potentials, in contrast to slow waves, show a significant increase in amplitude as a microeletrode is moved through tissue and approaches an active neuron.

Slow waves, on the other hand, exhibit a relative constancy of amplitude throughout large domains of tissue, with abrupt increases in amplitude occurring only upon penetration of the neuron (approximately 100- to 500-fold). This biophysical distinction between spikes and slow waves can be explained if it is assumed that: (a) the capacitative and resistive properties of the extracellular medium act as a low-pass filter (Humphrey, 1968; Pollen, 1969), and (b) the extracellular medium is nonhomogeneous to slow waves; that is, the extracellular mucopolysaccarides and mucoproteins provide a nonhomogeneous medium such that the amplitude of slow wave sources do not decrement with distance.

This second point is emphasized by the fact that EEG voltage follows Ohms law and is equal to the product of the resistance of the extracellular space and the local current density (i.e., $V = R \times I$). The fact that EEG voltage does not decrement means that this product, RI, is constant throughout large regions of tissue (see Elul, 1972).

In support of this point is a recent study by Abraham, Bryant, Mettler, Bergerson, Moore, Maderdrut, Gardiner, Walter, and Jennrich (1973). These workers analyzed the relative contribution of EEG generators near a recording electrode versus the contribution by generators great distances away. It was found that 20 to 50% of the EEG is produced by generators near the electrode. This determination was not based on measurement of overall amplitude, since the EEG amplitude was indeed constant throughout large domains of tissue, but rather it was based on measures of the coherence between frequencies of EEG at different points, which reflect synchronized behavior. Details about this aspect of EEG will be discussed in later sections.

E. Contributions by Summated Synaptic Potentials

It has been shown that EEG arises from slow waves generated within individual neurons. The question remains: "Exactly what produces intracellular slow waves?" The obvious candidates are synaptic potentials. Indeed, there is little doubt that synaptic potentials are a primary contributor to intracellular slow waves. This has been demonstrated in studies as those of Fujita and Sato (1964) who injected currents through intracellular electrodes and observed conductance changes caused by synaptic activity occurring near the tip of their electrodes (see Smith, Wuerker, & Frank, 1967 for details of this technique) during specific phases of the EEG. Similar studies, using injected currents to detect synaptic conductance changes, also demonstrate the contribution of synaptic potentials to the intracellular slow waves (Feldman & Purpura, 1970).

Chang (1951, 1952) studied what he termed "dendritic potentials" from neocortical systems. The dendritic potentials were slow waves produced by stimulating afferents running in the molecular layer and known to terminate on the dendrites of cortical pyramidal cells. Recent intracellular analyses from immature cortical neurons have demonstrated conclusively that dendritic slow waves are synaptic in origin and make a major contribution to the EEG (Purpura et al., 1965).

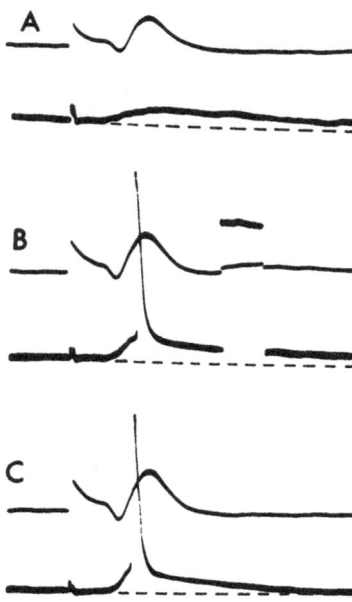

FIG. 2.4 Prolonged EPSPs (80–100 msec) evoked by ventrolateral thalamic stimulation in a sensory-motor cortex neuron from a 6-day-old kitten. Upper channel records indicate cortical surface activity, negativity upward. Amplitudes of cortical evoked responses ranged from 100–300 μV in this study. (A) Weak stimulation elicits an 18- to 20-msec latency EPSP with slow rise time and prolonged declining phase. (B) and (C) Stronger stimulation decreases the rise time and increases the amplitude of the EPSP. Cell discharge is secured at the crest of the EPSP. Calibration in (B) 50 mV; 20 msec. (From Purpura et al., 1965.)

The latter study was conducted on 2-week-old kittens in which basal dendritic and somatic synapses are not developed, with the majority of synapses occurring only on apical dendrites (Purpura et al., 1964). At this age, EEG is present and synaptic activity arising from apical dendrites, observed intracellularly, is capable of producing rather large extracellular field potentials (Purpura et al., 1965). Figure 2.4 shows an example of intracellular activity recorded from immature cortical pyramidal cells and the corresponding surface field potentials.

F. Contributions by Afterpotentials

Investigations using microelectrodes in fiber tracts have demonstrated rather large extracellular EEG-like slow waves (Verzeano, 1972). For instance, recordings from the internal capsule, pyramidal tract (Verzeano, 1955) and the optic tract and optic nerve (Doty & Kimura, 1963; Laufer & Verzeano, 1967) reveal slow waves with frequencies ranging between 3 and 55 Hz. Figure 2.5 shows a microelectrode record obtained from the optic tract of the cat. As Verzeano

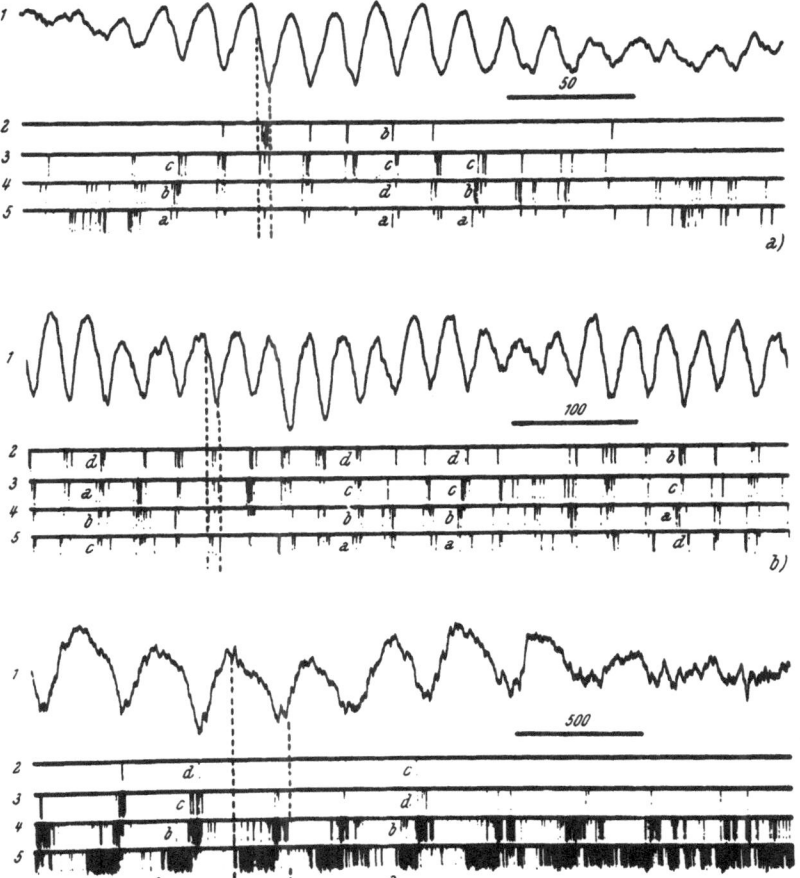

FIG. 2.5 Phase relations between gross waves and axonal action potentials in the optic tract. (A) Tracing 1 shows oscillations at a frequency of 55 per sec, recorded by means of a gross electrode in the optic tract of the cat, under maintained illumination. Tracings 2–5 show the pulses corresponding to the axonal action potentials, simultaneously recorded from the same location and separated by an amplitude discriminator into four amplitude ranges decreasing in equal steps, from 2 to 5. Thus, action potentials in tracing 2 occurred near the tip of the electrode, action potentials in tracing 5 occurred far from it. One vertical dotted line coincides with the first action potential of a group of action potentials, the other with the last action potential of the same group; this illustrates the phase relations between the action potentials and oscillations. The action potentials always occur during the initial part of the positive phase. The letters (a–d) illustrate, for some cases, the successive appearance of the groups of action potentials in the different amplitude ranges, indicating activity approaching or moving away from the tip of the electrode. (B) Similar relations, recorded after cessation of illumination, when the frequency of the oscillations has decreased to 35 per sec. (C) Similar relations, recorded a few minutes later, still in darkness, when the frequency of the oscillations has decreased to 3–6 per sec. Time in msec. (From Laufer & Verzeano, 1967.)

(1972) points out, these waves cannot be explained in terms of electrotonic spread or by field effects resulting from cortical activity. Note the phase relations between spike bursts and slow waves in Fig. 2.5. This suggests the two phenomena (spikes and slow waves) are closely linked and generated at the recording point. These data also show that the slow waves are not electrotonically transmitted down the axons from the cell bodies or soma located in the retina.

The accumulated findings of EEG in fiber tracts make it clear that afterpotentials or some slow wave process, closely associated with axon spikes, contribute to the EEG. This is of considerable significance since the cerebral cortex contains a large number of fiber bundles that interconnect intracortical and subcortical systems.

G. Contributions by Intrinsic Membrane Oscillations

Finally, the possibility exists that oscillations which are, to some extent, independent of synaptic input, but intrinsic to neurons, contribute to the EEG. Intrinsic oscillations are a basic feature of the invertebrate nervous systems (Kandel *et al.*, 1969; Klee & Hess, 1969; Strumwasser, 1968; Waziri *et al.*, 1969). A detailed analysis of the neural basis of behavior modification (habituation in aplysia) shows that the fundamental control operation involves modulation (by excitation or inhibition) of neural intrinsic oscillations (Kandel *et al.*, 1969; Kupferman *et al.*, 1970).

A convincing demonstration of intrinsic oscillations involves recording from neurons with all synaptic connections removed. Such a method is comparatively simple in the case of invertebrate cells (Strumwasser, 1968; Klee & Hess, 1969) but methodologically very difficult with vertebrate neurons. Nonetheless, some inferential evidence exists which is suggestive of intrinsic oscillations in vertebrate neurons. For instance, Gerard and Libet (1940) applied nicotine to isolated cortical slab preparations in sufficient concentrations to obliterate all synaptic activity, and yet EEG oscillations persisted. Smith and Smith (1964, 1965) investigated the properties of spontaneously discharging neurons in the isolated cat forebrain. They found that the "interburst" process recorded from cells anywhere in the cortical mantle can be affected by stimulation (flashing lights, direct stimulation of the brain by electrical pulses, and depth depolarization), whereas intraburst processes, that is, the control of the interval between spikes within a burst, was "extremely nonlabile" and most likely endogenous. Recent studies by Schwindt and Calvin (1973) add further credence to the notion of intrinsic oscillations by showing that the frequency of intracellular oscillations in motor ventral horn cells can be altered systematically by injecting DC currents directly into the cell. Also, Fidone and Preston (1971) demonstrated that IPSPs reset oscillatory activity in Fusimotor neurons. The latter two studies indicate that modulation of oscillations, as in invertebrates, is probably an important and perhaps basic mode of neural control in higher organisms.

The last line of evidence to be mentioned relevant to intrinsic oscillations comes from biological clock studies in which diurnal and circadian rhythms are reflected in discharge frequencies of single cells (Lickey, 1969; Moore, 1974; Strumwasser, 1965, 1967, 1968).

To summarize: afterpotentials and possibly intrinsic oscillations contribute to the EEG. However, the primary contributions come from summated synaptic potentials arising on the dendrites and soma of neurons. The summated synaptic currents, generated by "functional synaptic units" or by synchronous activation of synaptic ensembles, travel throughout the brain in small channels within the extracellular space.

III. STATISTICAL ORGANIZATION OF EEG GENERATORS

It is now necessary to direct attention to the problem of cooperative behavior among neurons. The previous section argued for individual unitary generators of EEG, with synaptic potentials, afterpotentials, and, possibly, intrinsic membrane oscillations as the major contributors to these rhythmic waves. However, it is a well-established anatomic fact that innumerable connections exist between any given neuron and any other neuron chosen nearly anywhere within the neural axis (Lorente de Nó, 1938). It has been conservatively estimated that each cortical neuron is connected to 600 other neurons, most of which are in the close proximity of that cell (Cragg, 1967). The density of cortical synapses is estimated to vary between approximately 7,000 and 13,000 synapses per neuron (Cragg, 1967), which provides an estimate of the amount of axonal branching.

Data by Globus and Scheibel (1967) helped to quantify the degree of interconnectivity within small cellular domains. Spatial fields of influence of a given cortical cell onto cells of other classifications are shown in Fig. 2.6. The dendritic–somatic receptive domains differ geometrically from one cell type to another (see Fig. 2.6). These figures help to visualize the type and distribution of synaptic drive which a given cell receives and delivers, and illustrate the extent to which neurons are interconnected both within small cellular domains and between domains.

The extent and complexity of neuronal interconnectivity is responsible, in part, for the development of statistical descriptors of EEG phenomena. Theories have been formulated that maintain that neural population behavior is inherently statistical in nature. These theories are based on the fact that there is considerable variability in the behavior of individual neurons and that many of the interactions between individuals of a population are nonlinear in nature. The brain performs many diverse functions. The question is how do individual neurons cooperate within groups to mediate those various functions?

Because this subject is of considerable theoretical importance, three lines of argument will be discussed in regard to these points. The first level of argument is largely inductive and is epitomized by the positions of Hebb (1949) and

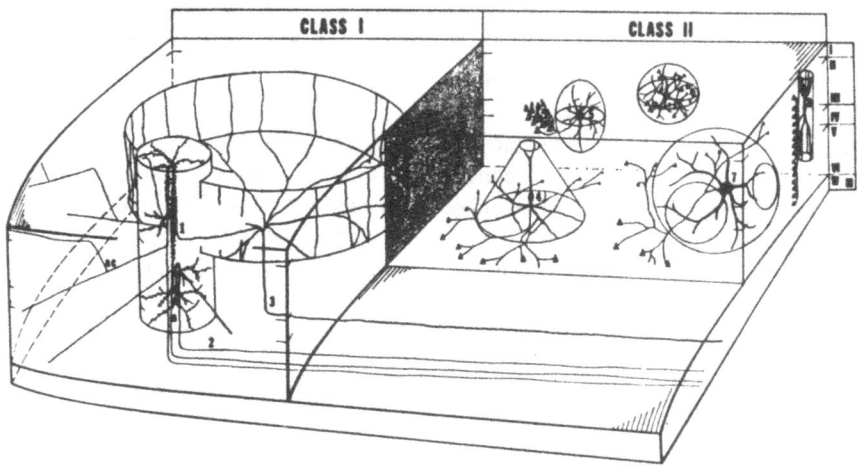

FIG. 2.6   A diagrammatic representation of a curved slab of cortex showing dendritic and axonic characteristics of the large class I neurons on the left, and smaller class II neurons on the right. "ac" is a collateral fiber connecting two columns Class I neurons have spines, a specific dendritic field, and axons that leave the cortex. Class II neurons have few or no spines, a variable dendritic field, and axons that do not leave the cortex. (1) The cylindrical dendritic module of cortical pyramidal cells; (2) descending axons leaving the cortex; (3) sensory afferent showing terminations on the apical dendrite of a pyramidal cell (1). The dendritic domains of class II neurons are described as a truncated cone in the case of a spindle cell (4); an ellipsoid for one stellate cell (5); a sphere for a second stellate (6); a large sphere for a third stellate (7); and a cylinder for a fusiform cell (8). The axonal aborizations of class II neurons synapsing on pyramidal cells (filled triangles) are drawn to represent the density and extent of branchings typical of these neurons. The various layers of the cortex are depicted on the right in roman numerals. (From Globus & Scheibel, 1967.)

Lashley (1942). Both workers agreed that CNS neurons are continuously active, exhibiting spontaneous discharges, and that information transmission must necessarily interact with this background activity. Both would agree that, at any instant in time, any individual neuron which was previously responsive to some particular sensory input may be refractory to that same input, or vice versa.

Statistical analyses of spontaneously active neurons show that the probability distributions of discharge are commonly Poisson in nature (Smith & Smith, 1964, 1965). This random background is the substrate on which information from the outside world is superimposed. Hebb (1949) argued that there must be parallel conduction of information such that on each of several parallel axons complete afferent information is available. In other words, spontaneous background activity contributes noise to the resolution of the total system but, nonetheless, at any moment the complete message is carried by some subset of parallel conducting axons. Lashley (1942) advocated a statistical theory, postulating that each axon carries only a part of the message and that the flow of information is distributed within networks whose elements fire in random order (see Freeman, 1972). According to Lashley, an observer cannot know what message was sent from a record of the pulse train from one neuron. He must repeat the message again and again until the averaged activity of the axon being observed converges to the behavior displayed by all the axons in the channel during a single transmission, or in other words, he must average across all axons simultaneously. Recently, John (1967a) has tried to integrate these two views by postulating that the information within an ensemble of cells consists of the deviation from randomness of the population's activity. In this formulation, the "noise" which Hebb recognized would interfere with transmission is dealt with by subtracting the random behavior of the population, while the distributed nature of the information postulated by Lashley is dealt with by averaging across the ensemble. Similar statistical theories of cooperative neural behavior have been offered by Freeman (1972a, 1975), Adey (1967), and Elul (1967b, 1969).

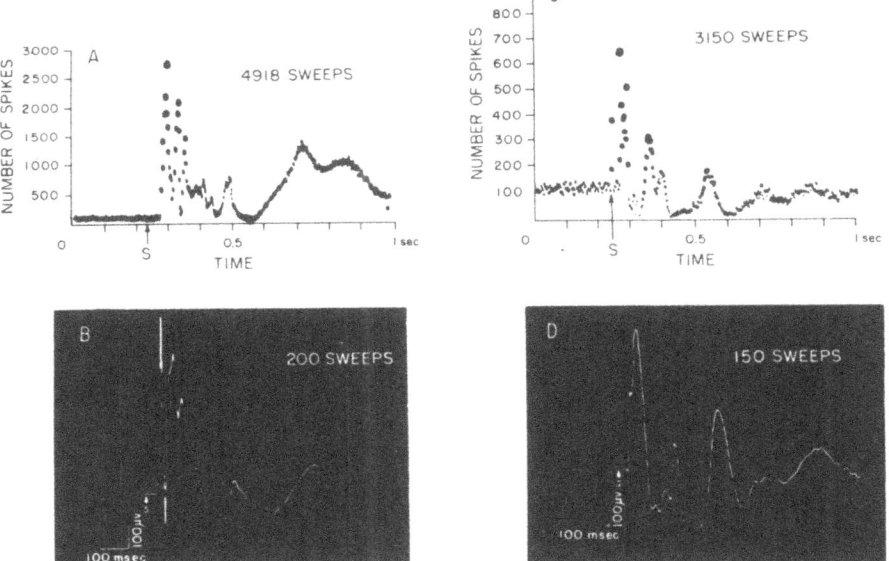

FIG. 2.7 Relation between probability of firing of a single cell and evoked potential waveform. (A) Frequency distribution of spikes from a single cell in the visual cortex of a cat after stimulation with 4918 flashes; (B) averaged evoked potential (200 oscilloscope sweeps) recorded from the same microelectrode, after cell death ($r = .60$ $p < .001$). Similarly, spike distribution for a single cell is shown in (C) (3150 sweeps) and the corresponding averaged evoked potential in (D) (150 sweeps) ($r = .51$; $p < .001$). Ordinate (for unit distributions): number of times the cell fired in response to light flash. Abscissa (for unit distributions): time, in 100-msec divisions. (From Fox & O'Brien, 1965. Copyright © 1965 by the American Association for the Advancement of Science.)

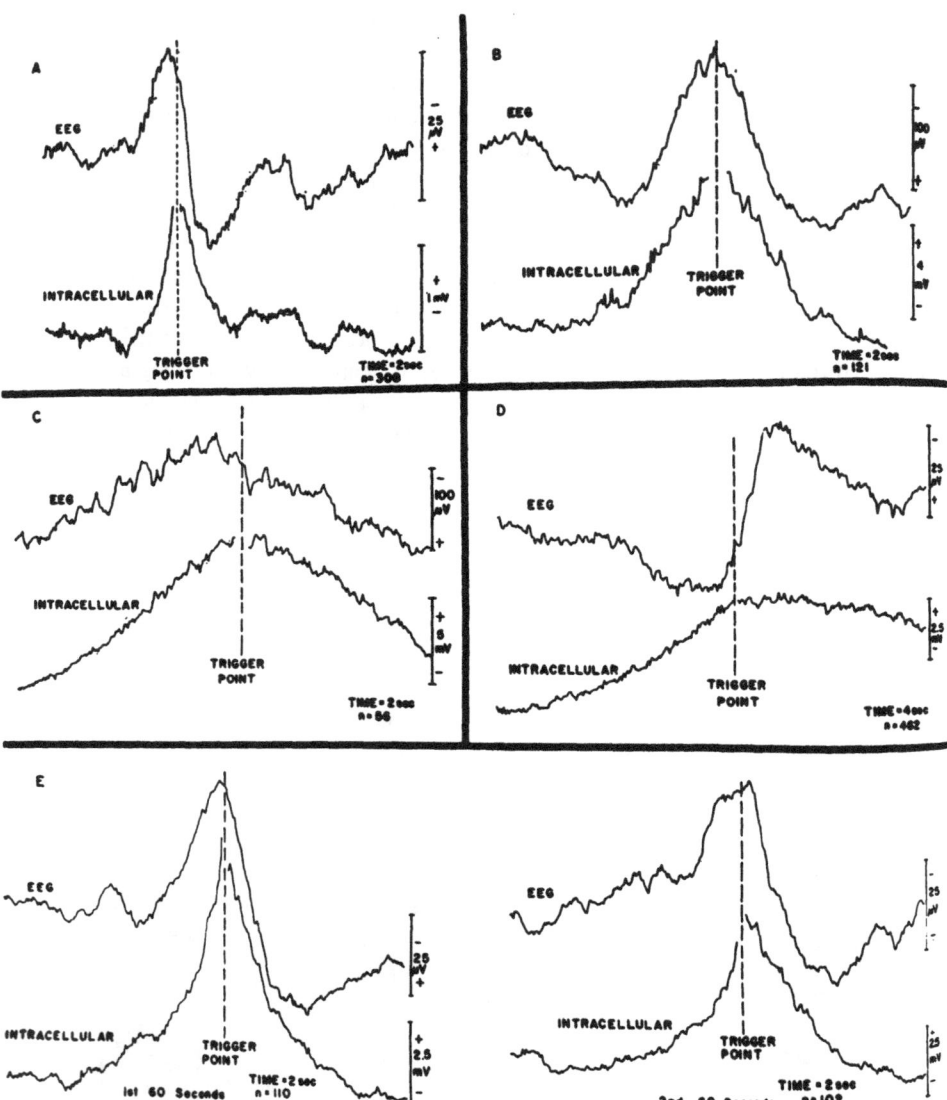

FIG. 2.8 Examples of averaged EEG and intracellular data timelocked to the spike discharges of different neurons (A–E). Total time of each trace is 2 sec. (E) The similarity of responses from the same neuron at different times. *N* equals the number of individual traces. The dashed line (trigger point) represents the time of spike discharge. All EEG records were obtained from the surface of isolated cortical slabs. The correlations between intracellular oscillations and surface EEG could not be seen by visual analysis and required electronic averaging before consistent relations were revealed. (From Frost, 1968.)

The second and more direct line of argument is based on studies that correlate single cell behavior with the slow wave EEG. The advent of the computer has contributed significantly to the quantification of this relationship. Fox and O'Brien (1965) were among the first to demonstrate a statistical relationship between the probability of discharge of a single cell and the slow wave response of the evoked potential. Frost (1968), Frost and Gol (1966), and Frost and Low (1967) were among the first to quantify similar relationships between the firing probability of single cells and the EEG. Figures 2.7 and 2.8 show some of the results of the Fox and O'Brien experiments and the investigations of Frost and co-workers. In Fig. 2.8, the probability of single cell discharge, plotted as a function of time following a flash of light (summed for 5000 flashes of light), is demonstrated to reproduce the waveshape of an evoked potential. In the Frost and Gol (1966) experiment, EEG waves that preceded and followed the occurrence of a spike were averaged. Figure 2.8 shows that spike discharge is correlated with certain phases of the EEG. Figure 2.9 demonstrates, using the same method as in Fig. 2.8, the correlation between the intracellular slow wave, spike discharge (dashed vertical line), and the surface EEG. The latter example further demonstrates the contribution of intracellular slow waves to the EEG discussed earlier. The relationships reproduced here are not readily apparent using visual inspection of the EEG, but require statistical methods. Further examples of statistical descriptions of the EEG (Elul, 1967a; Elul & Adey, 1965) and EEG-unit statistical relationships (Fox & Norman, 1968; John & Morgades, 1969b; Morrell, 1967; Verzeano et al., (1965) followed these initial observations.

The theoretical interpretation of EEG-unit statistical relationships was further illuminated by John and Morgades (1969b). Based upon the conflict between Hebb and Lashley discussed earlier, the following interpretation of the Fox and O'Brien (1965) data arose. The Fox and O'Brien study showed that the response of a single cell to thousands of light flashes approximated the shape of an averaged evoked potential. Studies by Lorente de Nó (1938) and others (see Scheibel & Scheibel, 1967b, 1970) have shown that neurons are anatomically organized in loops. As will be discussed in later sections (Sections IV, V, and VI) and in Chapter 7, the functional organization of the brain involves a complex network of loops. From this viewpoint, the Fox and O'Brien finding that related single responses to the averaged evoked potential can be explained if it is assumed that the neuron under observation is a "nodal" neuron, that is, a neuron shared by different sized neural loops. Data by Verzeano and co-workers (see Verzeano, 1972) have indicated that a given neuron may be a member of one loop at one instant and a member of a different loop at the next instant. If the loops are of different sizes involving different transit times, then a single light flash will activate a complex subset of loops, and the nodal neuron will exhibit a particular set of discharge latencies. The next light flash will activate a somewhat different subset of neurons and the nodal cell will exhibit a different

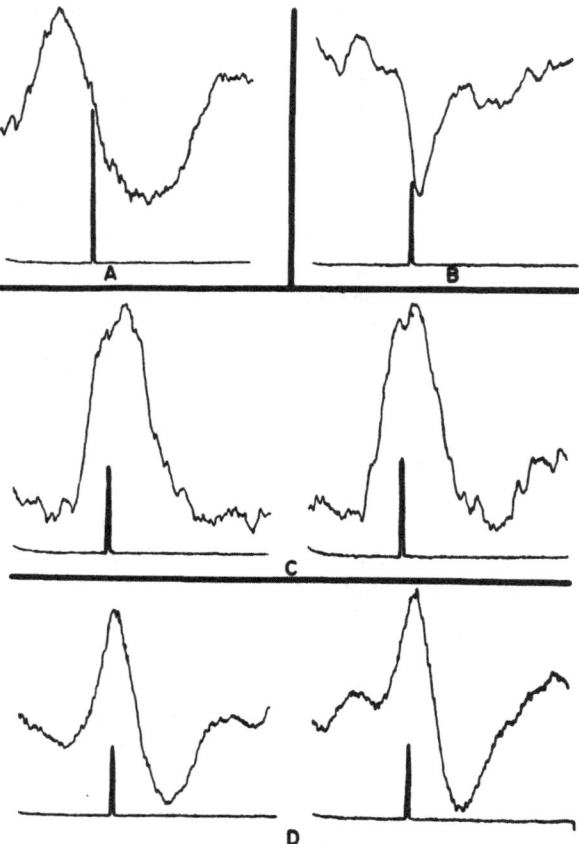

FIG. 2.9 Additional examples of averaged EEG waveforms showing a relationship between spike discharge (vertical arrow) and surface slow wave EEG as revealed by electronic averaging. The total duration of each average is 2 sec. All records are from isolated cortex. (From Frost & Gol, 1966.)

set of latencies. Using thousands of light flashes, a family of cellular response latencies will be generated which represent the mean or averaged oscillation of all subsets of neurons activated by the flashes. Thus, in the work of Fox and O'Brien, the responses of one cell to thousands of flashes is equivalent to ensembles of thousands of cells. It will be recalled that Lashley (1942) argued that partial information is carried by many axons, with the ensemble or constellation of such axons representing the totality of information at any one instant.

The phenomenon described by Fox and O'Brien offers a unique opportunity to test this notion. Since information is statistically distributed throughout the ensemble, the following ergodic relationship should hold: the response of a single cell to many thousands of light flashes must be statistically equivalent to the response of many thousands of cells to a single light flash.

The stochastic prediction was recently tested by John and Morgades (1969b). These workers recorded multiple units, using a spike height voltage discriminator to increase or decrease the size of the unit population being averaged. They found that as the size of the unit population *increased,* the *fewer* were the number of light flashes necessary to reproduce the Fox and O'Brien result. This is in conformity with the principle of ergodicity (a principle of great importance in statistical mechanics that states that a time average of a macroscopic quantity is the same as a single ensemble average).

The third line of argument regarding statistical relations comes from mathematical treatments of unit-EEG relations that show a nonlinear interaction between population elements (Freeman, 1972a, 1975). By "nonlinear" is meant that the principle of superposition does not hold and that the interaction of one element with another is important so that emergent properties appear in a population which are not discernible by observing the behavior of any individual element. In nonlinear systems, emphasis is placed on the population interaction or ensemble response in which "freedom" or variability of the individual is great even though there is relative stability of the whole (Goodwin, 1963; Weiss, 1967).

This purely mathematical treatment is supported by empirical observations demonstrating nonlinear relationships between probability of discharge and fluctuations in the membrane potential of single neurons (Elul, 1968; Elul & Adey, 1966). Figure 2.10 shows a continuous intracellular recording in which the threshold membrane voltage for spike discharge varies from moment to moment. Data by Burke (1967) and Nelson and Frank (1967) have demonstrated that the site of spike initiation is not fixed. Immature cortical neurons and mature hippocampal neurons, as well as other cell types, are capable of dendritic spike initiation (Purpura, 1967; Purpura & Malliani, 1966; Purpura & Shofer, 1965; Purpura et al., 1966) which demonstrates the degree to which the site of spike initiation is capable of changing with respect to the axon Hillock locus.

Intracellular recordings from thalamic neurons (Feldman & Purpura, 1970) during the recruiting response, which produces a sequence of intracellular inhibition (IPSPs) and excitation (IPSPs) suggests the possibility that the site of inhibitory synaptic drive may even vary from stimulus pulse to stimulus pulse though the site of stimulation (several millimeters distant) remains constant. These studies indicate the considerable degree of freedom of the control of spike discharge in both the spatial and temporal domains.

These various lines of evidence taken as a whole indicate that descriptors of cooperative neural behavior (information) must be largely statistical in nature. Thus, a certain amount of randomness is inherent to a description of neuronal population dynamics and that the degree of randomness may vary.

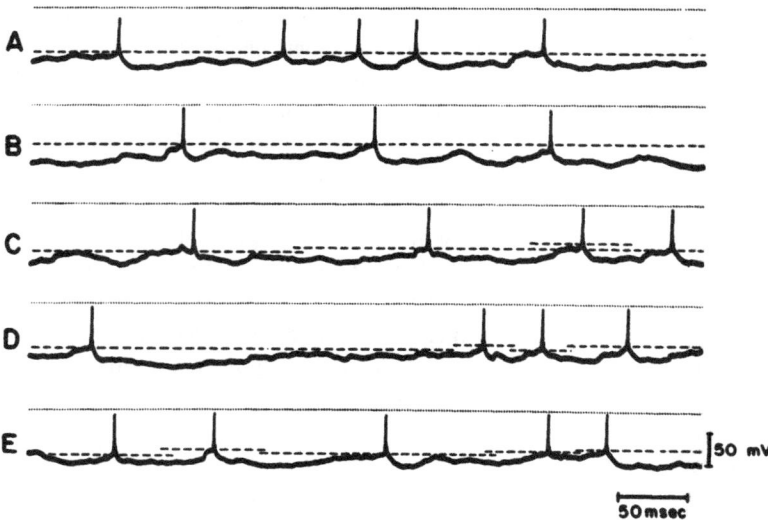

FIG. 2.10 Apparent fluctuations in the threshold for firing spikes in cortical neurons of unanesthetized cat. (A–E) From same neuron, sigmoid cortex, 500-μ depth. Depolarization preceding last spike in (A) exceeds threshold without provoking spike. (B) Third spike triggers at a level some 8 mV lower than for first two spikes. (C–E) Examples of variations in threshold for discharge of spikes. (From Elul, 1972.)

In fact, evidence indicates that the degree of randomness within a population of interacting neurons is itself a functional variable. For instance, a state of minimum randomness is present during convulsoid activity, characterized by low-frequency high-amplitude EEG. In this state, large populations of cells distributed over wide regions of the neural axis are synchronously active, exhibiting considerable coherence. In other states, such as arousal, which is characterized by low-amplitude high-frequency EEG, there is significantly less coherence among individual neurons. Thus, a wide spectrum of coherences can be observed among EEG generators.

IV. SIGNIFICANCE OF DEGREE OF COHERENCE
OF CELL POPULATIONS

The idea that the degree of coherence and/or randomness is a functional variable was first suggested by John and Killam in 1960. These workers observed that the similarity of electrical responses to a repetitive conditioned stimulus recorded from many areas of the brain increased during the course of learning; that is, the potentials evoked from wide areas of the brain by a conditioned stimulus become more similar as learning progressed.

Rafael Elul (1967a) devised a statistical measure of EEG amplitudes in which individual EEG generators were treated as Gaussian, allowing the construction of hypothetical amplitude histograms. Specific deviations from Gaussian probability deviations were found during the performance of mental tasks (Elul, 1969). This method distinguished between normal and mentally retarded children (Elul, 1972).

Similar findings showing valuable functional correlates to the variance of EEG generators were demonstrated by Adey *et al.* (1966a), Fox and Norman (1968), Gavrilova (1970), and Knipst (1970). In general, these studies suggest that the degree of cooperative activity of neuronal elements is an important variable correlated with mental performance.

Recently, Adey (1975) proposed a mechanism that can exercise control of the degree of coherence or randomness among a population of generators. According to this model, control resides in the interaction of neurons, neuroglia, and the extracellular space. Emphasis is placed on a mechanism similar to that known to operate in the interaction between neurons and hormones or neurohumoral substances (which are effective in minute amounts). In the Adey model, the macromolecular structure of the extracellular space, with its numerous fixed negative charges, is believed to operate as a "sensor" and "amplifier" of weak electric fields, influencing many of the neurons in the local domain.

This contention is supported by recent data which indicate that minute extracellular current flows can affect neural excitability and, consequently, behavior. For instance, low energy fields (modulated DC or VHF) can systematically alter the concentration of bound Ca^{2+} as well as certain amino acids (glutamic acid and gamma-aminobutyric acid) known to be involved in synaptic transmission (Adey, 1975; Kaczmarek & Adey, 1973a, b). The field strengths that were used (20 to 50 mV/cm) cause current flows that are smaller than synaptic currents which are known to flow in the extracellular compartment (Elul, 1967b). Experiments have shown that spinal motor neurons generate extracellular field gradients greater than 50 mV/cm and that these gradients can alter neuronal excitability (Nelson, 1966).

The functional effect of extracranial low energy fields on neural coherence has also been demonstrated in behaving monkeys and cats (Adey, 1975; Bawin, Gavalas-Medici, & Adey, 1973; Gavalas, Walter, Hamer, & Adey, 1970).

Figure 2.11 shows the acquisition and extinction performance of a cat in an operant avoidance task in the absence of a very high frequency (VHF) field. The task involved producing a characteristic brain rhythm to delay the delivery of a noxious stimulus. After initial training of five cats, the animals were divided into two groups; one was given reacquisition and extinction in the presence of the VHF field ($N = 3$), while the other ($N = 2$) did so in the absence of the field.

Figure 2.12 shows the performance of animals in the two field conditions. Acquisition occurred rapidly in cat C4 (field-on animal) as compared to cat C3 (field-off animal). It can be seen, however, that the most pronounced effect was

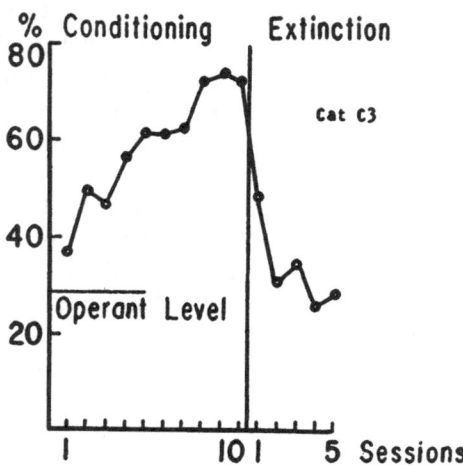

FIG. 2.11 Performances of a cat (C3) in conditioning and extinction of a 14-Hz rhythm in nucleus centrum medianum, without VHF fields. Data normalized over total number of CS presentations within session. (From Bawin *et al.*, 1973.)

on extinction performance, which was significantly prolonged in the field-on cat. In one cat, the operant response was a 4 Hz EEG spindlelike response. Spectral analysis showed that the field did not produce a 4 Hz burst in naive or extinguished animals, but it did enhance the frequency of occurrence of 4 Hz activity in trained animals (Bawin, Gavalas-Medici, & Adey, 1973).

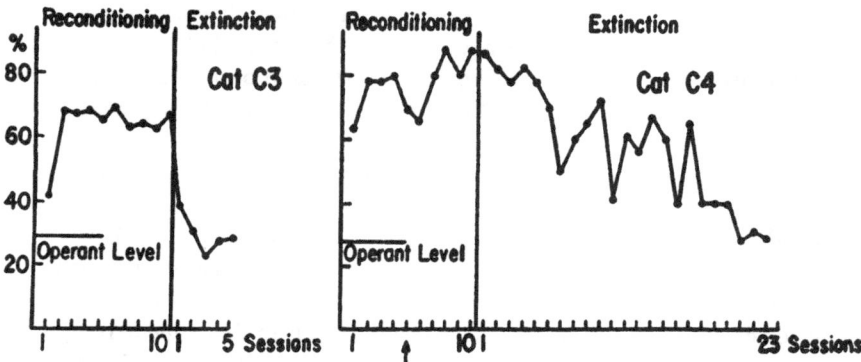

FIG. 2.12 Comparison of performance in two cats during reconditioning and extinction; cat C3 (centrum medianum, 14 Hz–no field) and cat C4 (hippocampus, 4.5 Hz–VHF field amplitude-modulated at 4.5 Hz). Data normalized over the total number of CS presentations within session. The arrow indicates the first day of exposure to the fields. (From Bawin *et al.*, 1973.)

Performing DRL task. 80 sec. segments near end of the 4 hr. runs. Combined data from 2 experimental-control runs		7 CPS On vs. Off	10 CPS On vs. Off
MONKEY J	L. HIPPOCAMPUS	p = .048	p = .025
	R. HIPPOCAMPUS	p = .001	p = .011
	R. AMYGDALA	p = .003	p = .001
(OTHER STRUCTURES OBSERVED: LMBRF , L.V. CX, R. M. CX, R.V. CX)			
MONKEY Z	R. HIPPOCAMPUS	p = .006	p = .020
	L. CENTRE MEDIAN	p = .001	p = .001
(OTHER STRUCTURES OBSERVED: AUD CX, R.V. CX, R. AMYG, L. HIPP)			
MONKEY A	R. HIPPOCAMPUS	p = .001	
	L. HIPPOCAMPUS	p = .001	No run
	L. CENTRE MEDIAN	p = .059	
(OTHER STRUCTURE OBSERVED: L. AMYG)			

Non-performing: Sitting quietly		7 CPS On vs. Off
MONKEY A	R. HIPPOCAMPUS	p = .001
	L. HIPPOCAMPUS	p = .036
	R. CENTRE MEDIAN	p = .045
	L. AMYGDALA	p = .003
(OTHER STRUCTURES OBSERVED: LCM, R. AMYG)		

FIG. 2.13 The EEG peak quotient is a statistical test comparing the EEG spectral plot to a logarithmic distribution over the frequency range 4–20 Hz. The test is based on the fact that the spectral curve of the EEG in the 4- to 20-Hz range closely approximates an exponential shape in the absence of fields (see Gavalas et al., 1970). Significant changes in EEG spectral power occurred in particular brain areas in the presence of a 7 V/m frequency modulated field. It has been calculated that the electric gradient produced in the brain by these fields is approximately .02 μV/cm (Adey, 1975). These extremely weak fields affect the EEG frequency distribution in both behaving (monkeys J, Z, A) and nonbehaving subjects (monkey A). (From Gavalas et al., 1970.)

In other experiments (Gavalas, Walter, Hamer, & Adey, 1970), it has been shown that the frequency of an external modulating field is an important variable in the control of EEG rhythms. Figure 2.13 shows the effects of two different electrical field frequencies on different brain structures in monkeys. Antenna effects from implanted electrodes or shielded leads cannot account for these changes since only specific structures were affected and the effects were not uniform. Also, systematic changes in time estimation (Gavalas et al., 1970), reaction time (Hamer, 1968), and circadian cycles (see Adey, 1975) occur in the presence of external fields in unimplanted subjects. Gavalas and his colleagues (1970) found that discriminant analysis of spectral parameters revealed increased coherence and raised total EEG power in the presence of fields. These data indicate that brain function can be influenced, to some extent, by properly controlled fields external to the head. They also strongly suggest that macromolecular–ionic interactions in the extracellular space may be part of the mechanism that controls neural coherence.

The idea that neuronal activity can be modulated by extracellular currents is actually quite old. Such influences were referred to as *ephaptic* transactions in the earlier neurophysiological literature. Ephaptic transmission played an important role in certain theoretical formulations of information processing (McCulloch, 1951), was directly demonstrated in experiments with single axons (Eccles, 1946), and was invoked to explain interaction between auditory and visual stimuli at the level of the superior and inferior colliculi (Gerard, Marshall, & Saul, 1936). More recently, direct evidence of the modulation of functional activity by extracellular fluids was demonstrated by Terzuolo and Bullock (1956), who showed that minute DC fields could alter the timing of the pacemaker neuron in the invertebrate cardiac ganglion. Nonetheless, because of the difficulty of establishing that synaptic action could be excluded from contributing to apparent ephaptic processes in the CNS, it has long been assumed that such transactions, if they existed at all, were of negligible functional relevance.

However, the belief that extracellular current flows are an important factor in neural control has been revived in view of this recent experimental support. Classic notions regarding neural control have long been based on the abundance of data showing that synaptic action, in particular coordinated phasing of EPSPs and IPSPs, has a very direct and powerful effect on neural ensemble behavior. It is clear that synaptically mediated control, in which groups of neurons are turned "off" or "on" in an organized sequence, occurs throughout the brain and must constitute a major part of any coherence controlling mechanism. The extracellular factors described in this discussion must be considered as supplemental, modulating basic anatomic and synaptic mechanisms. To further emphasize this point, the subject of thalamic regulation and control of cortical and related neural systems will be discussed in the next chapter.

3

The Genesis of Alpha Rythms
and EEG Synchronizing Mechanisms

I. INTRODUCTION

"Synchronization" means that populations of neural generators act in concert, that is, together in time. Such a process underlies all of EEG, whether delta (.5 to ≈ 3.5 Hz), theta (3.5 to ≈ 7 Hz), alpha (7 to ≈ 13 Hz), or beta (13 to ≈ 22 Hz) waves, since each of these rhythms involves the synchronous activation of generator populations. What is the process that organizes disparate cell groups into the same temporal patterns of activity; what are the mechanisms which achieve synchronization? The answers to these questions are crucial in understanding how disparate cell populations function in concert. In later chapters it will be argued that neural representational systems play a fundamental role in the processing of information in the CNS. In the section on perceptual framing (IV, B), data will be presented in support of the idea that synchronization and gating processes, whereby specific cells are excited and others inhibited, underlie the creation of neural representational systems. For background purposes, the following discussion is concerned with the general mechanisms of EEG synchronization.

Alpha rhythms are a dominant phenomenon in cortical EEG, occurring during the waking state throughout the neocortex and increasing developmentally until adulthood when 70% of the EEG is at alpha frequencies (7 to 13 Hz) (Matousek & Petersén, 1973). In this section we will discuss animal models of thalamic and cortical spindles. The reason for this is that no other model of global EEG slow wave phenomena has been studied in as much physiological detail. Over the past 30 years innumerable extracellular and intracellular analyses have been directed to elucidating the synaptic mechanisms underlying synchronized thalamocortical spindle activity. By spindle is meant a relatively brief period of rhythmic waves at a fairly constant frequency which wax and wane to produce a spindle-shaped

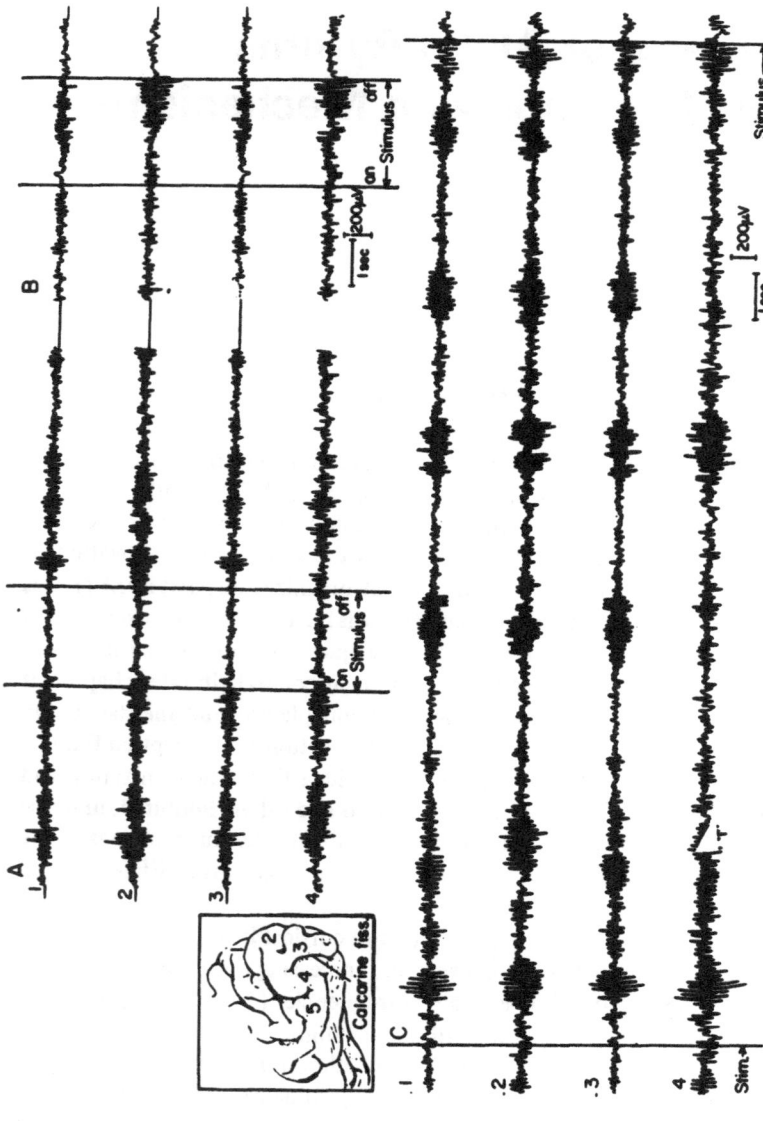

FIG. 3.1 EEG records from a human patient sitting quietly. Location of pia electrodes are shown in the insert. This figure demonstrates typical alpha spindles showing the envelope structure of synchronized 10 per sec rhythms. Alpha suppression to stimulus onset is shown in upper part of the figure (A and B). (C) Alpha spindles from a subject who has habituated to repetitive flash stimulation. (From Morrell, 1967.)

envelope of amplitudes. Understanding how this phenomenon arises should help us gain insight into the generation of more sustained synchronous processes.

Several reservations can be expressed about the use of a spindle model to understand alpha rhythms. One is that cortical spindles involve rostral thalamic structures and appear primarily upon the anterior cortex, whereas alpha waves occur most strongly in the occipital cortex of man. A second is that cortical spindles occur during Stage II sleep, whereas alpha is present during relaxed and even aroused states. And third, cats have been primarily used to elucidate spindle mechanisms, and this species does not exhibit strong alpha rhythms.

Nonetheless, in spite of these reservations, the general wave morphology of cortical spindles is similar to the alpha spindle (cf. Figs. 3.1 and 3.2). Both phenomena involve the synchronization of huge masses of neurons, and the thalamus, because of its unique anatomic position, most likely plays a fundamental role in the production of both. It seems plausible that the basic features of interneuronal interaction involved in the production of these synchronized slow cortical waves can be generalized to a wide variety of cortical slow wave phenomena.

When an animal or human begins to sleep, the cortical EEG exhibits frequent spindles, fusiform or ringing waves during the initial drowsy state, and an

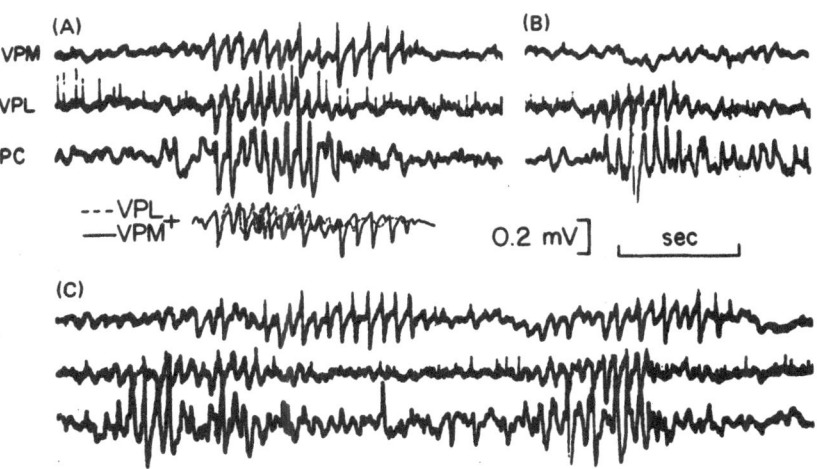

FIG. 3.2 Microelectrode recording from the VPM and the VPL and the corticogram from a point in the postcruciate gyrus (PC). (A) Compound spindle with perfect synchrony between the points in the VPL and PC. Below is a superimposition of the VPM (full line) and VPL spindles (broken line), showing the synchrony of the spindle at the two locations. (B) Local spindle involving the VPL and PC, but without concomitant activity in the VPM area. (C) Two compound spindles in which the VPM spindles lag markedly relative to the VPL and PC spindles. (From Andersen & Andersson, 1968; after Andersen et al., 1967b.)

increase in the frequency of such waves in subsequent Stage II sleep. Examples of cortical spindles are shown in Fig. 3.2.

Spindles are characterized by a preceding spike and wave complex that increases in amplitude with each successive wave and then decreases in amplitude, thus producing a spindlelike appearance. Alpha bursts also exhibit this pattern (see Fig. 3.1). The form of the waves that constitute a spindle varies greatly. The waves can be either purely surface positive, diphasic, or surface negative, with the latter two the most common. The form of the spindle may change markedly with small movements of the recording electrode. This suggests that spindle systems are anatomically heterogeneous in nature, involving specific distributions of cells. The peak-to-peak amplitude of spindles, when recorded from the surface of the brain, is from 100 to 800 μV; when recorded from the dura the amplitude is reduced, and when recorded from the scalp, the peak-to-peak amplitudes seldom exceed 200 μV.

There are two general forms of spindles. The Type I spindle pattern is largely surface positive, sometimes showing a diphasic pattern, whereas Type II spindles are simpler in morphology and are usually surface negative (Calvet, Calvet, & Scherrer, 1964; Spencer & Brookhart, 1961). Microelectrode recordings show that the Type I spindles have a relatively simple source, namely, depolarization of cell bodies in cortical layers. The sources of Type II spindles are more complex involving somatic and dendritic interactions (Andersen & Andersson, 1968; Spencer & Brookhart, 1961).

Bremer, in 1935, described the effects of isolating the forebrain structures of the cat, which basically involves separating the neocortex, diencephalon, and basal ganglia from the midbrain. In this preparation (called a cerveau isole), the cortical EEG was dominated by spindles interspersed with irregular activity. Thus, Bremer's experiment showed that the reticular formation did not play an important role in the production of spindles since connections to it had been severed and that either one or a combination of the remaining forebrain structures were principally involved. Adrian (1941) helped identify the critical structures by removing the cortex of a cat and recording spindlelike waves from the killed ends of thalamocortical fibers. Other workers (Andersen & Andersson, 1968) subsequently recorded spindles from within the thalamus in decorticate preparations. These findings, which indicated an intrathalamic origin of cortical spindles, are supported by lesion experiments in which selected thalamic nuclei were ablated with the consequent abolition of cortical spindles (Andersen *et al.*, 1967a; Kristiansen & Courtois, 1949).

The lesion experiment showed that removal of sensory-specific thalamic relay nuclei abolished cortical spindles, whereas destruction of nonspecific midline thalamic regions did not (Andersen *et al.*, 1967a). Also, ablation of the anterior third of the thalamus was most effective in abolishing spindles (Kristiansen &

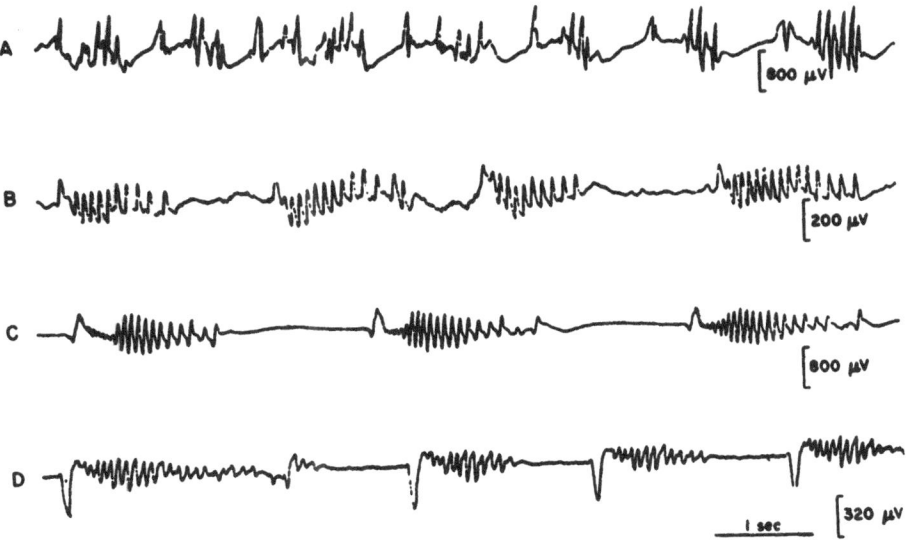

FIG. 3.3 Results of corticoelectrical stimulation after application of 1% eserine and .5% acetylcholine. (A) Intact cortex; (B) thalamectomized hemisphere; (C) undercut cortex; (D) completely isolated cortex. This shows that rhythmical 10 to 12 per sec waves can occur in the cortex, independent of the thalamus. (From Kristiansen & Courtois, 1949.)

Courtois, 1949). Furthermore, cooling of the inferior thalamic peduncle (a fiber system communicating from the mediodorsal thalamus to the frontal cortex) abolishes cortical spindles as well as recruiting responses which are a type of spindle phenomenon (Skinner, 1971) produced by electrical stimulation.

These studies demonstrated that the thalamus plays a fundamental role in the production of spindles and that the midline nonspecific thalamus is not a necessary prerequisite for spindle activity. This is not to say that the midline thalamus is unimportant. On the contrary, through the diffuse midline system a powerful influence is exerted upon rhythmic generating mechanisms probably at both thalamic and cortical levels. The latter point has been established strongly by studies involving the recruiting response (Dempsey & Morison, 1942; Morison & Dempsey, 1942).

However, it must also be pointed out that the cortex, by itself, is capable of producing rhythmic spindle waves. This is shown in Fig. 3.3, showing cortical activity recorded following progressive isolation of the cortex. In Fig. 3.3 (D), the isolated cortex is seen to be capable of exhibiting spindle envelopes, indicating that there exist independent thalamic and cortical spindle systems which, in the intact organism, are coupled to form a functional unit.

II. THALAMOCORTICAL COORDINATION
IN THE PRODUCTION OF SPINDLES

Such lesion and stimulation studies provided indirect evidence for an intimate coordination between the thalamus and cortex in the production of rhythmic EEG phenomena. Direct evidence of thalamo-cortical concordance was provided by Andersen and co-workers in studies involving simultaneous thalamic and cortical recordings (Andersen *et al.*, 1967b). A particularly good example of simultaneous thalamic and cortical recordings which provides information about the size of a coupled thalamocortical projection system is shown in Fig. 3.4. The striped inset shows the cortical area exhibiting spindles and the cortical site of monopolar recording points in and near the area (a–c). In Fig. 3.4 (A–D) there is considerable concordance between the cortical spindles (lower lines) and simultaneously recorded thalamic spindles. Note that the point-to-point correspondence is absent in (E) and (F), where the cortical recording electrode (position c from inset) is outside the thalamocortical spindle projection region. This, to-

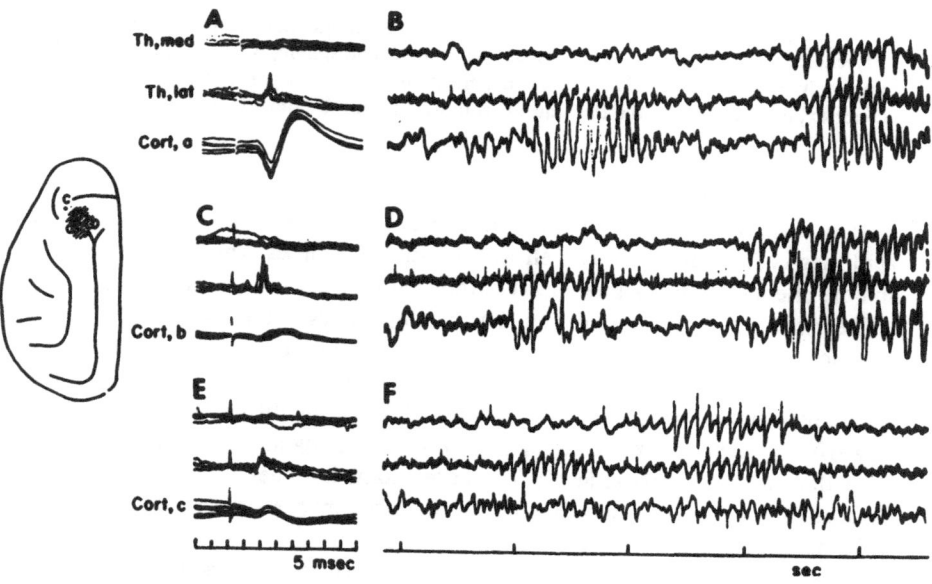

FIG. 3.4   In each section, the three traces are, from above: microelectrode recordings from a medial (VL) and lateral (VPL) electrode in the thalamus (Th, med and Th, lat), and the corticogram taken from the points a, b, and c of the inset, respectively. (A, C, and E) Responses to stimulation of the contralateral ulnar nerve with the cortical electrode recording from points a, b, and c, respectively. (B, D, and F) Spontaneous spindle activity from the corresponding locations. (From Andersen & Andersson, 1968; after Andersen *et al.*, 1967b.)

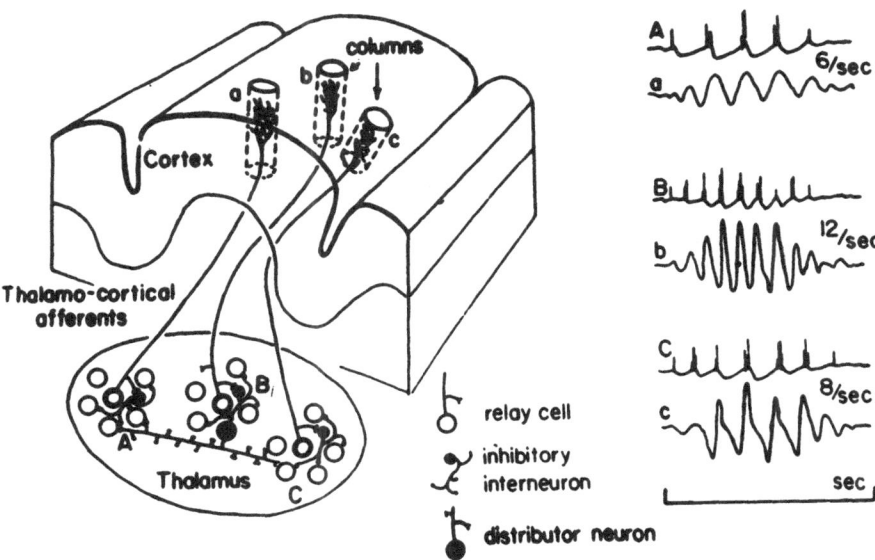

FIG. 3.5 A model of the thalamocortical rhythmic correspondence. The neurons of three thalamic nuclear groups, A, B, and C, send their axons to the appropriate part of the cerebral cortex, activating the columns a, b, and c. Collaterals of these axons excite inhibitory interneurons (black), which have profusely ramifying axons that can initiate postsynaptic inhibition simultaneously in a large number of thalamic neurons. The different thalamic nuclear groups have been given different intraspindle frequencies, times of onset and stop, and interspindle periods. The corresponding features of the cortical spindles vary accordingly, as illustrated in the right-hand column. The upper lines are imaginary spindles from the thalamic groups A, B, and C, whereas the lower lines show the corresponding cortical spindles as they would appear at points a, b, and c, respectively. (From Andersen & Andersson, 1968.)

gether with stimulation studies, shows that rhythmic activity involves relatively discrete locus-to-locus thalamocortical interactions.

Andersen and Andersson (1968) estimate that the cortical projection region is at most .8 mm in diameter (which is slightly larger than the size of a cortical column of cells) and that the thalamic pacemaker or rhythm generator is approximately 200 μ in diameter. Figures 3.5 and 3.6 show a model of Andersen and Andersson's proposed thalamocortical spindle mechanism

Studies by Verzeano and colleagues (Verzeano & Negishi, 1960; Verzeano, Laufer, Spear, & McDonald, 1965) indicate that there is a mosaic of thalamic spindle producers, with some overlap but nonetheless with definable borders approximately 200 μ in diameter. Within these domains thalamic neurons beat in full synchrony.

Multiple microelectrode recordings, as well as recordings involving the successive movements of a single electrode, indicate that there exist complex networks

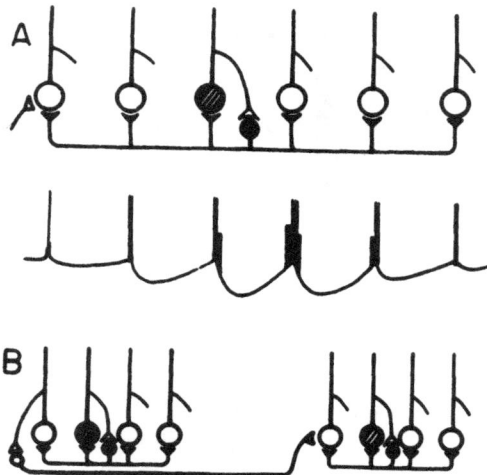

FIG. 3.6 Schematic representation of the proposed mechanism giving rhythmic discharges of thalamic neurons. (A) A discharge of the cell (hatched) causes recurrent inhibition via an interneuron (black) that hyperpolarizes many neighboring projection cells. During the postinhibitory rebound, many of these cells discharge action potentials and an increasing number of cells participate during the successive cycles (A, lower part). Alternatively, intrathalamic connections via distributor neurons (B) spread the rhythmic activity from one group to other parts of the thalamus (from left-to-right group). (From Andersen & Andersson, 1968.)

of cells constituting spindle-generating groups. These groups have been called "facilitative pacemakers" by Andersen and Andersson (1968) and are independent to some extent, although capable of interacting and thus becoming synchronized, even across widely distributed regions (Andersson and Manson, 1971; Andersson *et al.*, 1971). It is believed that there is a profuse multiplication of these rhythm-generating neuron groups, each involving recurrent inhibitory feedback.

The evidence presented thus far should not be construed to suggest that thalamocortical relations are a one-way street. There are abundant cortico-thalamic connections capable of driving thalamic neurons, as demonstrated most clearly by Frigyesi and Schwartz (1972). Small inhibitory feedback loops within both the thalamus and the cortex plus larger thalamocortical and cortico-thalamic connections represent a constellation of interconnected systems, capable of cooperative behavior, which form the heart of the mechanism producing rhythmic spindle waves. It appears likely that the proposed facilitative pacemaker systems are distributed in many regions of the neural axis, particularly in the thalamus, cortex, septum, and hippocampus. Various degrees of cross-coupling and cooperative behavior are exhibited by these systems.

III. MECHANISMS OF SYNCHRONIZATION
AS REVEALED BY INTRACELLULAR ANALYSIS

Thalamic mechanisms of synchronization are perhaps best demonstrated by low-frequency (6 to 12 sec) stimulation of the anterior midline thalamus. Such stimulation induces the classical recruiting response, first discovered by Dempsey and Morison in 1942. The recruiting response involves a progressive buildup of cortical-evoked potentials which wax and wane in amplitude in a manner illustrated in Fig. 3.7. The fusiform pattern of the recruiting response is markedly similar to that seen in spontaneously occurring spindles and alpha rhythms (see Figs. 3.1 and 3.2).

Intracellular recordings from thalamic neurons during electrically evoked EEG synchronization reveal a sequence of excitatory and inhibitory postsynaptic potentials (Fig. 3.8). The salient features of the synchronizing process are (1) long duration (100 to 180 msec) IPSPs, and (2) evoked EPSPs which are of short latency (5 to 10 msec), typically leading to spike discharge. In Fig. 3.8 the EPSP is followed (within 10 to 30 msec) by a powerful and prolonged hyperpolarizing response which suppresses further discharge. In the recruiting response, repetitive stimulation produced a long duration inhibitory process that is distributed throughout extensive regions of the thalamus (Purpura, 1970; Schlag & Villablanca, 1968). The EPSP–IPSP sequences shown in Fig. 3.8 are typical of recordings from many thalamic neurons. A study by Schlag and Villablanca (1968) showed that there are clear differences between EPSPs and IPSPs in terms of the intensity and distribution of these two postsynaptic processes following midline thalamic stimulation. For example, EPSPs are shorter in duration, more intense, and spatially more confined than IPSPs. The spatial distribution of IPSPs increases incrementally following each successive stimulus in the recruiting train (Schlag & Villablanca, 1968). This incremental increase in the distribution and intensity of IPSPs underlies the recruitment of neighboring cells and is, in part, responsible for the envelope structure characteristic of spontaneous spindles and alpha rhythms (Andersen & Andersson, 1968; Purpura, 1969, 1970).

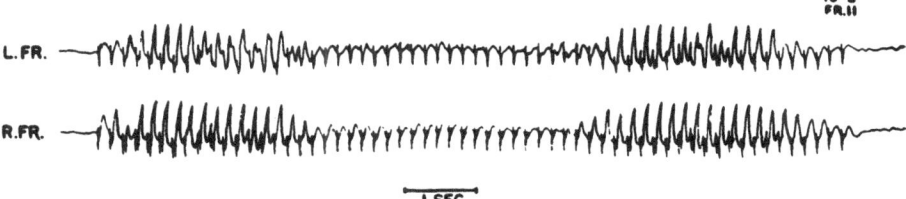

FIG. 3.7 Cortical recruiting responses from the left and right gyrus proreus of the cat under barbiturate anesthesia. Bipolar stimulating electrodes situated in the nucleus centralis medialis of the thalamus. Repetitive stimulation at 7 per sec. (From Jasper, 1949.)

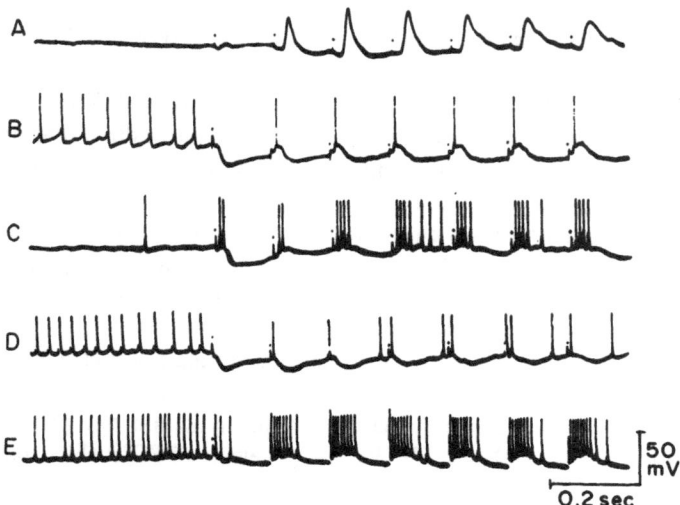

FIG. 3.8 Cortical surface (A) and intracellular potentials in thalamic neurons (B–E) during centralis medialis stimulation at 7 per sec. (A) Surface negative recruiting responses elicited in the motor cortex. (B) Neuron in nucleus ventralis anterior showing long-lasting IPSP to the first stimulus and an EPSP–IPSP sequences to successive stimuli. (C) Neuron in nucleus ventralis lateralis discharging double spikes to the first stimulus and an ensuing IPSP. The following stimuli elicit two or several spikes, succeeded by an IPSP of varying amplitude. (D) Neuron with a discharge pattern similar to that shown in (B). (E) Neuron in the intralaminar region showing a long-lasting IPSP to the first stimulus and an EPSP with a prolonged discharge terminated by an ensuing IPSP in response to the successive stimuli. (From Purpura & Shofer, 1963.)

Several workers agree that the intensity and distribution of IPSPs is important in synchronizing widely distributed cells, particularly since the time course of the IPSP is essentially uniform across large domains of tissue; that is, many cells are nearly simultaneously inhibited and then released from inhibition and thus synchronized. A major difficulty in understanding the mechanisms of synchronization lies in visualizing how excitation and inhibition are phased so precisely throughout a heterogeneous mixture of cells with complex interconnections.

Andersen and Andersson (1968) and Andersen and Eccles (1962) postulated a very simple mechanism to explain these phenomena. Their hypothesis is based upon the observation that a large depolarizing wave (excitation) occurs after termination of inhibition (Andersen & Sears, 1964; Coombs et al., 1959). In other words, the membrane potential does not always return to the prestimulus level after an IPSP but sometimes exhibits an excitatory overshoot, called "postinhibitory rebound." In view of this phenomenon, the Andersen and Andersson (1968) model for the genesis of alpha rhythms was proposed (see Figs. 3.5 and 3.6). An afferent volley or spontaneous excitation results in the discharge of a group of "principal" thalamic cells (Fig. 3.6A). Following discharge, recurrent collaterals activate nearby inhibitory interneurons (in black). The inhibitory neurons feed back onto the principal cell and turn it off for the duration of the IPSP. The subsequent termination of the IPSP results in an excitatory rebound discharge. This discharge in turn excites inhibitory interneurons which again produce an IPSP in the principal cell and the cycle is repeated again and again. In this way, cells are phased between excitation and inhibition, with the interval between discharges determined by the duration of the IPSP. Functional groupings of such inhibitory feedback systems can recruit as well as interfere with other groupings of cells.

The inhibitory interneurons are probably not located within the immediate vicinity of the principal cells. Electrophysiological recordings, in the midline thalamus, only rarely reveal spike discharges possessing characteristics of an inhibitory interneuron (Marco et al., 1967). Inhibitory interneuron burst discharges at latencies and durations consistent with the intracellular IPSPs are found most commonly in the nucleus reticularis. This nucleus is a complex matrix of cells arranged as a pear-shaped mantle overlying the lateral and anterior borders of the thalamus and located perhaps a centimeter distant from the midline (Scheibel & Scheibel, 1967b, 1970).

The Andersen and Andersson (1968) model was seductively simple and consequently was not seriously challenged until recently. Purpura and co-workers (Purpura, 1969, 1970; Purpura & Cohen, 1962; Purpura & Shofer, 1964; Purpura et al., 1966) argued from the outset that the thalamic synchronizing mechanisms involved a complex interplay of inhibitory as well as excitatory loops (of an unspecified nature), and did not emphasize the importance of inhibitory rebound. A similar view has recently been advocated by Andersson and Manson (1971) which is in opposition to the model of Andersen and Andersson (1968). However, experimental evidence clearly opposing the Andersen and Andersson (1968) model has only recently been reported.

These data are derived from intracellular analyses of the developing thalamus of the immature kitten (Thatcher & Purpura, 1972, 1973). For example, thalamic synchronizing mechanisms do not develop until the second week of postnatal life (Scheibel & Scheibel, 1965; Thatcher & Purpura, 1973). The Thatcher and Purpura (1973) study involved the systematic investigation of thalamic synaptic mechanisms in 1- and 3-day-old kittens as well as in 2- and 3-week-old kittens. Figure 3.9 shows typical intracellular responses to electrical stimulation in kittens one week old or less while Fig. 3.10 shows the effects of electrical stimulation in 3-week-old kittens. Two important features of synaptic development underlying evoked synchronization were observed; (1) excitatory synaptic drives were poorly developed in kittens less than ten days old, and (2) IPSP durations were very short (<60 msec) in 3-day and 1-week-old kittens but became progressively longer as kittens matured ($\approx$ 125 msec in 2- and 3-week-olds. Maturational changes in IPSP duration are shown clearly in Fig. 3.11.

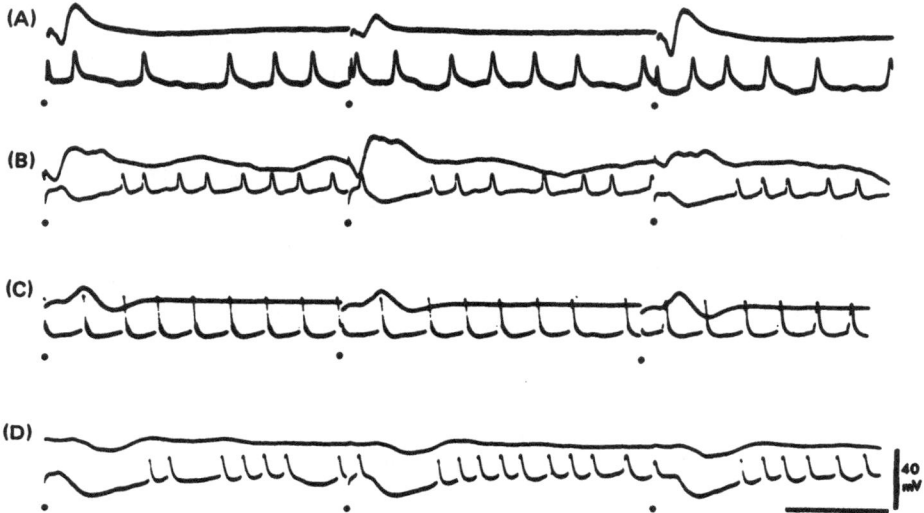

FIG. 3.9 Examples of characteristics of IPSPs evoked in four thalamic neurons from 7-day-old kittens in response to 3.3 per sec medial-thalamic (MTh) stimulation. Upper channel records are cortical surface responses, lower channel records are the intracellular responses. (A) A slight increase in the interval between discharges immediately after the MTH stimulus is the only effect detectable. (B and D) Short duration IPSPs are elicited by MTh stimulation in cells with partial spike potentials. (C) Cell unaffected by MTH stimulation. Membrane potentials are all greater than 30 mV. (From Thatcher & Purpura, 1973.)

Two lines of evidence from this study of the development of synchronizing mechanisms fail to support the Andersen and Andersson model. First, none of the neurons in the immature kitten exhibited postinhibitory burst discharge, and, second and most important, the duration of the IPSP was *clearly dissociated* from intraspindle and recruiting frequencies. At one developmental stage (one week or less) when short duration IPSPs were present, spindles were either absent or the intraspindle frequency was very low. As kittens mature (two weeks) and IPSP duration lengthens, intraspindle frequency increases; this relationship between frequency and IPSP durations is precisely opposite from that expected from the Andersen and Andersson model; that is, as IPSP duration increased the interstimulus interval necessary to elicit recruitment decreased.

The Thatcher and Purpura (1972, 1973) studies support a different model of thalamocortical synchronization. This model relies on the findings of Schlag (1966) and Johnson and Hanna (1970) regarding the *thalamic delayed response*. This response (also called a *thalamic echo*) can be observed when a single electrical pulse is delivered to the thalamus and recordings are either taken from

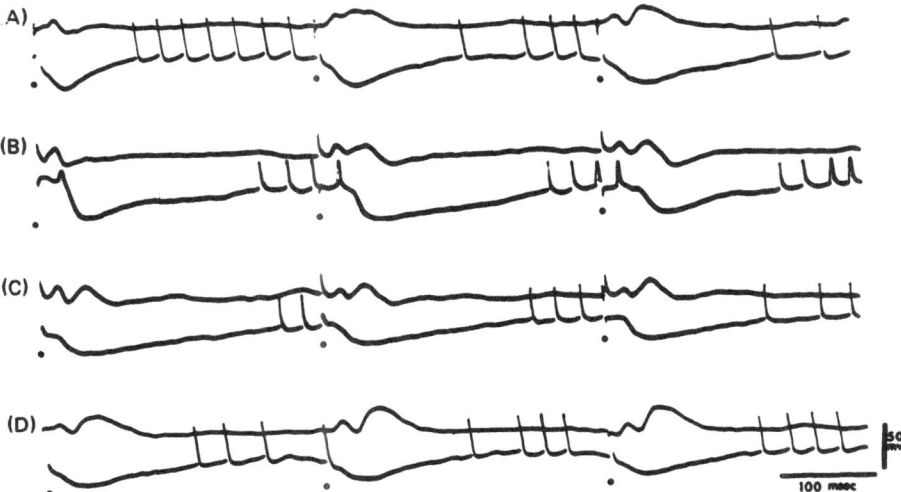

FIG. 3.10 Examples of IPSP characteristics observed in four different thalamic neurons (A–D) from a 3-week-old kitten during low-frequency MTh stimulation. Spike potential amplitude and duration are similar to spikes of thalamic neurons in adult animals (see text). This display emphasizes the synchronizing features of the long-duration IPSPs that limit spike discharges to the brief periods after IPSPs and just prior to the succeeding stimulus of the repetitive train. Note in particular that IPSPs in (B) attain a duration of nearly 200 msec. In the other cells the IPSPs range from 100 to 150 msec. (From Thatcher & Purpura, 1973.)

the stimulating electrode or the surface of the motor (pericruciate) cortex. The thalamic delayed response is a long latency (110–250 msec) response which can sum with the effects of a second stimulus pulse delivered near the time of its occurrence. If a succession of pulses are delivered at the same interstimulus interval, the recruiting response is produced.

Broggi and Margnelli (1971) and others (Hanberry & Jasper, 1953; Morison & Dempsey, 1942) have shown that, in the adult, the recruiting response is "frequency tuned," showing no recruitment at 3 Hz, a maximum response at 5.8 Hz, and very low amplitude responses at 15 Hz. The time course of the thalamic delayed response or echo approximately corresponds with the frequency tuning feature of the recruiting response and is considered a fundamental process underlying recruitment. In 1- and 3-day-old kittens, the thalamic delayed response has a latency of one or two seconds. In a 1-week-old kitten the delayed response latency is approximately 700 msec, while in the 2- and 3-week-old kitten latencies of 200 to 300 msec are common. Maximal recruiting responses in the kitten are produced when the thalamic delayed response latency ap-

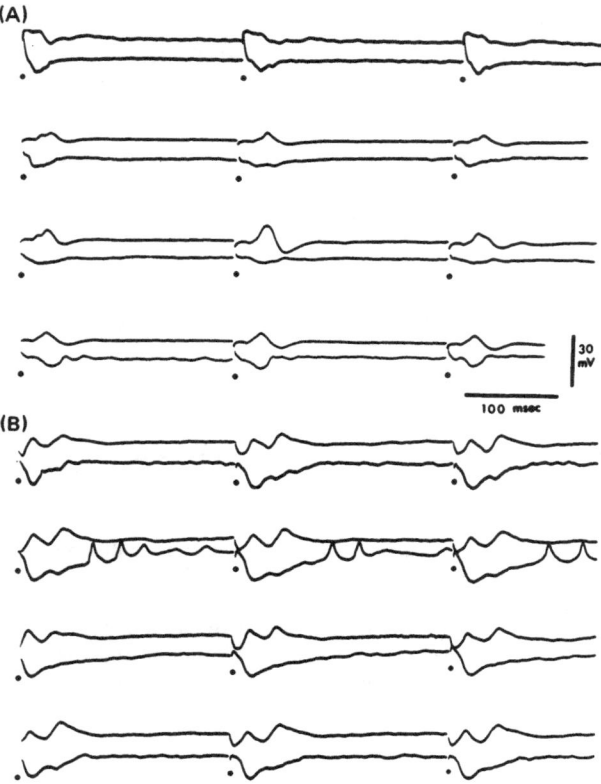

FIG. 3.11 Intracellular records of IPSPs in eight different thalamic neurons with membrane potentials greater than 30 mV that failed to exhibit spikes following medial-thalamic stimulation (3.3 per sec). Upper channel is from cortical surface, lower channel is intracellular record. (A) From kittens 1-week-old or less. (B) From kittens 2 weeks old. Although cells failed to produce spikes, IPSPs were consistently elicited. In (A) IPSPs are of very short duration (<50 msec). (From Thatcher & Purpura, unpublished observations.)

proaches the evoking interstimulus interval. Also, spontaneous spindles appear at the same time (≈ two weeks postnatal).

These observations suggest that thalamic synchronizing mechanisms depend on the development of two parallel loop processes, one excitatory and the other inhibitory. These processes are dissociated in the developing brain and mature at different rates, but converge as two interacting processes in the adult. Thus, the Purpura (1969) and Andersson and Manson (1971) models are supported by these data (Thatcher & Purpura, 1973), with the emphasis placed on the cooperative action of excitatory and inhibitory loops rather than upon inhibitory rebound.

IV. THALAMOCORTICAL GATING FUNCTIONS

The transition from sleep to wakefulness is accompanied by a dramatic change in the frequency of the EEG. Specifically, low-frequency high-amplitude waves are replaced by an activated pattern consisting of high frequencies and low amplitudes. A similar shift to desynchronized frequencies occurs when attention is altered in the awake subject. The discovery of the role of midline core structures (e.g., the reticular formation and its rostral thalamic extensions) in the control of generalized arousal (Moruzzi & Magoun, 1949) led to intensive efforts to elucidate the interneuronal mechanisms of EEG desynchronization. The most intensely explored variant of desynchronization is that which follows high-frequency stimulation (20 to 100 Hz) of the midline thalamus.

It will be recalled that low-frequency midline thalamic stimulation produced synchronous cortical slow waves, with the accompanying appearance of an EPSP–IPSP sequence in thalamic cells. Attenuation and blockage of IPSPs occur

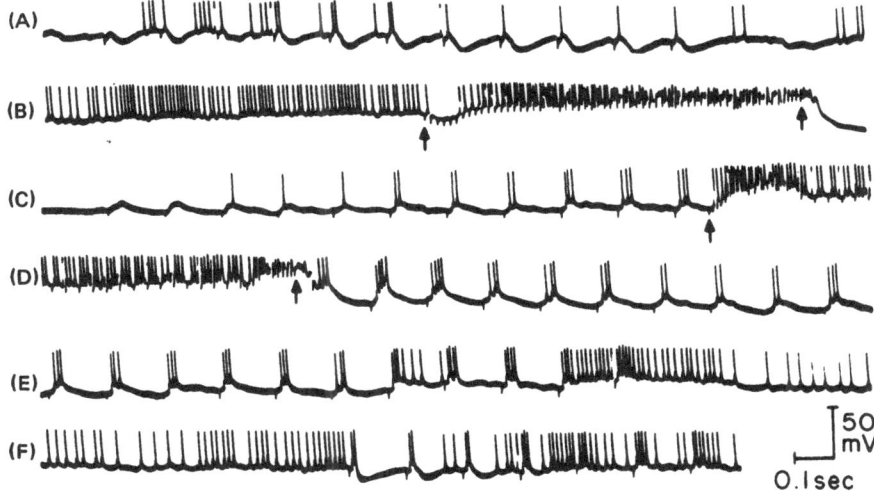

FIG. 3.12 Effects of repeated high-frequency medial-thalamic stimulation on a ventromedial cell that exhibited a synchronization pattern characterized by short-latency EPSPs and prolonged IPSPs prior to activation. (A–E) Continuous record. (A) Low-frequency medial-thalamic stimulation succeeded by a phase of hyperexcitability (B). At first arrow in (B), a prolonged IPSP is initiated by the first stimulus of the high-frequency (60 per sec) repetitive train. Successive stimuli after the IPSP evoke summating EPSPs associated with high-frequency spike attenuation. (D) Second period of low-frequency medial-thalamic stimulation after repolarization initiates only prolonged, slowly augmenting EPSPs. Changes in stimulus frequency between arrows, in (C) and (D), induce high-frequency repetitive discharges superimposed on depolarization, the magnitude of which is related to stimulus frequency. (F) Several seconds later. Note reappearance of IPSPs during low-frequency medial-thalamic stimulation. (From Purpura & Shofer, 1963.)

when high frequencies of midline thalamic stimulation are used. Figure 3.12 shows intracellular records from a thalamic neuron which was activated by a high-frequency midline stimulus. Note (Fig. 3.12) that in addition to the blockage of IPSPs considerable increase in excitatory synaptic drive follows the onset of high-frequency stimulation (at the arrow). The blockage of the IPSPs was thought initially to involve an actual inhibition of inhibitory cells (Purpura, 1970). However, intracellular analyses from hippocampal cells involved in a pencillin focus demonstrate that inhibition is simply overridden by excitatory drives produced by high-frequency fornix stimulation (Dichter & Spencer, 1969). A similar phenomenon may operate in thalamic interneuronal systems.

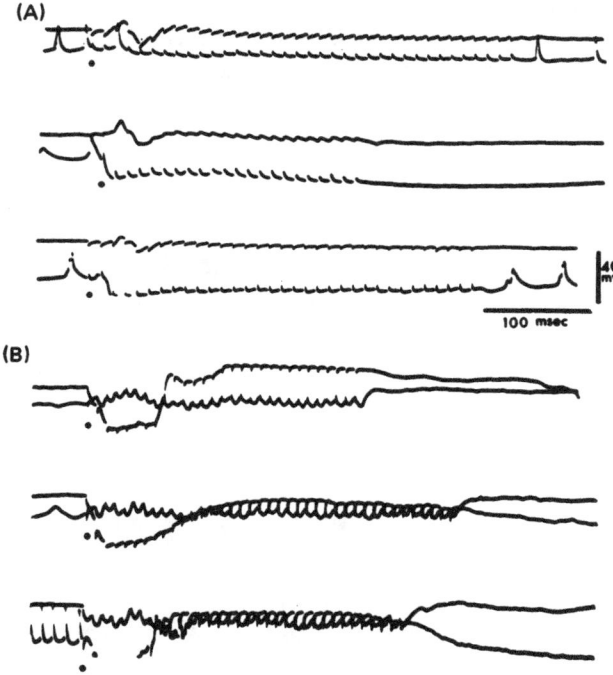

FIG. 3.13 Changing effects of high-frequency (80 per sec) MTh stimulation on thalamic neurons as a function of postnatal age. (A) Recordings from three different neurons from 3-day-old kittens. Onset of high-frequency MTh stimulation indicated by dots. IPSP summation is prominent though variable in the three neurons. In the upper and lower sets of records spike discharges occur shortly after the last stimulus of the repetitive train. In the middle set the increase in membrane potential associated with IPSP summation persists after cessation of MTh stimulation. Electrocortical activation is not seen at this age during high-frequency MTh stimulation. (B) Neurons from 14-day-old kittens. High-frequency MTh stimulation initially evokes 60–80 msec duration IPSPs which are terminated by sustained depolarizing shifts in membrane potential. Recovery of membrane potential occurs slowly after cessation of stimulation. (From Thatcher & Purpura, 1973.)

Figure 3.13 shows examples of intracellular recordings from immature thalamic neurons in 1-week-old and 2-week-old kittens (Thatcher & Purpura, 1973). It can be seen that in 1-week-old kittens high frequencies of stimulation (arrow) produce a maintained hyperpolarization. At this age, strong excitatory drives have not developed. In 2-week-old thalamic neurons powerful excitatory drives have developed and, as can be seen in Fig. 3.13, overpower the initial IPSP but apparently do not inhibit it. In any case, activation of midline core structures results in an increase in corticopetal excitatory drive and a blockage of the synchronizing IPSP. This sequence of events is believed to underlie the production of desynchronization of corticothalamic EEG. The blockage of IPSPs, however, may or may not involve inhibition of inhibition.

Evidence exists to suggest that the thalamus is capable of gating on or off the afferent drives projected onto the cortex. Figure 3.14 provides a diagrammatic representation of the anatomic interrelations between the thalamus and reticular formation. The heavy black arrows (1, 2, and 3) represent the rostral flow of

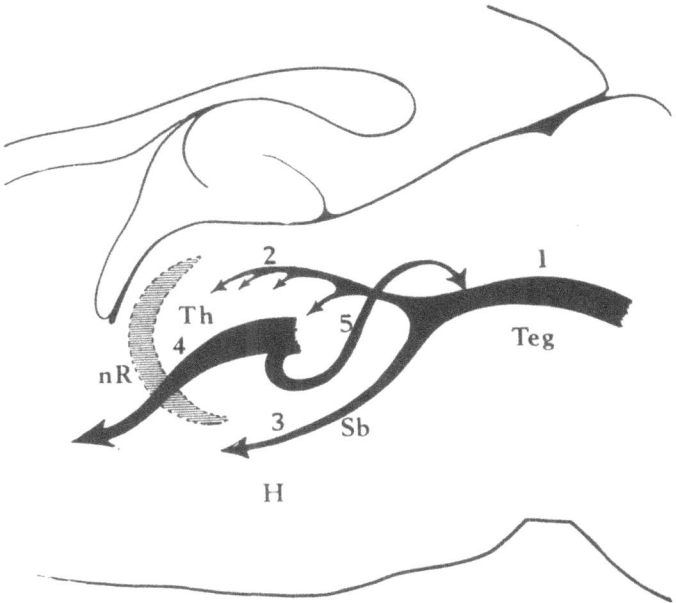

FIG. 3.14 Schematic representation of the projection paths of the two nonspecific systems. The brain stem reticular core (1) ascends through the mesencephalic tegmentum (Teg) and divides into a smaller dorsal lamina (2) which projects into several nuclei of the thalamus, Th, and a larger central lamina (3) which projects into subthalamus (Sb) and hypothalamus (H). The thalamic nonspecific (intralaminar) system (4) sends a major projection rostrally through the nucleus reticularis thalami, nR, and a secondary projection (5) caudally into tegmentum. (From Scheibel & Scheibel, 1967a.)

input from the reticular formation. Arrow 3 involves a rostral reticular projection that bypasses the thalamus.

An interesting experiment by Schlag and Chaillet (1963) helped to elucidate the position of the thalamus in the control of desynchronizing influences. These workers sectioned the rostral reticular formation pathway just anterior to the mesencephalic junction leaving intact arrow 3. It was observed that high-frequency stimulation of the midline thalamus failed to desynchronize the EEG even though high-frequency reticular formation stimulation was still successful as a desynchronizing force. As Scheibel and Scheibel (1970) point out, this suggests that the rostral thalamus, particularly the encapsulating nucleus reticularis through which the rostral coursing reticular output must perforate (arrow 4), serves a gating function by blocking high frequencies but allowing the passage of low frequencies. The Schlag and Chaillet (1963) finding can be explained if it is assumed that high-frequency thalamic stimulation exerts a desynchronizing influence on the neocortex via the secondary projection system which projects caudally into the tegmentum (arrow 5, Fig. 3.14). Pathways 2 and 3, involving reticulocortical control, are thought to play an important role in attention gating mechanisms discussed in Chapter 4. As will be discussed later (Chapter 4), parallel reticulocortical and reticulohippocampal influences operate in a reciprocal but coordinated manner to control orienting and attention responses.

Other studies, using intracellular recordings, support the view that the thalamus serves a gating and a multiplexing function in regard to the flow of ascending sensory influences to the cortex (Desiraju *et al.*, 1969; Purpura, 1970). Reciprocal interactions between the more lateral thalamic relay nuclei and the midline core structures of the diffuse projection system reveal a complex functional interplay between sensory-specific and sensory-nonspecific influences (Desiraju *et al.*, 1969). A clear example of the sensory gating capabilities of the thalamus is shown in Fig. 3.15. In this figure the arrows represent the time of stimulation of the medial lemniscus (somatosensory fiber pathway) and show the subsequent excitatory response from neurons in the lemniscal relay nucleus of the thalamus (Fig. 3.15A and E). The dots in Fig. 3.15 (B–D) represent a stimulus to the basal ganglia which produces a subsequent inhibitory potential (IPSP) in cells of the thalamic relay nucleus. Note in Fig. 3.15 (C and D) that the excitatory sensory response (arrow) is "gated" off by basal ganglia stimulation. Data such as these (see Desiraju *et al.*, 1969; Maekawa & Purpura, 1967) have led Purpura (1970) to conclude that the thalamus serves an "interface" function involving gating of sensory and motor influences to and from the cortex.

Thus, the anatomy and physiology of the thalamus suggest this structure to be the prime candidate for the role of a master synchronizer, involved in determining information distribution and time parsing of information flow. This proposed role implicates the thalamus in the phenomenon of perceptual framing, which will be discussed in a later section.

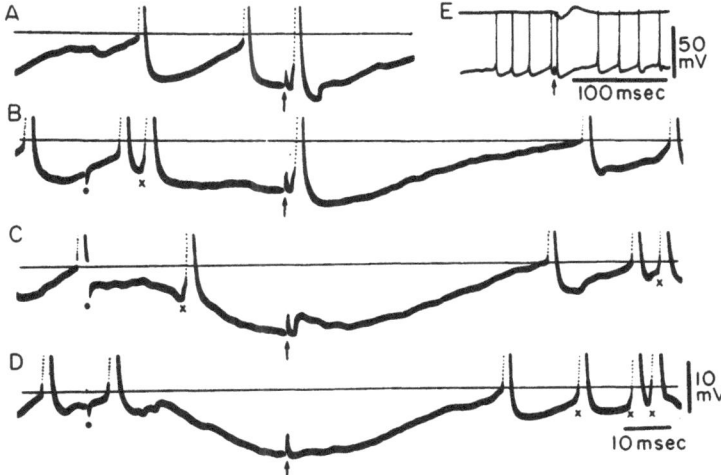

FIG. 3.15 Characteristics of excitatory postsynaptic potentials (EPSPs) and variations in firing level of a ventrobasal (VB) thalamic neuron activated by medial lemniscal stimulation (at arrows). (A–D) Photographically enlarged segments of records from continuous recording. Spikes are truncated to show the IPSP action. (E) Example to show full spike amplitude and primary response (upper channel) at cortical surface to lemniscal stimulation. (A) Lemniscal stimulation elicits a fast-rising EPSP which initiates a spike discharge at a higher level of membrane potential than do preceding spontaneous spikes. (B, C) Lemniscal stimulation is preceded by internal capsule stimulation (at dots), which evokes long-latency inhibitory postsynaptic potentials (IPSPs) in the VB relay neuron. In (C) spike generation is suppressed by the IPSP, whereas in (D) the EPSP fails in an "all-or-none" fashion. Variations in firing level are indicated by "×," which also identifies spikes preceded by fast depolarizing prepotentials. (From Maekawa & Purpura, 1967.)

A. EEG Scanning Mechanisms

The concept of scanning was originally introduced by Pitts and McCulloch (1947) in relation to perception in humans. Their idea of a periodic subcortical scanning pulse that probes the momentary state of neural excitability was consistent with evoked potential recordings demonstrating excitability fluctuations coincident with background EEG. For instance, Bartley in 1942 demonstrated a periodic alteration in the excitability cycle of the sensory evoked potential coincident with the frequency and phase of the alpha rhythm. Similarly, Dustman and Beck (1965) and Morrell (1966) demonstrated that behavioral reaction time was minimal when the triggering stimuli occur on the rising phase of the cortical alpha rhythm. Studies by Lilly and Cherry (1954, 1955) and Lilly (1954) involving simultaneous recordings from large arrays of cortical surface electrodes demonstrate that EEG rhythms are in constant motion, shifting across the cortical surface in complex but recurring patterns. Experiments by Elul (1969, 1972) show that the EEG recorded from the cortical surface represents intermittent synchronization of relatively small groups of neurons forming mosaics that shift in as yet indiscernible patterns across the cortical surface.

Although the functional significance of alpha rhythms is still unclear, these results suggest that the alpha rhythm and possibly other rhythms may reflect the action of a scanning mechanism that alternately excites and inhibits adjacent domains of cortical neurons.

This possibility is attractive, particularly since alpha rhythms are not solely confined to the visual cortex (Case, 1942), and even persist following complete removal of the striate cortex (Masland et al., 1949). It is a common finding that alpha spindles are most prominent when a subject's eyes are closed or when the subject has no visual imagery (Cobb, 1963a). This is not to say that when "alpha spindles" are absent there is an absence of EEG energy in the 7 to 13 Hz range. On the contrary, Morrell (1966b) has demonstrated that a flash of light may elicit large amplitude alpha spindles and Matousek and Petersén (1973) have shown that considerable power in the EEG is continuously present in the alpha

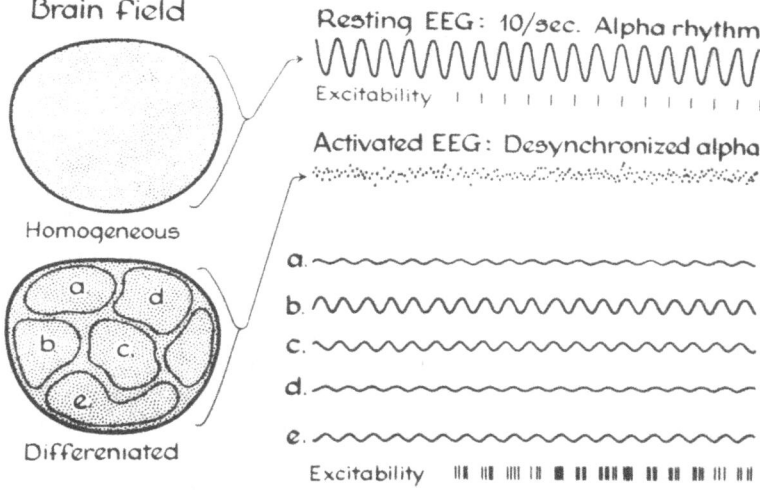

FIG. 3.16 Hypothetical brain fields, illustrating type of EEG assumed to be associated with each. Resting alpha at 10 per sec characterizes the homogeneous, or relaxed, condition. A desynchronized activated EEG is associated with a differentiated state of brain function (independent states a–e). The activated EEG is characteristic of attention and problem-solving and, in general, more efficient perception and performance. According to the concept that an excitability cycle is associated with the waxing and waning phases of the waves, it is evident that alternating periods of excitability and unexcitability could occur only 10 times per sec in the case of the resting EEG, whereas almost continuous excitability is represented in the case of the desynchronized EEG resulting from a differentiated brain field. (From Pribram, 1971; redrawn after Lindsley, 1961.)

range in awake subjects. Further, the proportion of energy at alpha frequencies (7 to 13 Hz) increases significantly as a function of postnatal maturation (Matousek & Petersén, 1973). The proportion of power at alpha frequencies also increases as a function of primate phylogeny (Jurko *et al.*, 1974).

It is important to distinguish between *alpha spindles* and *alpha rhythms*. Power spectral analyses show that in the waking state alpha rhythms ($\approx$ 10 Hz) are usually present (Bickford, 1975; Bickford, Brimm, Berger, & Aung, 1973; Matousek & Petersén, 1973). The rhythm generators appeared to be desynchronized, for the most part, and do not form spindles. However, when excitability conditions permit, alpha rhythms synchronize in space and time, creating the alpha spindle. This phenomenon can be visualized as a distributed or spatially differentiated set of alpha generators that suddenly come into common phase to produce synchronized waves of alpha activity. This relationship between continuously present alpha generators and momentary alpha spindles (alpha synchronization) is represented in Fig. 3.16. Andersen and Andersson (1968) have presented data indicating that the smallest alpha generator involves a thalamocortical projection of 200 μ in diameter (see Fig. 3.3). Work by Verzeano and co-workers (Verzeano, 1972) indicates that expansion, contraction, and differentiation in space and time occurs among thalamocortical loop (alpha) generators. Multiple interactive factors control the size and distribution of cooperative alpha generators. However, in aroused states differentiated alpha rhythms may represent multiple and quasi-independent scanning systems.

The strongest line of evidence favoring an EEG scanning mechanism comes from the work of Verzeano and co-workers who have used arrays of multiple microelectrodes to detect circulating neural activity (Verzeano, 1955, 1963; Verzeano & Negishi, 1960). Figure 3.17 shows some of the conditions under which circulating cortical patterns appear. In this figure an array of microelectrodes E_1, E_2, E_3, E_4, were placed in a line with tip separations from 100 to 150 μ. Figure 3.17 shows various degrees of synchronization and demonstrates a highly ordered spatial delay of unit firing that follows a cyclical motion. The geometric complexity of a circulating pattern is shown in Fig. 3.18 in which three microelectrodes recorded different thalamic units simultaneously. Figure 3.19 shows that the velocity of circulation varies in correspondence with the rhythmic frequency of the EEG, shown in line d. Verzeano and co-workers first discovered circulating patterns of unit discharge, which correlated with EEG, in the 1950s and have explored this phenomenon extensively in recent years (Verzeano, 1955, 1963, 1970, 1972; Verzeano & Negishi, 1960, 1961; Verzeano *et al.*, 1965). Based on these studies the following summarizing conclusions have been drawn (Verzeano, 1972): (1) the circulating activity proceeds along curved lines and frequently exhibits reversals of direction; (2) the velocity of circulation, estimated by the time it takes to cover the distance between two successive tips, ranges from .5 to 8 mm sec^{-1}, a rather low value indicating that many synapses and/or long axons are transversed; (3) high correlations are found with

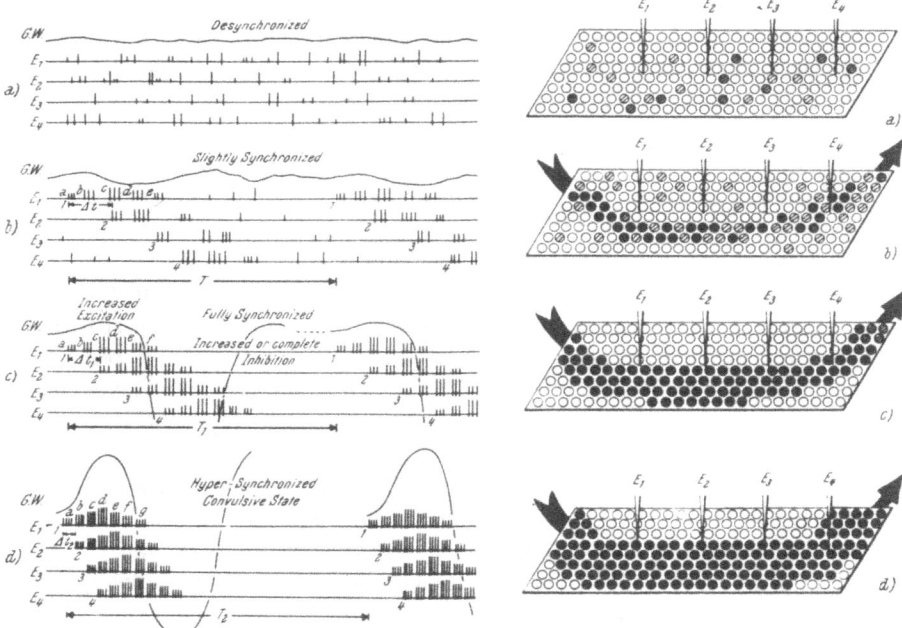

FIG. 3.17 Relations between neuronal discharge, circulation of neuronal activity, and synchronization of the gross waves, as they appear when recorded simultaneously with four microelectrodes with tips separated by 100 to 150 μ. Left: Diagrammatic representation of oscilloscope tracings, showing the progressive changes that take place from the desynchronized to the hypersynchronized state. Right: Diagrammatic two-dimensional representation of hypothetical neuronal networks, showing the neuronal activity corresponding to each successive state, E_1, E_2, E_3, and E_4 represent the microelectrodes through which oscilloscope tracings E_1, E_2, E_3, E_4, (at left) would be obtained; the circles represent neurons in the two-dimensional networks; the degree of darkness in each circle represents the degree of excitation of that particular neuron; the arrow represents the direction of the circulation of neuronal activity. (A) Oscilloscope: infrequent clustering of the neuronal action potentials, no circulation of neuronal activity, no synchronization of the gross waves. Neuronal network: scattered, sporadic neuronal activity, at low level of excitation; no circulation of neuronal activity. (B) Oscilloscope: increased clustering of action potentials; occurrence of neuronal activity in regular succession at each one of the tips of the microelectrodes, indicating circulation through the neuronal network; decreased activity in the interval (T) between successive passages of circulating activity through the network; gross waves slightly synchronized. Neuronal network: neuronal activity at higher level of excitation, concentrated mostly in the pathway of circulation (arrow); decreased activity outside this pathway. (C) Oscilloscope: high degree of clustering of action potentials; increase in the velocity of circulating activity ($\Delta t_1 < \Delta t$; activity abolished in an increased interval (T_1) between successive passages of circulating activity through the network; gross waves fully synchronized. Neuronal network: neuronal activity at high level of excitation in the pathway of circulation; no activity outside the pathway. (D) Oscilloscope: extreme degree of clustering of action potentials; high velocity of circulation ($\Delta t_2 < \Delta t_1$); further increase in the

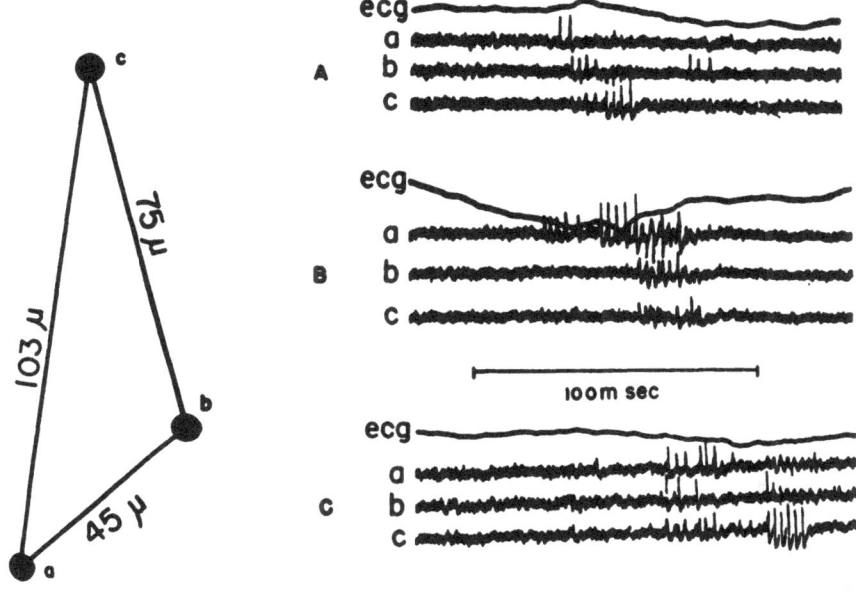

FIG. 3.18 Pathway of circulating neuronal activity in a tridimensional network. Simultaneous tracings obtained out with three microelectrodes whose tips (a, b, c) were arranged as shown in the diagram at the left; location, nucleus lateralis posterior of the cat. In (A), activity circulates from a to b then to c and back to b. In (B), activity arrives first at a, then progresses toward the center of the triangle (all three microelectrodes show action potentials simultaneously). In (C), activity appears first near the center of the triangle (all three microelectrodes show action potentials simultaneously), then moves toward the vicinity of c (high-amplitude action potentials on tracing). Anesthesia: sodium pentobarbital 6 mg/kg. Negativity up. (From Verzeano & Negishi, 1960.)

the frequency of rhythmic EEG; and (4) the pathway of circulation follows a series of loops, whose "locus" advances spatially through the neuronal networks. Some of the loop dynamics are illustrated in Fig. 3.20. In this figure three loops are represented as a', b' and c'. Electrodes E_1, E_2, E_3, and E_4 are positioned so that successive circulations through loops appear as progressive bursts of cells at the different electrode locations. In Fig. 3.19 a simpler circulation was involved, where the tips of microelectrodes were dispersed along a single loop of the pathway. In this case, activity traveled directly from one tip to the next with a

interval (T_2) between successive passages of circulating activity through the network; neuronal activity in this interval abolished; hypersynchronized gross waves of high amplitude. Neuronal network: neuronal activity at very high level of excitation concentrated exclusively in an enlarged, multilane pathway of circulation; completely abolished outside this pathway. G.W.: gross waves recorded within the same networks, by the same microelectrodes. (From Verzeano, 1963.)

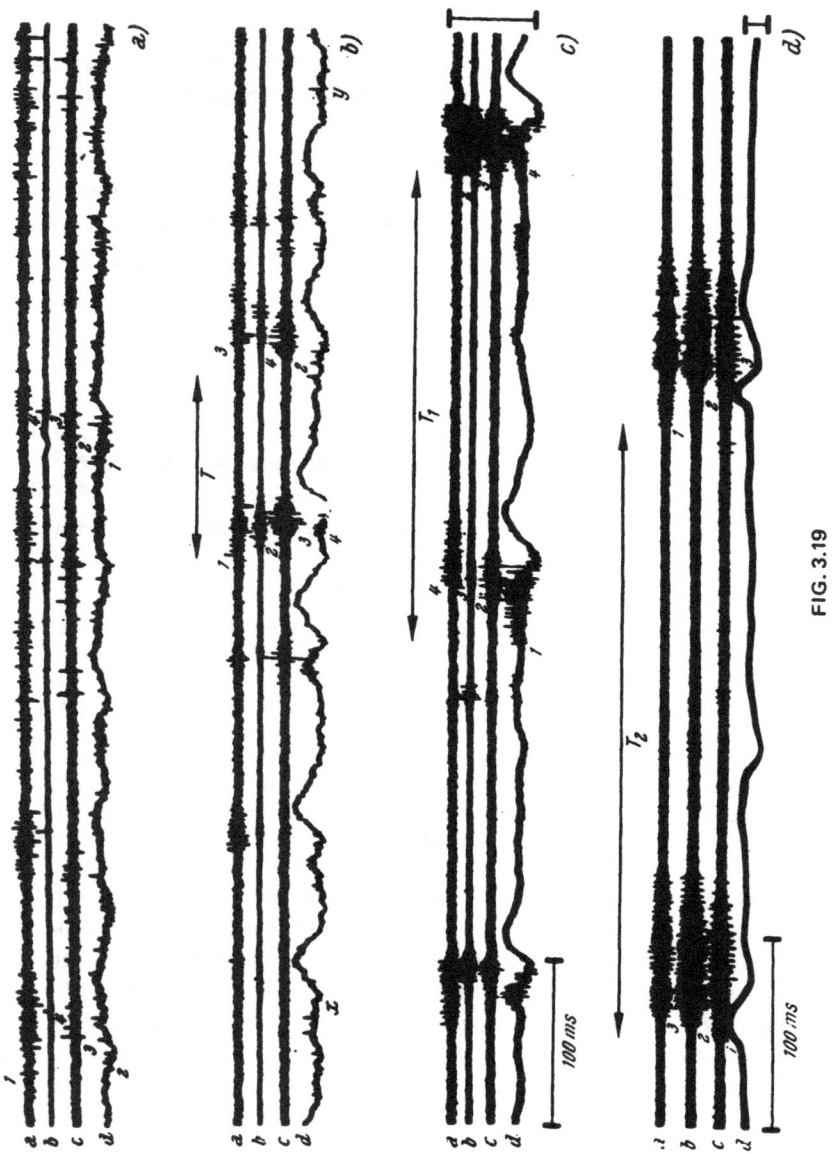

FIG. 3.19

FIG. 3.19 Relations between neuronal discharge, circulation of neuronal activity, and synchronization of gross waves, in thalamic neuronal networks, shown by recordings obtained, simultaneously, with four microelectrodes (a–d). Channels a, b, and c show neuronal action potentials; channel d shows action potentials as well as gross waves. (A) Recordings obtained from the nucleus ventralis medialis of the waking cat under gallamine. (B) Another section of the same recording as in (A), in which some "synchronization" of the gross waves occurs in wakefulness (x to y) and in which clustering of neuronal action potentials and circulation of neuronal activity can be seen (at 1–2–3–4), in association with the gross waves. (C) Recordings obtained from the same animal in the same experiment, with the same array of microelectrodes at the same thalamic location, in sleep induced by sodium pentobarbital; the increase in the synchronization of the gross waves is accompanied by an increase in the clustering of neuronal action potentials, an increase in the regularity of the circulation of neuronal activity (at 1–2–3–4), an increase in the number of neurons involved, and an increase in the duration of the period of silence (T_1) between successive passages of circulating activity through the network. (D) Recordings obtained, in the "hypersynchronized" preconvulsive state, from another animal, under gallamine, 15 min after the administration of .9 mg/kg of picrotoxin. The clustering of neuronal action potentials is very marked, the circulation of neuronal activity is highly regular (1–2–3), a large number of neurons is involved in it, and the period of silence between successive passages of activity (T_2) through the network is much longer. These changes correspond to a great increase in the amplitude of the gross waves. The time line and amplitude calibration (.5 mV) under (C) apply to (A), (B), and (C); the time line and amplitude calibration (.5 mV) under (D) apply only to (D). Distances between the microelectrodes; in (A)–(C): ab = 120 μ; bc = 100 μ; cd = 155 μ; in (D): ab = 50 μ; bc = 80 μ; cd = 90 μ; microelectrode d in (D) shows only gross waves. Negative up in all tracings. Note the frequent reversals in the direction of circulation (at 1, 2, 3, 4): in (B) from a–b–c–d to d–c–b–a; in (C) and (D) from d–c–b–a to a–b–c–d. (From Verzeano, 1963.)

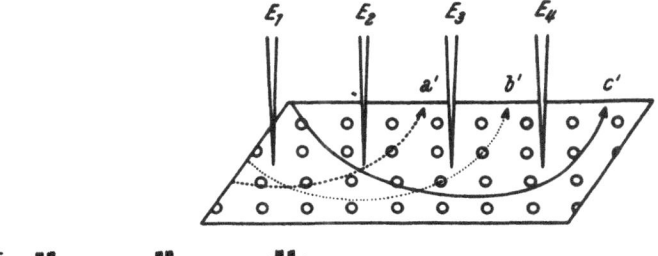

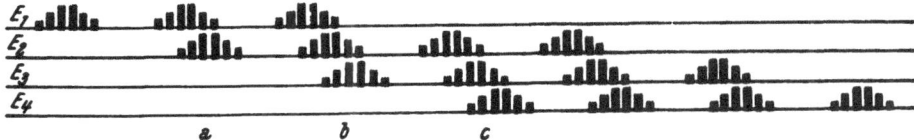

FIG. 3.20 Two-dimensional diagrammatic representation of the configuration of the pathway of circulation of neuronal activity as determined with four microelectrodes (E_1, E_2, E_3, E_4) with tips displayed along a straight line. Activity appears at one extremity of the line (E_1) and proceeds by successive steps (a, b, c, etc.) toward the other extremity (E_4), suggesting a pathway of circulation which follows along a series of loops (a, b, c) whose "locus" advances progressively through the neuronal network. (From Verzeano & Negishi, 1961.)

high apparent velocity. In Fig. 3.17 successive loops were involved, thus the apparent velocity of circulation was lower. The evidence accrued by Verzeano and co-workers over the years strongly suggests that there are networks in neural systems through which circulation appears as moving waves capable of sweeping across large domains of tissue. These data are consistent with the phenomena of EEG waves apparently moving across surfaces, as mentioned earlier (Lilly & Cherry, 1954, 1955), as well as with the findings of other workers (Elul, 1972; Gavrilova, 1970; Knipst, 1970). As discussed in Chapter 7, the slow movement of neural activity in a loop system may also underlie the neural representation of time.

The phased loop systems demonstrated by Verzeano and co-workers are for the most part thalamocortical, cortical-cortical, and thalamic-thalamic. The vertical thalamocortical loop can activate a single cortical column (Andersen & Andersson, 1968; see Fig. 3.4). In Figs. 3.17 and 3.19 the size of the cortical loop system can be seen to change as a function of the degree of synchrony in the cortical mantle. This suggests that the cortical EEG is produced by the activity of a large number of vertical loop systems that sweep across the surface of the cortex, with recruitment of adjacent loop systems occurring during EEG synchrony. During EEG desynchrony, these loop systems are fragmented and broken into a large number of independent oscillators. This suggests that the cortical EEG may reflect a multitude of scanners, whose extent may vary depending on levels of arousal.

The question must be asked: What might be the function of such a scanning system? Is it related to memory, to cognitive "set," or to perception? The answer is probably that scanning is related to all of these phenomena. The circulation times of 1 mm sec^{-1} to 8 mm sec^{-1} show that the scanners are slow. However, the microelectrode studies of Verzeano and co-workers reveal activity on only one surface of a three-dimensional system. The slow scanning times represent a movement in space via a sequence of thalamocortical loops, producing activity traveling within a spiral (see Fig. 2.33). Single vertical thalamocortical and corticothalamic connections conduct rapidly (10 mm sec^{-1} to 20 mm sec^{-1}; Hursh, 1939). Thus, there exists a wide spectrum of conduction and circulation times. This spectrum of rapid pulsatile and slow scanning movements may be involved in a wide spectrum of temporal processes. For instance, stimulus recognition, as reflected by evoked potential wave forms, occurs in less than 120 msec (John *et al.*, 1973). Stimulus matching processes, as reflected by evoked potential wave forms, occur in less than 400 msec (Thatcher, 1974b; Thatcher & John, 1975). Change of cognitive set sometimes requires many seconds (Lashley, 1958). Memory retrieval and search processes, with which we are all familiar, may involve minutes. Finally, the "specious present," which is the duration of the consciousness of an event belonging to the present, after which occurs a distinct perception of the event as belonging to the past, has been estimated by William James (1890) to be 8 to 16 sec long. Thus, there are hierarchical levels of temporal organization (see Chapter 7). EEG scanning mechanisms, which encompass a wide spectrum of "time," may be involved in any or all of these phenomena.

B. The Phenomenon of Perceptual Framing

The electrophysiological data reviewed thus far can be summarized as evidence that the brain is composed of a large number of complexly interacting loop systems. In the previous sections thalamocortical and corticothalamic loops as well as intrathalamic loops were discussed. It was emphasized that the thalamus exerted a powerful and periodic influence upon cortical projection regions and most likely operates as a sensory and motor integrative gating network. In this section, the role of the thalamus in the phenomenon of perceptual framing will be discussed. From the outset it is important to emphasize that the reticular formation and the more rostral midline core thalamus are believed to form a complex integrative system (Scheibel & Sheibel, 1967b, 1970). The extreme diversity of cell type and the nearly isotropic organization of cell connections renders the midline core among the most difficult to dissect anatomically and the most difficult to understand in terms of its pluripotentium of functional roles. In light of the complexity of thalamic midline core structures, it is indeed remarkable that comparatively simple electrophysiological patterns appear redundantly in electrophysiological studies (Purpura 1970, 1972). Dominick Purpura (1970, 1972) recently surveyed the entire field of intracellular analyses of thalamic, cortical, cerebellar, and basal ganglia structures; reviewing studies spanning some 30 years. For the purposes of this discussion two rather general conclusions from Purpura's (1972) analysis will be considered: (1) "The most common PSP (postsynaptic potential) pattern observed in neurons of the mammalian brain consists of an EPSP–IPSP sequence" and, (2) "Long duration PSPs (postsynaptic potentials) are characteristically observed in mammalian forebrain neurons, irrespective of the morphology or location of the elements in a particular organization." It is important to emphasize that the duration of these omnipresent EPSP–IPSP sequences in mammalian forebrain neurons commonly range from 80 to 200 msec. The question one must ask is: Is there anything functionally important about this highly pervasive interval of time?

Studies by Efron (1963b, 1967, 1970a, b) and others (Fraisse, 1963; Shallice, 1964; White, 1963) suggest that this particular interval may be of considerable importance. These workers maintain that the neurophysiological processes underlying perception are temporally discontinuous involving a sensory sampling interval or frame estimated to be approximately 100 msec in duration. In conjunction with the scanning properties of EEG (Pitts & McColloch, 1947) it has been speculated that neural sensory processes operate by successively integrating discrete temporal samples of incoming information (Allport, 1968). This continual process of integrating discrete sensory samples is called the *traveling moment hypothesis* of perception (Allport, 1968; Sanford, 1971). Efron (1967) maintains that the onset of a sensory event initiates an integrative process which has a duration less than 66 msec and organized such that any other events occurring within the integration span are added together or integrated to represent a perceptual entity or frame. In support of this, Efron (1967) reports an experiment in which it is demonstrated that if two 20-msec duration visual stimuli, one blue the other red, are presented in rapid succession, then the human observer commonly perceives a single green flash of light. A similar auditory integration was reported in which two different 20-msec tone bursts (1000 Hz and 4000 Hz) were presented successively. In this case subjects perceived a 2500-Hz tone (Efron, 1967).

The most convincing evidence for the existence of a perceptual frame is provided by well-controlled psychophysical experiments (Efron, 1970a, b). In these experiments, human subjects are instructed to match the onset of one sensory event (event B) with the offset of another (event A). The experimenter could vary the duration of event A, in a counterbalanced manner, and the subject could adjust the onset of event B with a Rheostat. With event A durations of 120 msec or greater, subjects could accurately match onset with offset. However, with durations of event A less than 120 msec the accuracy of matching decreased rapidly with errors of estimate increasing as a function of decreased duration of event A. This experiment, in conjunction with others (Shallice, 1964; White, 1963), indicates that there is a minimum duration of a perception and that the limiting temporal value for visual stimuli lies within the range of 120 and 240 msec and for auditory stimuli between 120 and 170 msec (Efron, 1970a).

Studies reported by Efron (1963a) indicate that the temporal discrimination of simultaneity and temporal order are performed in the speech dominant hemisphere. These studies, plus those cited earlier, are of clinical relevance. For instance, Efron (1963b) has shown that aphasic patients with unequivocal damage to the left hemisphere are impaired in their perceptions of simultaneity and temporal order. In a paradigm that required subjects to judge which of two stimuli appear first it was found that aphasic patients required interstimulus intervals sometimes as great as 600 msec before an accurate discrimination could be performed (Efron, 1963b). The significance of this can be understood by considering that a word is a sequence of auditory events (phonemes). Normal English speech contains 120–150 words per minute and an average of 5 phonemes per word. This is a rate of approximately 80 msec per phoneme. If the concept of perceptual framing is correct and is functionally related to the time parsing of sensory events, then it is understandable that damage to this process, particularly injury that reduces the capacity to "time label" or "sequence" phonemes, might effect the comprehension of speech.

In earlier discussions it was argued that the thalamus plays a fundamental role in synchronizing processes by gating and directing the flow of sensory information. The gating and synchronizing processes occur via the interaction of inhibitory synaptic potentials which turn certain systems off while allowing others to be excited. In line with these data is a recent intracellular investigation which directly implicates the IPSP in the segregation of perceptual frames (Kuhnt & Creutzfeldt, 1971). The latter study has shown that IPSPs in visual cortical neurons become severely attenuated at stimulus frequencies near flicker fusion. EPSPs, on the other hand, were largely unchanged. This finding, in conjunction with the studies of Desiraju *et al.* (1969) showing thalamic gating of sensory flow, indicates that inhibition isolates or segregates events in space and time. In later chapters neural representational systems will be defined in terms of information in the CNS (Chapters 6 and 7). The coordinated activity of IPSPs and EPSPs in thalamocortical networks will be argued to form the basis of a "sculpturing" process which is fundamental to the creation of representational systems. From this perspective, the phenomenon of perceptual framing can be understood as the first level of sensory packaging through which representational systems are created.

C. Control Systems and Thalamocortical Loop Processes

The importance of understanding the basis of synchronization and thalamocortical time parsing warrents further speculative remarks concerning how these processes might take place. In Section III evidence was presented showing that thalamocortical synchronization involved the maturational convergence of two loop processes, one excitatory, the other inhibitory (Thatcher & Purpura, 1972, 1973). These loops are differentiated at birth, mature at different rates, and converge to form a functional system in adulthood. It appears that these balanced inhibitory and excitatory loop processes function on at least two interrelated but clearly different levels. On one level the loops appear to be part of a regulatory or control system involving inhibitory (negative) and excitatory (positive) feedback. This balanced feedback system operates in accordance with homeostatic control systems seen throughout the body (Guyton, 1971) which keep dynamic operations within specified limits. In the case of neural control systems interactive loop processes keep excitatory and inhibitory drives in bounds, which prevents epilepsy and regulates the levels of excitation such as in sleep–wakefulness and attention (see Chapter 4). The second set of functional operations of the loop systems appears to directly involve information processing. For instance, many studies show that in sensory systems *lateral inhibition* (where one neuron inhibits a neighboring neuron) enhances contrast by focusing excitation within local cellular domains (Hubel & Wiesel, 1959, 1963; Lettvin *et al.*, 1960; Ratliff, 1960). As discussed in the previous section evidence indicates that, in contrast to lateral inhibition, *recurrent inhibition* (where a neuron inhibits itself) operates to localize or segregate events in time. On the other hand, excitatory feedback, which in control systems is kept in bounds by inhibition, functions to select and ensure the preservation of excitatory flow within certain loop systems. In the case of information processing, inhibition serves to *sculpture* the flow of excitation in space and time. Thus, the regulatory or control properties of negative and positive feedback can be visualized as a creative process in which inhibition shapes excitation within certain loop systems. In Chapter 6 more detailed information will be presented in support of the idea that representational systems have specific shapes in space and time and that the morphology or geometry of representational systems is itself of informational significance. In any case, the role of the thalamus and its dynamic inhibitory–excitatory systems is believed to be of primary importance in the formation of representational systems.

4
Neurophysiology of Arousal and Attention

I. INTRODUCTION

Now that the foundations have been laid, consisting of the overviews of basic neurophysiology, functional and neurochemical neuroanatomy, and slow wave electrophysiology, we are ready to begin our examination of the performance of complex functions by the brain. A logical entry point to these problems is the question of how the levels of arousal of the brain are regulated.

Changes in general "brain state" in a normal individual range from deep sleep, including the dream or rapid eye movement (REM) state, to lighter stages of sleep, through drowsiness to high levels of arousal, vigilance, and attention. Coma caused by damage to arousal systems, narcolepsy, catalepsy, agitation, and hyperactivity are a few of the pathological conditions involving disruption of the normal regulation of brain states. What are the mechanisms that control brain state? How do various neural subsystems interact to characterize a specific brain state? What are some of the dynamics involved in changes in levels of arousal?

Partial but important answers to these questions have developed since the 1930s, when Hans Berger (the "father" of EEG) first described EEG changes that parallel behavioral arousal. A considerable advancement in our understanding of arousal systems followed the discovery of the reticular formation and the diffuse thalamic projection system and the subsequent elucidation of the control these structures exert on the excitability of forebrain and spinal systems.

II. THE RETICULAR ACTIVATING SYSTEM

Figure 4.1 shows in schematic form some of the ascending and descending connections of the reticular formation. In this scheme, the reticular formation is represented by the dark central core extending from the medullary regions of

FIG. 4.1 Schematic representation of the relationship of the reticular core of the brain stem (black) with other systems of the brain. Collaterals pour in from long sensory and motor tracts (thin lines) and from the cerebral hemispheres (arrows directed downward into reticular core). The core acts in turn on cerebral and cerebellular cortices (upward-directed arrows) and associated structures, on spinal cord (long downward-directed arrow), and on central sensory relays (striped arrows). (From Worden & Livingston, 1961. Copyright © 1961 by the Hogg Foundation for Mental Health.)

the lower brain stem up through the region of the pons, midbrain, and diencephalon. Extensive reticulocortical and corticoreticular connections, as well as descending reticulospinal influences, are represented by the thick arrows. Ascending sensory fibers are shown which send collaterals to the reticular formation on the ventral surface. Corticofugal motor output fibers (shown on the dorsal surface) also give off collaterals.

The integrative properties of the reticular formation can be demonstrated anatomically and electrophysiologically. For example, unlike specific sensory relay systems, many reticular formation cells respond to all modalities of sensory stimulation (Palestini *et al.*, 1957; Scheibel & Scheibel, 1958). Multimodal cells are present throughout the rostral–caudal dimension of the reticular formation, with the greatest concentration in the medial two-thirds.

The most dramatic picture of the reticular formation's integrative properties is provided by anatomic studies revealing interneuronal relations such as those seen in Fig. 4.2. In this figure, a single neuron (R) located in the nucleus reticularis gigantocellularis gives rise to an axon that courses dorsally a few millimeters,

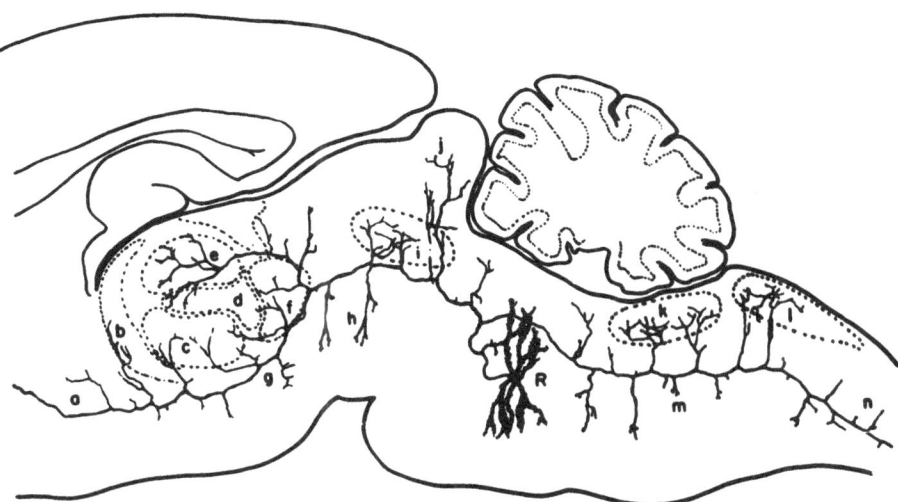

FIG. 4.2 Sagittal section through the brain of a young rat showing the axonal trajectory of a single neuron of the nucleus reticularis gigantocellularis (R). The rostral-coursing axonal component supplies collaterals to inferior colliculus (j); (i) the region of the III and IV nerve nuclei; (h) mesencephalic tegmentum; (f) posterior nuclear complex of thalamus; (e, d, and c) dorsal, intralaminar, and ventral thalamic nuclei, respectively; (g) zona incerta of hypothalamus; (b) nucleus reticularis thalami; and (a) basal forebrain area. The posterior-directed component seeds collaterals into the substance of the reticular core (m); the hypoglossal nucleus (XII) (k); the nucleus gracilis (l); and the intermediate gray matter of the spinal cord (n). (From Scheibel & Scheibel, 1958.)

then bifurcates to send a long descending branch to the intermediate gray matter of the spinal cord with axonal collaterals synapsing with both motor and sensory nuclei along the way. The ascending branch courses rostrally, synapsing with many neurons in the pons, mesencephalon, diencephalon, and even basal forebrain structures. This demonstrates the extensive rostral–caudal influence which a single midline core reticular neuron exerts, extending from the spinal cord to the forebrain.

Note the rather extensive dendritic branchings of neuron R. The dendritic domain of such a neuron samples input from many sources. Figure 4.3 shows the dendritic organization that is typical of midline core neurons. In general, the dendrites of neurons comprising the medial two-thirds of the reticular formation are compressed along the rostral–caudal axis, with extensive branchings in the lateral or dorsal–ventral plane. As a consequence, the dendritic domains of these neurons produce a set of poker chips stacked in succession. Such a neural organization operates by sampling, at each poker chip, the ascending and descending confluences of afferent sensory-specific and efferent motor fibers coursing in the lateral one-third of the core structure (see insert, Fig. 4.3). These

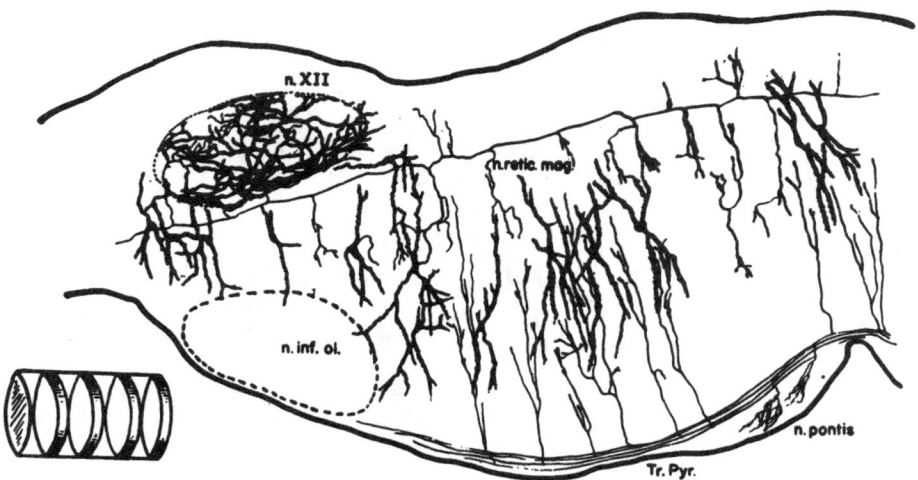

FIG. 4.3 Sagittal section through the lower half of the brain stem of a 10-day-old rat. Most of the dendrite mass of reticular core cells is organized along the dorsoventral axis as seen in this type of section, with marked compression along the rostrocaudal axis. This orientation places the dendrites parallel to the terminal presynaptic components which in this case arise from pyramidal tract (Tr. Pyr.) and from a single axon of a magnocellular reticular neuron (n. retic. mag.). This type of dendrite organization, which is especially characteristic of reticular cells of the medial two-thirds of the core, produces sets of essentially two-dimensional modular neuropil fields leading to the stack of chips analogy (see text and inset diagram at lower left). (From Scheibel & Scheibel, 1958.)

figures help to emphasize that sensory input and motor input are continuously integrated along the midline core in both thalamic and reticular formation systems. This lends credence to the idea that perception is an active process that entails coordinated sensory–motor interaction.

III. RECIPROCAL INTERRELATIONS IN THE CONTROL OF AROUSAL

Relatively low levels of high-frequency stimulation in the midline thalamus and reticular formation can transform a sleeping animal, whose EEG is dominated by slow waves, to one displaying an activated EEG and behavioral arousal. This effect is mediated largely by excitatory drives exerted on cortical, limbic, and basal ganglia structures.

Lesions of the mesencephalic and pontine reticular formation result in a comatose state, characterized by a slow wave EEG which resembles sleep. The symptoms of the sleeping disease, encephalitis lethargica (Economo, 1918) are largely the result of lesions in the central gray of the brain stem. Similarly, tumors in this region produce prolonged states of somnolence (Fulton & Bailey, 1929).

These classical findings, which constituted an important contribution in the development of contemporary neuroscience, led to the formulation of the concept of an Ascending Reticular Activating System involved in the control of sleep–wakefulness, (Lindsley *et al.*, 1949a; Moruzzi & Magoun, 1949). Damage to the sensory-specific systems had little effect on sleep–wakefulness, whereas lesions placed a few centimeters more medially would produce a depressed, somnolent state.

The possibility of reciprocal corticoreticular influences on the control of arousal was demonstrated through stimulation studies by French, Hernández-Peón, and Livingston in 1955. Lindsley (1950) was one of the first to emphasize the role of descending cortical influences in the control of arousal and attention, as well as in emotion and motivation. The recognition that limbic system and cortex made contributions to the reticular formation made it possible to conceive of how internal states such as thoughts, worries, and apprehensions might generate arousal activity in the reticular formation.

Cortical as well as reticular control of afferent sensory information processes at peripheral levels have been demonstrated at the first synapse (Granit, 1955; Hagbarth & Kerr, 1954; Hernández-Peón, 1955; Kerr & Hagbarth, 1955; Spinelli & Pribram, 1967), as well as at more central thalamic and cortical levels of sensory systems (Bremer & Stoupel, 1959; Dumont & Dell, 1960; Hernández-Peón, Scherrer, & Velasco, 1956b). Although corticofugal feedback is largely inhibitory, it has also been shown that reticular formation stimulation can enhance evoked potential amplitude in thalamic relays and cortex (Bremer & Stoupel, 1959; Dumont & Dell, 1960) as well as facilitate temporal resolution of evoked responses to paired flashes (Lindsley, 1958). Fuster (1958) showed that reticular formation stimulation in monkeys can reduce reaction time in a visual discrimination task, and Lansing *et al.* (1959) showed that an arousing preparatory signal can produce decreased reaction times.

Thus, reciprocal reticulocortical interactions as well as reticulo and corticofugal (descending) influences on sensory relay nuclei and spinal motor systems have been demonstrated.

A. The Control of Sleep–Wakefulness

Some of the complex reciprocal reticular formation interactions involved in the control of sleep–wakefulness have been shown in both lesion and stimulation studies. For instance, it appears that sleep can be produced not only by the reduction of activity in a tonic activating system, but also by the influence of an active sleep-producing system. Hess (1944) and others (Akimoto *et al.*, 1956; Hess *et al.*, 1953) demonstrated that low-frequency stimulation of diencephalic structures would cause otherwise alert cats to select a likely place, after which they would curl up and then proceed to go to sleep much as a normal animal. In a study of the hypothalamic region of rats, Nauta (1946) found that lesions of the posterior thalamus produced prolonged somnolence whereas lesions of more anterior hypothalamic regions produced rats incapable of sleeping. These results fit in with the findings of Economo (1918) who found, in some encephalitic patients, prolonged sleeplessness associated with lesions in the anterior hypothalamus (Ochs, 1965).

More recent investigations have mapped out the areas within the hypothalamus, limbic system, and brain stem involved in the control of sleep and wakefulness (Hernández-Peón & Ibarra, 1963; Hernández-Peón et al., 1963; Jouvet 1969). Hypnogenic regions have been found throughout the pons, upper mesencephalon, diencephalon, and limbic system, as well as the orbital frontal cortex (Akert et al., 1952; Hess, 1954). A reciprocal activation–suppression relationship exists within the pontine reticular formation. Batini et al. (1959) demonstrated that lesions of the rostral pons gave rise to an EEG picture of arousal, while lesions a few millimeters lower, at the midpointine level, resulted in an EEG pattern typical of sleep. The inference drawn from these studies was that a mechanism present in the lower part of the reticular formation brings about sleep by inhibiting the upper reticular formation (Moruzzi, 1960). This inference was supported by experiments that blocked the upper and lower brain stem structures separately through the perfusion of barbiturates in the carotid and vertebral arteries, respectively (Magni et al., 1959). Perfusion of the caudal half of the brain stem through the vertebral arteries resulted in activation of the EEG, while barbiturate perfusion of the upper portion via the carotids resulted in sleeplike EEG.

Thus, two systems are involved and distributed within both diencephalic–limbic and reticular formation structures, one a sleep producing or hypnotic system, the other an awakening or arousal system. These systems appear to be reciprocally connected so that activity in one will suppress the other and vice versa, thus giving rise to the sleep–wakefulness cycle (Hernández-Peón & Ibarra, 1963).

Reciprocal biogenic amine interactions also play a role in sleep production. For instance, deposition of drugs such as p-chlorophenylalanine, which blocks serotonin production without affecting noradrenaline or dopamine, causes insomnia and significant reductions in both paradoxical and slow wave sleep (Jouvet, 1969). A normal state can be restored by the administration of 5-hydroxytryptophan, the immediate precursor of serotonin. On the other hand, deposition of chemicals that inhibit the production of noradrenaline cause a selective decrease in paradoxical sleep (Jouvet, 1969).

These biogenic amines are located in different regions of the brain. Serotonin is manufactured largely in the nucleus raphe in the midline reticular substance, while noradrenaline is manufactured largely in the locus coeruleus of the mesencephalic reticular formation (Kety, 1967, 1970; Moore et al., 1973, see pp. 26–30). Lesions of the nucleus raphe cause insomnia in cats, while lesions of the locus coeruleus have little effect on slow wave sleep but abolish REM or paradoxical sleep (Jouvet, 1969).

Jouvet believes there is an intimate relationship between these systems, such that a serotonin mechanism in the caudal raphe "primes" the noradrenaline system in the locus coeruleus to "trigger" paradoxical sleep.

B. Rostral and Caudal Influence of Reticular Formation

Some other dynamic reciprocal (facilitatory–inhibitory) relations in the diencephalon and brain stem are shown diagrammatically in Fig. 4.4. This set of relations (pluses representing facilitatory outflows and minuses representing inhibitory influences) was discovered by Magoun and collaborators (Bach & Magoun, 1947; Lindsley et al., 1949b; Magoun & Rhines, 1948). These workers demonstrated that when stimulated, the rostral portion of the reticular formation (involving mesencephalon and pontine regions labeled as structure 5) caused facilitation or enhancement of spinal reflexes and cortically initiated movement. A second, more dorsal facilitatory region (labeled 6) is related to the vestibulospinal system. Stimulation of this system affects tonic facilitatory mechanisms involved in the control of posture.

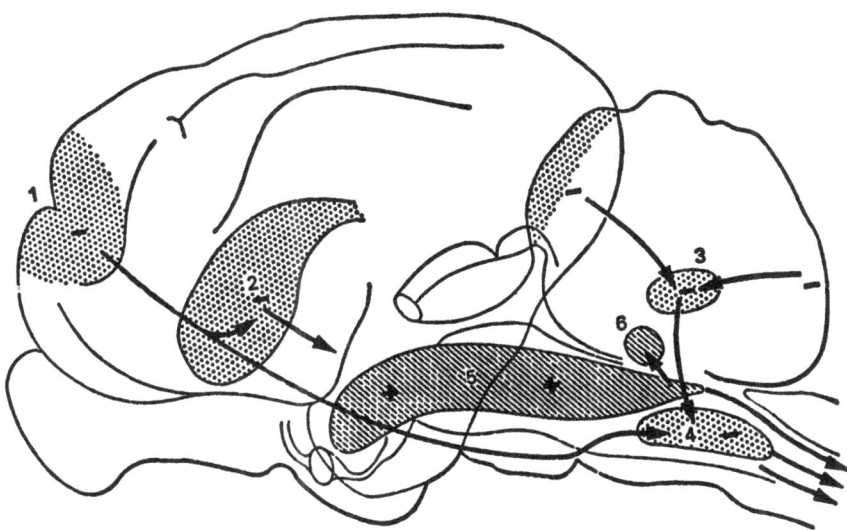

FIG. 4.4 Excitatory and inhibitory influence of stimulation of the nuclei of the reticular formation; (5) and (6) facilatory, and (4) inhibitory zones of the brain stem reticular formation and connections running to it from the cortex (1) thalamus (2) and cerebellum (3). (From Luria, 1973.)

In contrast, stimulation of area 4 (caudal reticular formation) results in an inhibitory effect on spinal motor outflow and has been shown to reduce spasticity and hyperreflexia. Stimulation of the other inhibitory regions (1, 2, or 3) reinforces the suppressor or inhibitory role of the caudal reticular formation. Lesions of the inhibitory regions result in increased spasticity or hyperreflexia due to the dominance of the facilitatory reticular formation system. Thus, it appears that an appropriate balance between inhibition and facilitation can be maintained by such a mechanism, with the normal state of a waking animal favoring facilitation of spinal reflexes and motor outflows. In relaxation, drowsiness, or sleep, there is a progressive tendency for the inhibitory system to dominate, resulting in sluggish reflexes and loss of postural control.

This complex set of interrelationships reveals a system with enormous capabilities for the regulation of levels of arousal. First of all, the system possesses wide-ranging ascending influences so that its outflow can modulate the level of excitation of extensive regions of the cortex. In the case of absent or diminished outflow, the tonic level of cortical excitability will drop. Second, the inputs to this system from relay nuclei, limbic system, and cortex can act upon it so as to cause alterations in the level of arousal. Finally, outputs from this system to structures which are possible sources of such inputs provide the capability of fine adjustments, increasing the effectiveness of particular inputs so small changes in afferent intensity will cause major changes in arousal level, or decreasing the effectiveness of other inputs so they are temporarily excluded from acting upon the system. Such a system is ideally suited for differential monitoring of the major sources of environmental information, as will be discussed in Chapter 9.

IV. CHANGES IN SINGLE UNIT DISCHARGE DURING AROUSAL AND ATTENTION

Our examination in the previous sections of the mechanisms which regulate arousal make it apparent that a complex functional organization exists which can modulate the effectiveness of afferent input. Not only can cortical and reticular formation influences facilitate certain thalamic relays and inhibit others, but the reactivity of the cortical and reticular regions themselves can be differentially enhanced to certain inputs. One important consequence of this capability, of course, is that the system can exclude repetitious stimuli of low information value from exerting a continuous stimulating effect, while permitting other novel stimuli to impinge upon it, leading to adaptive arousal. Such a mechanism provides a great increase in the efficiency of environmental monitoring for sources of danger or potential food. Another important consequence, perhaps less obvious, is that by modulating the reactivity of afferent systems in a differential way, the system can selectively enhance the signal-to-noise ratio of events in a particular sensory modality. Potentially, this specification of configurations of stimuli to be differentially admitted or excluded might even operate within a sensory modality. When the level of arousal of the brain increases, an animal appears to become more vigilant. It becomes restless and often displays searching head movements which orient its olfactory, auditory, and visual sensory receptors toward all regions of the environment. Such behavior is described as an increased level of vigilance or attentiveness. By "attention," we mean the focusing of perceptual mechanisms upon inputs in particular sensory modalities or upon the specific configuration of stimuli which correspond to a unique event in the environment. This process enhances the signal-to-noise ratio of the event brought into focus, facilitating the processing of information about the event. To the extent that attentiveness facilitates adaptive behavior, it can be considered as a rudimentary class of cognitive process.

It is instructive to consider the various neural interactions that characterize different brain states. Figure 4.5 shows neural discharge in two simultaneously recorded neurons during waking, slow wave sleep, and paradoxical sleep (S-LVE). It is important to note that the change from sleep to wakefulness does not necessarily involve reduced neural output but rather a modification in the pattern or organization of the discharge (Evarts, 1964). During waking, the correlation of discharge between two simultaneously recorded *cortical* neurons is nearly zero (i.e., random with respect to each other) with a wide range of discharge frequencies and interspike intervals present in both trains (Fig. 4.5). The transition from waking to sleep involves a tendency of cells to exhibit burst discharges and for two simultaneously recorded trains to become more correlated. With deeper levels of slow wave sleep, intercorrelations between neurons further increase and cells exhibit more pronounced burst discharges interspersed with periods of relative inactivity. As can be seen in (Fig. 4.5), unit discharge (bursts) during paradoxical sleep contrasts markedly with that manifested during the waking state. Thus, although the EEG is desynchronized during paradoxical sleep and resembles the waking EEG, unit discharge is not paradoxical. Thus, the two states can be distinguished on the level of unit activity, but only poorly on the more macrolevel of the spontaneous EEG. During paradoxical sleep, the burst duration and burst discharge frequency of units increases while the intervening periods of inactivity lengthen. This is merely a continuation of the trend toward bursting activity observed during slow wave sleep. However, *unit discharge during paradoxical activity resembles neural discharges during waking in terms of a markedly reduced intercorrelation between neuron pairs* (Evarts, 1964; Noda & Adey, 1970, 1973). Reduced intercorrelation between neuron pairs is therefore a common factor underlying EEG desynchronization.

Now, what does this tell us about information processing during the awake, desynchronized state? As pointed out by Spinelli and Pribram (1966), neurons during slow wave sleep exhibit maximum redundancy, that is, not only are a large number of neurons bursting, but they are also bursting in synchrony (Fig.

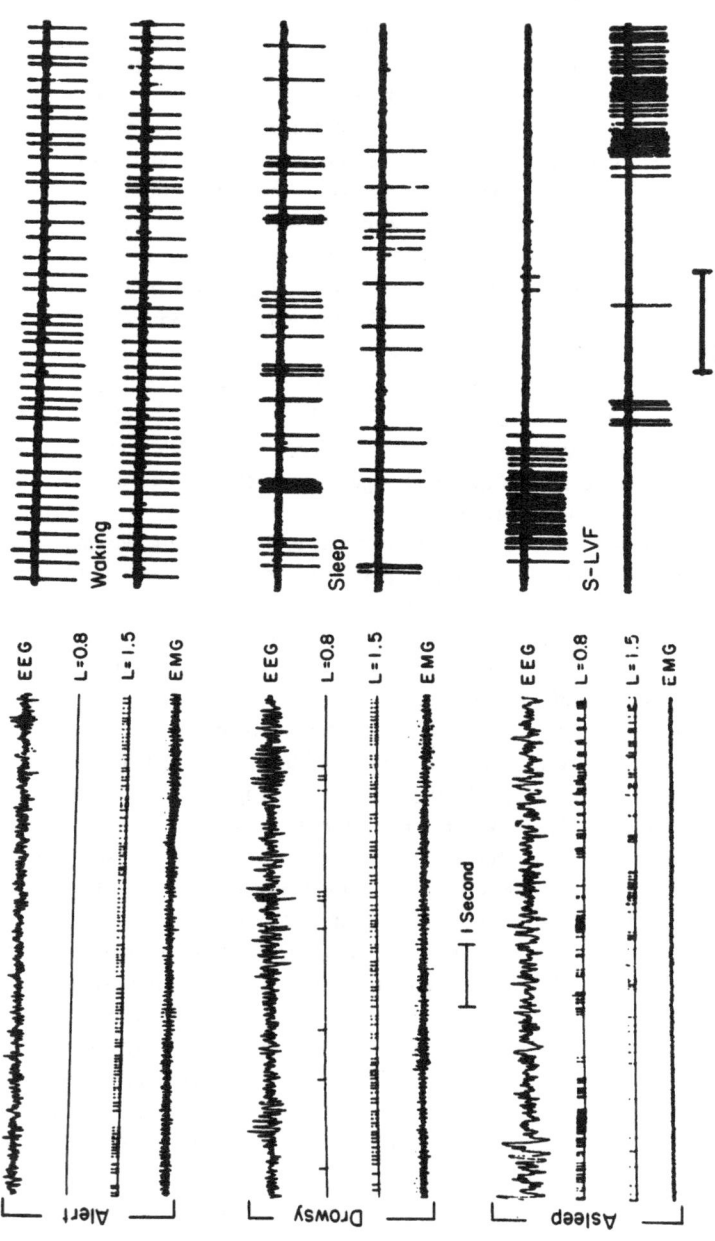

FIG. 4.5 Patterns of unit discharge during waking, sleep (with EEG slow waves), and during paradoxical sleep (S-LVF). During waking, discharge tends to be regular with an absence of both short and long interspike intervals. During slow wave sleep (middle pair of traces), there are bursts interspersed with periods of relative inactivity. With S-LVF, burst duration increases, intervening periods of inactivity become longer, and discharge frequency rises. The EEG is similar (low-voltage high-frequency activity) during waking and S-LVF. Note, however, that the pattern of discharge is markedly different during these two states. (From Evarts, 1964.)

4.5). This represents a state of minimum information processing (maximum redundancy) since a large number of neurons are doing the same thing and interneuronal predictability is high. During waking, these periodic cellular groupings break into more independent systems and thereby exhibit lower interneuronal predictability. This is a state where uncertainty is high and thus a state constituting high information transmission. It can be argued that during intense information processing neuron groupings differentiate, forming sets of cooperative or synchronized cell systems which process different aspects of the information. During slow wave EEG states neural systems can be considered to be in an equilibrium state in which the motion of the system is maintained by a regulative IPSP–EPSP balance (see Chapter 3, Section IV). The transition from sleep to wakefulness involves a more precise and active inhibitory and excitatory state in which the mechanisms of lateral and recurrent inhibition operate to differentiate and focus excitatory drives.

V. CHANGES IN NEURAL COHERENCE DURING ATTENTION

The foregoing discussion is germane to the phenomenon of attention, in that increased vigilance or attention is correlated with a desynchronized EEG. Figure 4.6 shows the results of averaging a series of evoked potentials to light flashes and to clicks in subjects who were instructed to alternately attend to the visual or auditory stimuli. It can be seen that flash responses during the visual task exhibited an increased amplitude compared with those present during the auditory task. Similarly, modality specificity of attention is evidenced by the fact that click evoked responses exhibited greater amplitude during the auditory task than during the visual task. This finding of increased amplitude of evoked potentials in a specific modality during arousal and attention is consistent with the results of many experiments (see Regan, 1972).

Such increased amplitude of the evoked response during the aroused or attending states appears to be due to two factors: First, there is a tendency for more generators to be available, to be responsive to the afferent stimulus. This may be due to desynchronization coupled with increased diffuse excitation during arousal. Second, there is a reduced variance in the latency of response so that average evoked responses exhibit greater amplitude (i.e., the neural generators are readily synchronized; see Fig. 4.7). This may be due to an increased tendency for neural generators in the aroused state, which are presumably desynchronized, to become synchronized by exteroceptive stimuli.

Further, it seems quite clear that these facilitations of particular afferent inputs are closely coordinated with inhibition of other simultaneous input. This was dramatically illustrated in the early work of Hernández-Peón (1955), showing that centrifugal influences could almost completely suppress irrelevant auditory input (click), at the level of a relay nucleus, while a cat attended to a

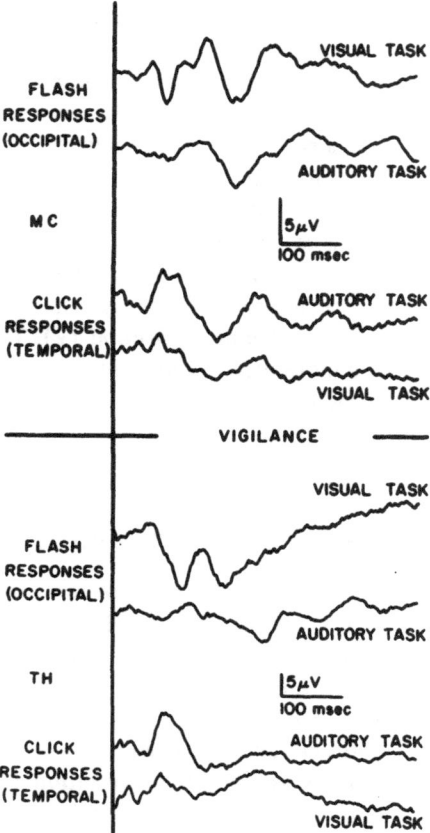

FIG. 4.6 Averaged cortical evoked potentials obtained from two subjects (MC and TH) in response to flashes and clicks. The potentials were recorded from the occipital and temporal areas for half of the experiment. The subjects were instructed to attend to the flashes and to ignore the clicks; for the other half of the experiment, subjects attended to the clicks and ignored the flashes. Note that there is an enhancement of response in the attended modality. (Modified from Spong, Haider, & Lindsley, 1965. Copyright © 1965 by the American Association for the Advancement of Science.)

meaningful visual input (mouse). This classical experimental finding is shown in Fig. 4.8.

A. Sensory–Motor Interaction

A complex sequence of changes in posture, orientation, and modulation of excitability takes place as an animal attends to a particular feature of the environment.

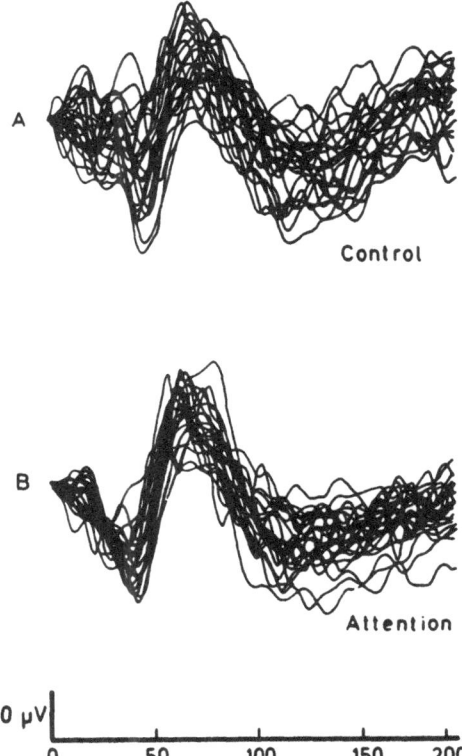

FIG. 4.7 Twenty-five individual evoked responses to a series of light pulses separated by 1 sec recorded from the middle sylvian gyrus of a cat. In one instance (A), the animal is indifferent to the stimulus; in (B) the cat has been trained to associate a foot shock with the flash. Note that in (B) the amplitude of the responses increases while the variance decreases. (From Beck, Dustman, & Sakai, 1969.)

The role of coordinated sensory–motor interactions in attention and perception has been infrequently investigated. However, this area is vitally important in understanding the focusing of attention, as well as such perceptual dysfunctions as aphasia and agnosia. Damage to cortical and brain stem regions can produce a multitude of effects related to attentional and perceptual, particularly visual, control processes. A visual stimulus is both seen and looked at (attended). It is clear that visual perception requires an integration of sensory and motor processes involving the whole of the visual apparatus, that is, fixation, lens accommodation, tracking, saccadic and vergence movements, and control of pupillary diameter.

This active process is demonstrated dramatically in Figs. 4.9 and 4.10 which show, with a small mirror attached to contact lenses, the pattern of fixation of

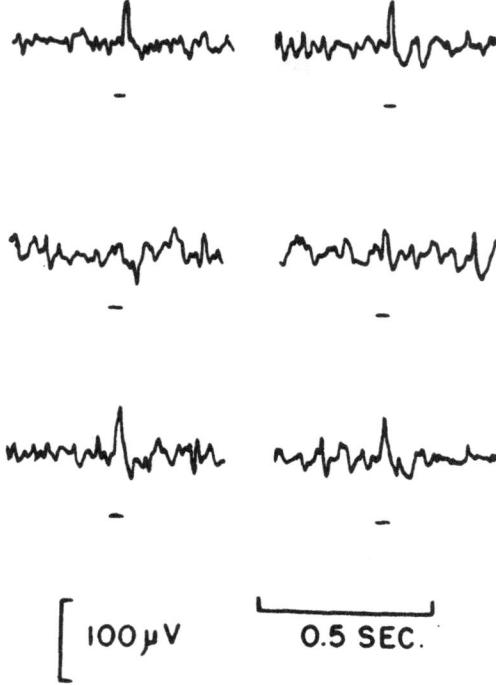

FIG. 4.8 Click evoked responses from the cochlear nucleus of the cat. (Top) Cat is relaxed; (middle) cat is attentively sniffing an olfactory stimulus; (bottom) cat is relaxed again. Note the attenuated amplitude of the click responses when the animal is sniffing. (From Hernández-Peón, Scherrer, & Jouvet, 1956.)

gaze in a normal and a cortically damaged individual. Note that the normal individual gazes at the angles and the relevant lines of the image, with saccadic movements sweeping across certain critical points in an information-seeking process which coincides with salient features of the image, while the cortically damaged individual scans the image in a relatively random, aimless fashion.

B. Changes in State

Not only must changes in brain state ensue as an animal attends to different aspects of the environment by modulation of excitability of afferent pathways and by differential orientation and tracking movements, but changes in state must also ensue as a result of internal processes related to the evaluation of incoming information. The latter factor has been indicated by several studies (Callaway, 1975; Callaway & Halliday, 1973; John *et al.*, 1973). In John's study (John *et al.*, 1973), individual evoked potentials were classified in terms of

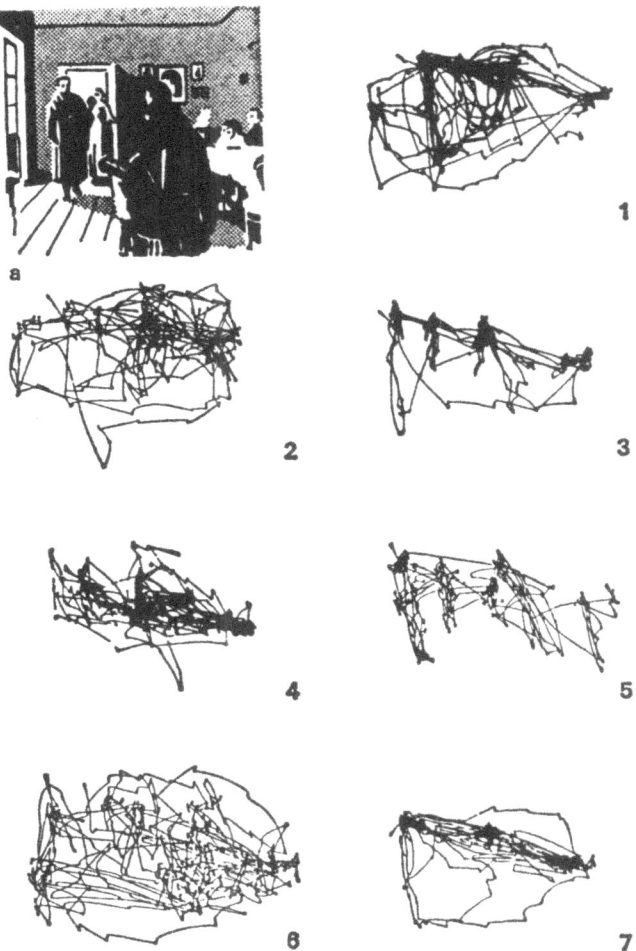

FIG. 4.9 Eye movements from a normal subject during observation of a picture (A) with different instructions. (1) Free observation (no instruction). (2) "Tell me, is the family poor or wealthy?" (3) "How old are the people in the picture?" (4) "What were they doing before the man entered the room?" (5) "Try to memorize the clothing the people are wearing." (6) "Try to memorize the placement of the furniture." (7) "How long had the man been away from his family?" Note that very different eye movements occur depending on the task requirement. (From Luria, 1966.)

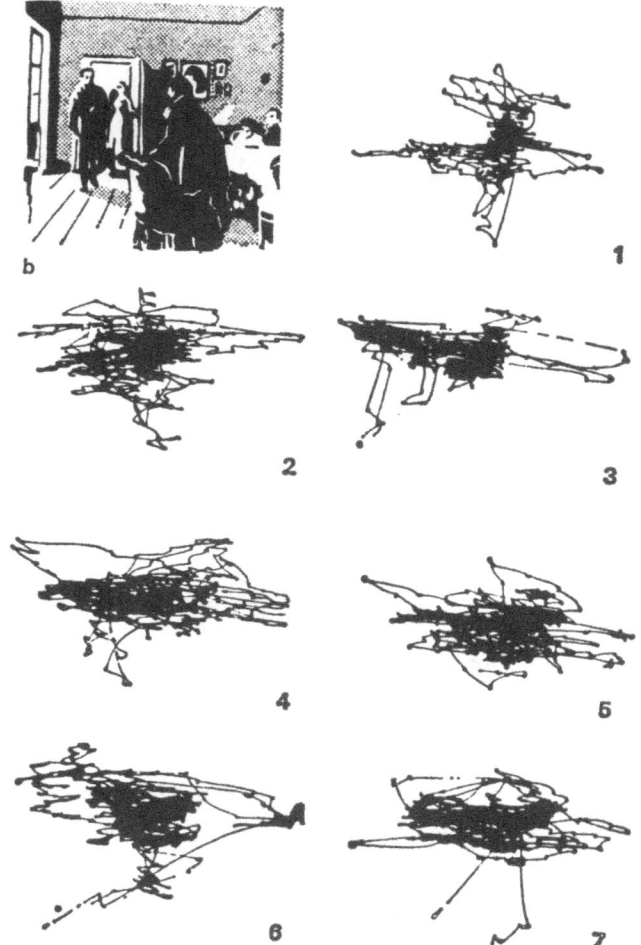

FIG. 4.10 Same as in Fig. 4.5 except the subject has suffered an intracerebral tumor of the right frontal lobe. Note that eye movement patterns change only slightly for the different instruction sets (1–7). In contrast to the normal subject in Fig. 4.5, the eye movements of this subject are erratic and lack information-resolving patterns. (From Luria, 1966.)

discrete modes. It was shown that a finite number of different modes are present throughout the course of a discrimination trial, with variations occurring in a specific manner. Changes from mode to mode took place without any change in posture or orientation. Thus, it appears that neural systems can assume a particular mode at one instant, and shift from that mode at a later instant, with specific modes showing an ordered pattern and being highly predictive of behavior (John et al., 1973).

CAT 10

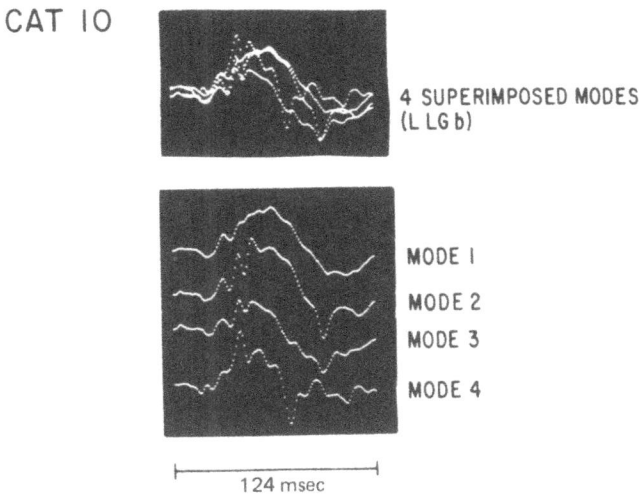

4 SUPERIMPOSED MODES
(L LG b)

MODE 1

MODE 2

MODE 3

MODE 4

|— 124 msec —|

TEMPORAL DISTRIBUTION OF EVOKED RESPONSE

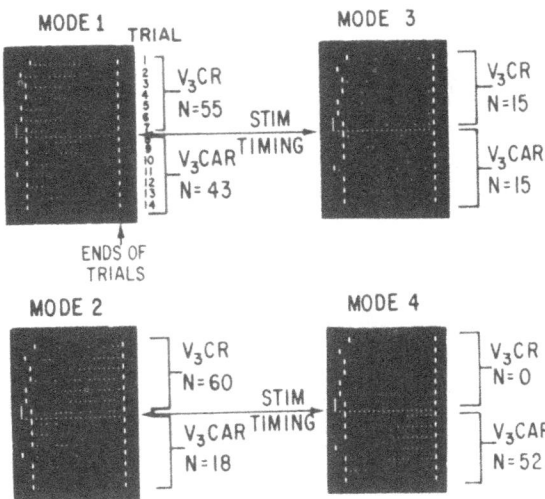

FIG. 4.11   Distribution of four evoked response modes recorded from the lateral geniculate of a cat exhibiting generalization responses to an ambiguous stimulus. The dark rectangles in the lower part of the figure show exactly where in a trial evoked potentials of a given mode occur. In this case, evoked potentials from mode 2 and mode 4 distinguish the behavioral outcome (CR versus CAR). The beginning of a trial is on the left and the end of a trial; when the animal responds is on the right. Note that the number of mode 2 and mode 4 evoked potentials increases toward the end of the trials. (From John *et al.*, 1973.)

CAT 14

LLGM

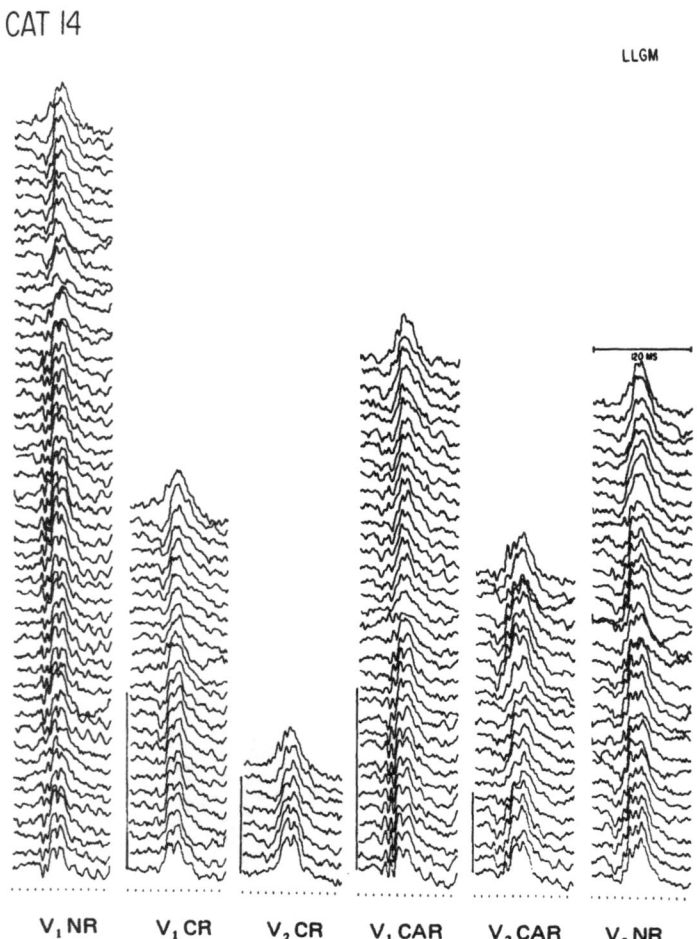

V₁NR   V₁CR   V₂CR   V₁CAR   V₂CAR   V₂NR

FIG. 4.12   Individual sequential evoked potentials recorded from a cat during trials result-ing in different behaviors (CR and CAR) to the same stimulus. Compare, for instance, V_1 CR to V_1 CAR and V_2 CR to V_2 CAR. As explained in the text V_1 and V_2 are different frequencies of flicker. Note that the waveshapes of the evoked potentials (marked by vertical lines) are invariant as a function of behavioral outcome and not stimulus frequency. Computer pattern recognition techniques such as shown in Fig. 4.7 (discussed in Chapters 10 and 15) show that invariant waveforms predict behavioral outcome nearly 100% of the time. (From John et al., 1973.)

These data, in conjunction with other studies (Hudspeth & Jones, 1975; John, 1972; Thatcher & John, 1975), demonstrate the need for a redefinition of brain state. Arousal, attention, and orientation are not sufficiently precise. Figure 4.11 is an example of a pattern recognition analysis of evoked potentials produced by a meaningless flickering light presented to a cat in a "differential generalization" paradigm (John *et al.*, 1973). In half of the trials, cats interpreted the flicker to mean "press the bar on the left" (trials 1–7) and in the other half of the trials the animal interpreted the flicker to mean "press the bar on the right" (trials 8–14). Four separate evoked potential modes (see top of Fig. 4.11) were distributed throughout the trials. These four evoked potential waveshapes are considered to reflect state variables. They exhibit definite invariances over time, correlate with specific variables in the experiment, and are highly predictive of behavior.

Further, evoked potential mode changes can occur very rapidly. They can occur within 250 msec, remain in that state for several seconds, and then rapidly change to a different mode (see Fig. 4.12).

In Fig. 4.12, the evoked potential reflects the momentary excitability of neural generators. These data suggest that the brain state of the animal shifted from one set of descriptors to another set in a succession of related but changing brain states.

One can define "brain state" in terms of the mathematics of information theory (see Chapter 6, Section I). In information theory, an information space is a coordinate framework or manifold in which the elements of a representation are ordered (Mackay, 1969a). Using this description of brain state implies that each unit of information (i.e., a representational system, see Chapter 6) is a different brain state and that the dynamics of brain function involve a discrete but continuous succession of such brain states. This subject is covered in more detail in Chapters 6 and 9.

C. Further Changes in Coherence during Attention

Figure 4.13 shows a series of averaged evoked potentials recorded during a vigilance task in which the subject was required to detect interspersed dim visual stimuli (Haider *et al.*, 1964). Performance is indicated in the third column from the left while reaction time is in the last column. This figure shows how vigilance and evoked potential amplitude wax and wane in correlation with performance and reaction time. Increased performance is correlated with decreased reaction time and increased amplitude of the evoked response.

Another example of the dynamic variation of evoked potentials during attention and arousal is shown in Fig. 4.14. In this study rabbits were classically conditioned to a tone and a painful ear shock. During conditioning, a noncontingent background flickering light was present to produce evoked potentials. The evoked potential averages from even and odd trials were compared in terms of their intercorrelations for four structures. The relative value of the square of the intercorrelation coefficient (r^2) indicates the percentage change in predictability of the waveshape. The first three successive averages (control responses before CS onset) show large variability and low correlation values. However, CS onset (arrow 1) results in an increase in evoked potential amplitude and a decrease in

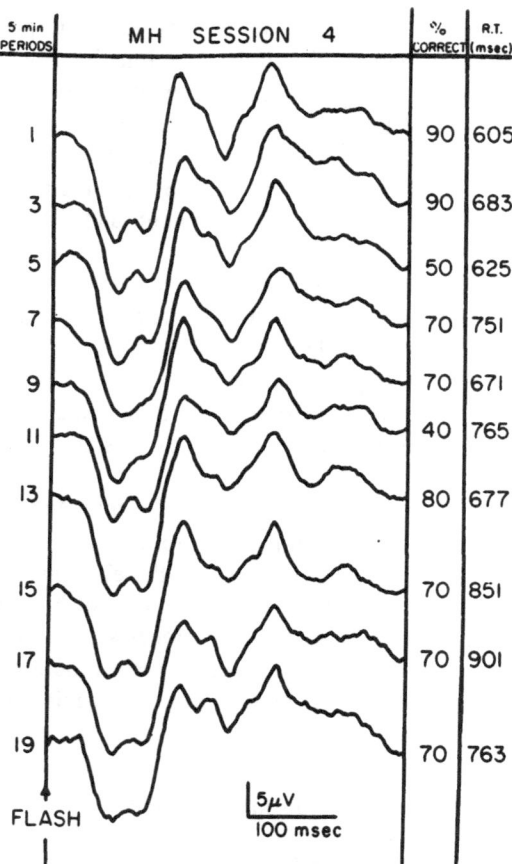

FIG. 4.13 Averaged occipital evoked potentials to nonsignal bright flash stimuli, presented during successive 5-min periods of a visual vigilance task, together with the percentages of randomly interspersed dim signal stimuli, correctly detected during the time periods. It can be seen that as detection efficiency diminished, the amplitude of evoked responses to nonsignal stimuli decreased and latency increased. It was shown that specific lapses of attention, revealed by detection failures, resulted in averaged evoked responses of lower amplitude to missed as compared with detected dim flashes. (From Haider *et al.,* 1964. Copyright © 1964 by the American Association for the Advancement of Science.)

variability, with a dynamic time course that reflects the main variables of the experiment.

These data offer support for the notion, discussed in Chapter 2, that evoked potential variability or the degree of nonstationarity in a system is itself a functional variable. Conceivably, neural ensembles activated during attention and learning exhibit greater global stability.

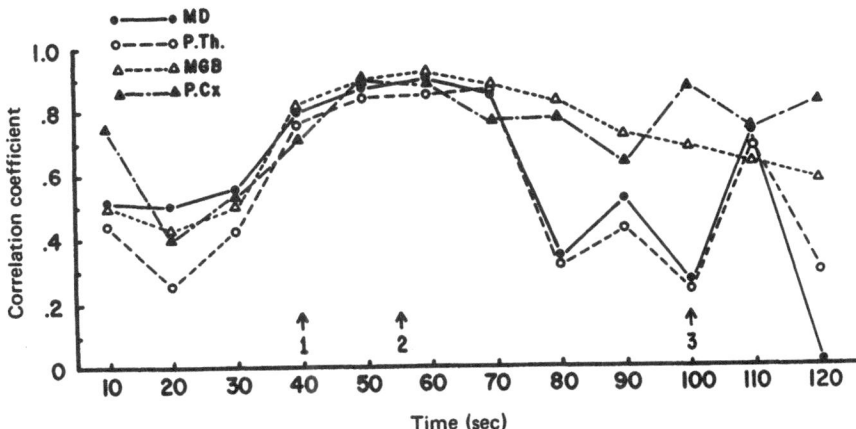

FIG. 4.14 Correlation coefficients for odd and even trialed evoked potentials recorded during a classic conditioning experiment in a rabbit. Evoked potentials were averaged at the beginning of successive 10-sec intervals starting with the onset of a background (noncontingent) flashing light. Tone onset (CS) occurred at 40 sec (arrow 1), UCS (ear shock) occurred at 60 sec (arrow 2), and tone offset occurred at 100 sec (arrow 3). It can be seen that correlation coefficients for each of the four structures (medialis dorsalis, MD; posterior thalamus, P.Th.; medial geniculate, MGB; and, posterior limbic cortex, P.Cx) increased during the CS with the MD and P.Th. showing a decrease following the UCS. The correlation coefficients reflect the degree of stability of evoked response over trials. Since the flicker is a noncontingent probe of excitability, the correlation coefficients indicate that the brain enters stable (i.e., replicable) modes of oscillation during critical phases of the experiment. (Printed by permission of W. H. Hudspeth, 1974.)

Such changes in neural dynamics during arousal and attention may be attributable to factors involved in controlling excitability and interneuronal coherence. For instance, recovery cycle measures of evoked potentials during various states indicate that inhibition in neural ensembles is more powerful and of a longer duration ($\approx$ 100 msec) in wakefulness than during either slow wave sleep or paradoxical sleep (Palestini et al., 1965). This indicates that there is both greater excitation and greater inhibition during arousal. Coherence measures during sleep–wakefulness (Holmes & Houchin, 1966; Noda et al., 1969) suggest that while paradoxical sleep states and wakefulness share comparatively low coherence, neurons in the awake state are more ready to respond to an exteroceptive stimulus and consequently show a greater tendency to be synchronized by the stimulus. In contrast, somatic evoked potentials elicited during slow wave sleep, while of large amplitude (Allison et al., 1966; Goff et al., 1966), show variable and long latencies. This is expected, since the bursting but coherent discharge of neurons during slow wave sleep (Evarts, 1967) provides a fluctuating background for exteroceptive stimulation in which afferent volleys arrive in phase with these periodic bursts some of the time, but out-of-phase at other times. Thus, it

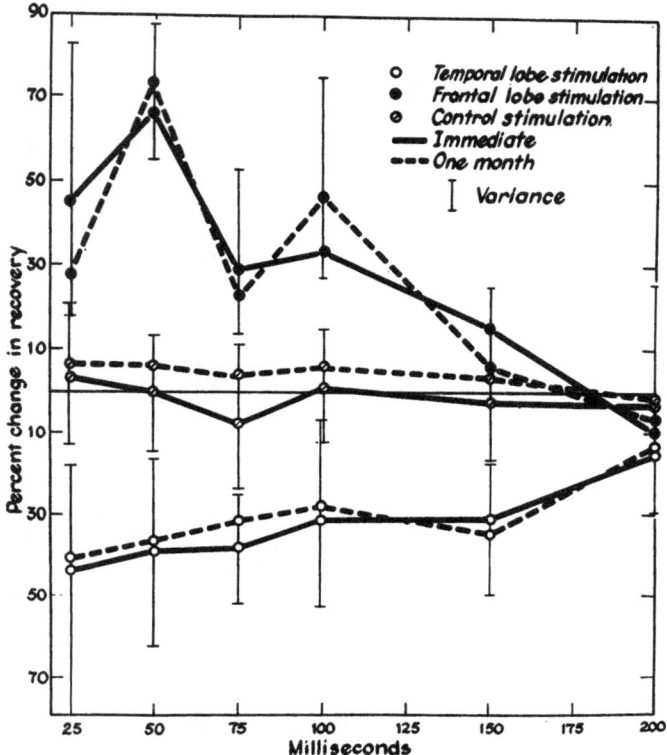

FIG. 4.15 The change in recovery of the primary components of the evoked response in a monkey to the second of a pair of flashes delivered at various intervals following the first. All curves represent percent change in recovery during cortical stimulation (temporal lobe, frontal lobe, and parietal lobe as a control) as compared to the prestimulation recovery function. Response curves were obtained immediately after the onset of stimulation and after one month of stimulation. Note that the recovery functions deviate in opposite directions depending on the cortical site of stimulation. (From Spinelli & Pribram, 1967.)

follows that since cells during slow wave sleep exhibit coherent tendencies, evoked potential amplitudes during such sleep will be large but of variable latency.

Further evidence indicating that the coherence of a neural ensemble is itself a functional variable is provided in Fig. 4.15. In this experiment, the amplitude of the evoked response to the second of a pair of light flashes was analyzed as the interval between members of the pair increased (i.e., the *recovery function*). The lines at the top of the figure are replicates of the recovery function from the visual cortex during frontal lobe stimulation. The lower pair of lines represent the recovery function from the visual cortex during stimulation of the adjacent inferotemporal lobe. It can be seen that a pronounced reduction in the rate of recovery is produced by stimulating the temporal lobes and an opposite enhancement of the rate of recovery is produced by frontal lobe stimulation. The authors (Spinelli & Pribram, 1966) interpreted these findings in terms of attention-gating mechanisms. The excitability within primary visual receiving areas is modulated by the association cortices, providing for a dynamic reciprocally related system that adjusts temporal and spatial coherence within the primary receiving stations.

These conclusions were supported further by single unit analyses of visual receptive fields in the relay nucleus of the thalamus (Spinelli & Pribram, 1967). Figure 4.16 shows a computer display of the receptive field of a lateral geniculate neuron in which each dot represents an excitatory response to a stimulus in the receptive field. The displays at the top left (n) and bottom right (m) are

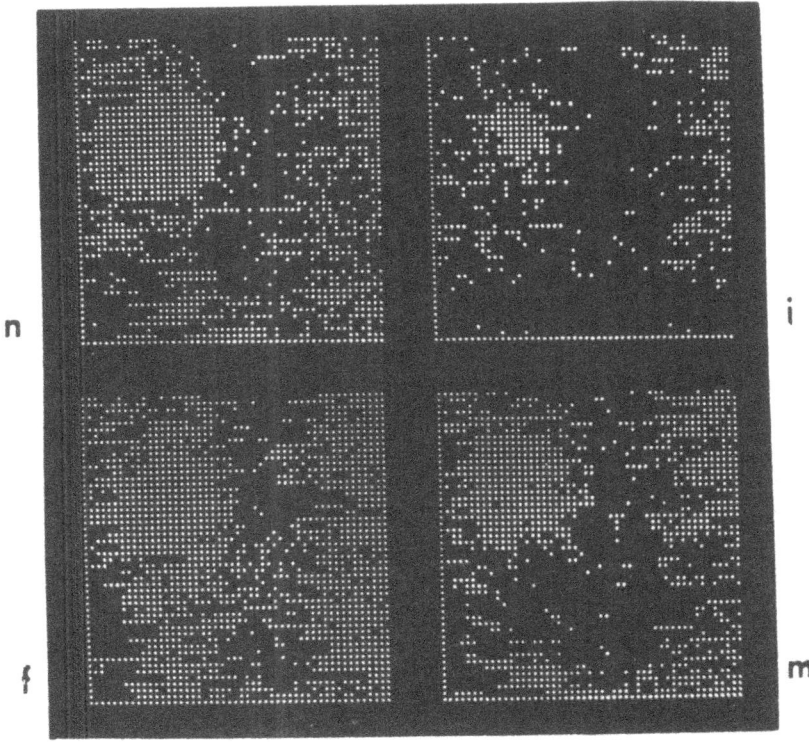

FIG. 4.16 Receptive field maps from a lateral geniculate unit of a monkey. Top left, n: control; i: mapped while inferotemporal cortex was being stimulated; f: mapped during frontal cortex stimulation; m: final control. Note that inferotemporal stimulation (i) decreases the size of the on center; frontal cortex stimulation (f), while not really changing the circular part of the receptive field (see Chapter 6 for a discussion of receptive field), brings out another region below it. The level of activity is three standard deviations above the normal background for this unit. (From Spinelli & Pribram, 1967.)

control displays obtained before and after stimulation of cortical regions. Display i is the receptive field plotted during inferotemporal cortical stimulation and demonstrates reduction of the area from which response was elicited. Display f is the receptive field from the same unit during frontal lobe stimulation, which shows an increase in the spatial domain within which visual stimuli were effective. This study indicates that activity in cortical association regions plays a fundamental role in attention (and perception) by influencing the dynamics of sensory processing at the primary sensory level.

VI. ANATOMIC SUBSTRATES OF ATTENTION FOCUSING

The presentation of a novel stimulus invariably elicits an attentive response. In Russian psychology this response is called an *orienting* response, since animals will orient toward the source of the novel stimulus. The animal's response seems equivalent to asking: "What is it?" This response is accompanied by widespread cortical EEG desynchronization.

Following continued presentation of the stimulus, the orienting response *habituates*, that is, gradually diminishes or even ceases to occur. The phenomenon of habituation has been demonstrated at many different levels of the central nervous system. An extensive literature exists on this topic, which has been reviewed elsewhere (John, 1961, 1967a; Morrell, 1961a). If the now irrelevant stimulus is made a CS for a conditioned response, then diffusely activated EEG reappears and subsequent changes occur that characterize learning (John & Killam, 1959). During learning, some of the changes in EEG activation (desynchronization) involve a dynamic spatial reorganization.

For example, Gastaut and his colleagues (1957) demonstrated that the widespread cortical desynchronization elicited in the early stages of conditioning gradually becomes restricted to the primary sensory projection area and the motor cortical region subserving the conditioned response. This phenomenon was interpreted as a concentration of excitatory processes in the sensory and motor analyzer systems and was believed to underly "connection–formation." Such spatial reorganization of excitatory desynchronization was observed in many early experiments on conditioning (Lindsley *et al.*, 1950; Morrell & Jasper, 1956; Moruzzi & Magoun, 1949; Rheinberger & Jasper, 1937).

In Section II some aspects of the anatomic organization of the reticular formation are discussed. In Figs. 4.2 and 4.3 a diffuse sensory system is shown in which multimodal interactions on reticular neurons take place throughout the rostrocaudal extent of the midline core system. This organization is likened to a stack of poker chips (Scheibel & Scheibel, 1958) or modular domains that sample activity in various afferent pathways coursing through the more lateral regions of the brain stem. This rather diffuse or nonspecific organization of the

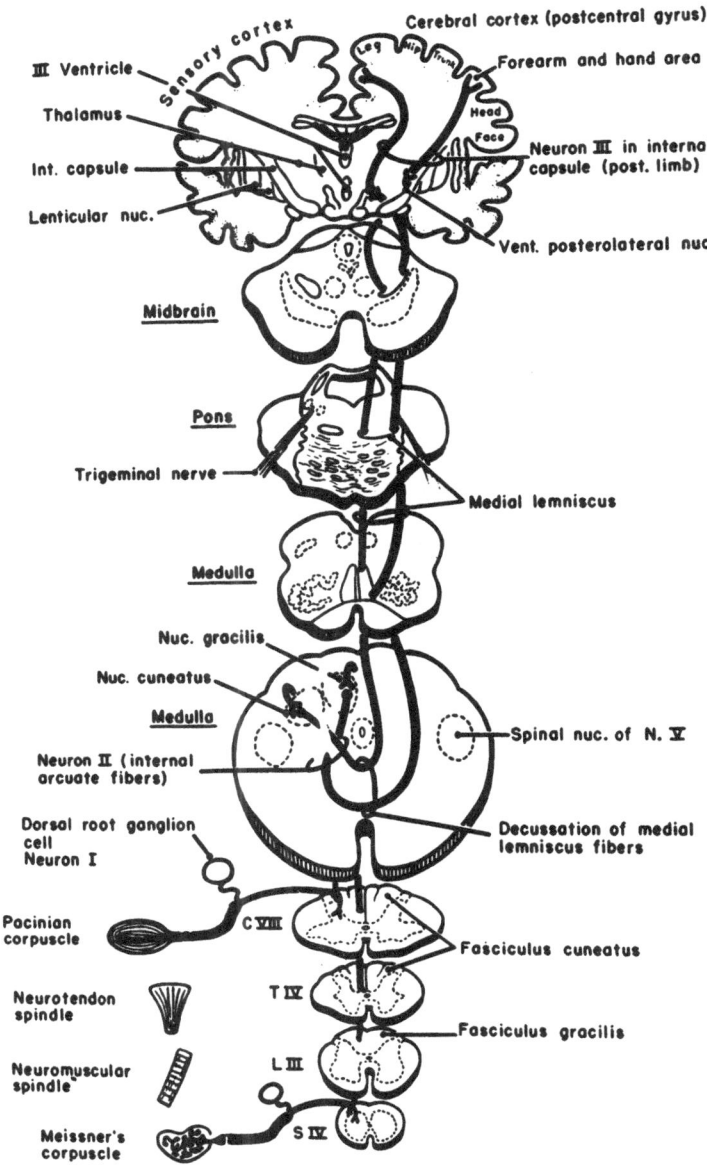

FIG. 4.17 Gross anatomy of the human somatosensory system. Dark lines show ascending pathways of leg and forearm somatosensory fibers. (From Truex & Carpenter, 1973. Copyright © 1973 the Williams & Wilkins Co., Baltimore.)

midline core is in marked contrast to the organization in specific sensory systems.

Figure 4.17 diagrammatically illustrates the anatomic pathways subserving the somatosensory system. These pathways can be considered a part of a specific sensory system conducting information from the skin and subcutaneous receptor tissues. The anatomic organization of this system insures maintenance of modality and topographic specificity. A similar anatomic and physiological maintenance of specificity occurs in other sense modalities (see Chapter 1, Section II). Of the various sense modalities, however, only the ascending somatosensory fibers do not give off collaterals to the reticular formation in passage. There is a definite separation of the specific and nonspecific system throughout both the

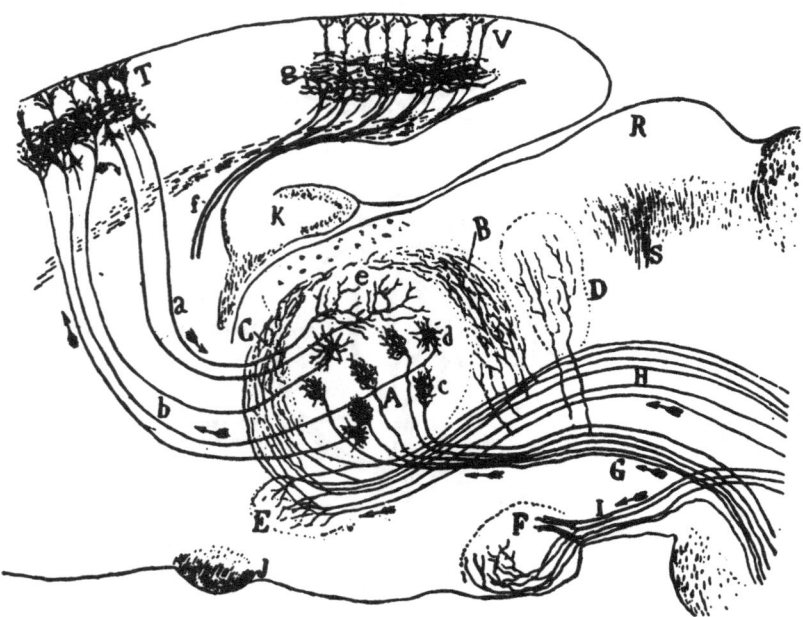

FIG. 4.18 Summarizing illustration showing some of the ascending thalamopetal pathways and the main axonal elements interconnecting thalamus and cortex. (A) Principal sensory nucleus; (B, C) accessory sensory or trigeminal nuclei; (D) posterior thalamic nucleus; (E) nucleus of the zona incerta; (F) lateral mammillary nucleus; (G) medial lemniscus; (H) central path for the trigeminal nuclei and other systems; (I) mammillary peduncle; (J) optic chiasm; (K) hippocampus; (R) superior colliculus; (S) elements of posterior commissure; (T) motor cortex; (V) visual cortex; (a) corticothalamic fibers; (b) thalamocortical fibers; (c) bushy arbor generated by terminating medial lemniscus fiber; (d) thalamocortical projection cell; (e) terminal plexus of corticothalamic fiber; (f) fibers of the optic thalamocortical radiation; (g) terminal afferent plexus of the optic radiation. (From Ramon y Cajal, 1911, Fig. 323; reproduced by permission of Consejo Superior de Investigaciones Científicas, Madrid.)

thalamus and reticular formation, with convergence occurring primarily in the cortex.

Conduction of somatosensory information through the relay nucleus of the thalamus (ventral posterolateral thalamus) occurs on a high priority, modality-specific basis. By "high priority" is meant that afferent input to thalamic relay neurons, even of moderate intensity, is usually sufficient to fire the relay cells and insure conduction to the cortex. This high priority of conduction is virtually guaranteed by bush axonal terminals that form a conelike structure that encapsulates the dendrites of thalamic relay cells. Each medial lemniscal axon produces a particularly dense arborization by repeated branchings of the terminal axon, providing a manyfold increase in the axonal surface area and thus increasing the synaptic drive onto the relay cell (Scheibel & Scheibel, 1970). This organization can be seen in Fig. 4.18, in which fiber tract G (medical lemniscus) gives rise to bushy arbors in the ventrobasal complex of the thalamus (A). This spatially localized, intensely arborized organization is not present in the nonspecific thalamus and reticular formation.

In general, the sensory-specific and nonspecific systems involve different anatomic substrates. Electrical stimulation of the specific systems results in replicable and short latency cortical responses. Electrical stimulation of the nonspecific sensory system produces variable long-latency responses which have the capability of summating and affecting behavior for long periods following stimulation (Morrell, 1961a).

An understanding of the possible anatomic substrate of attention is provided by examining the nature of specific—nonspecific convergence on cortical neurons. Scheibel and Scheibel (1970) demonstrated that afferents from the nonspecific sensory systems terminate in long climbing sequences along the entire length of the apical shaft of each pyramidal cell (Fig. 4.19). The rather diffuse distribution of these afferents results in relatively low synaptic densities, with consequently low synaptic drives, in comparison to the specific sensory afferents (Fig. 4.19) which terminate on the middle one-third of the apical dendritic shaft in dense configurations. The close packing of the specific sensory afferents near the spike-initiating zones of the cortical pyramidal cells insures a relatively high priority of transmission and modality specificity.

It is clear, on both electrophysiological and anatomic grounds, that nonspecific elements do not exert a significant phasic (i.e., strong and momentary) control over the output of the pyramidal cell. Rather, the pattern of distribution of nonspecific afferents results in low levels of tonic control or a gradual biasing of excitability. This differential organization of afferent drives is suggestive of an attention-focusing mechanism (see Scheibel & Scheibel, 1967b), in which a biasing of pyramidal cells by the nonsensory-specific system alters interactions between cortical cells and affects the routing of specific sensory control. The complexity of intracortical connections, which involves profuse collaterals and feedback from inhibitory interneurons, and inputs from the opposite hemisphere

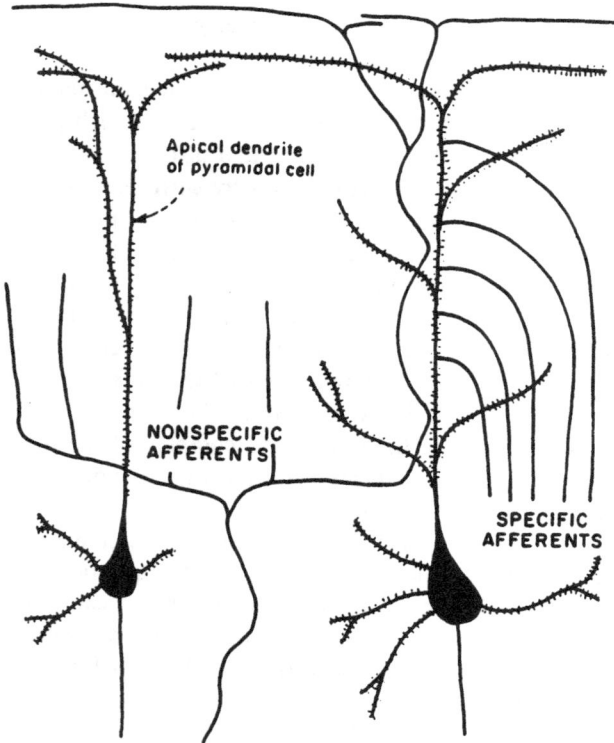

FIG. 4.19 Diagrammatic representation (from the work of Scheibel & Scheibel, 1967a) of the associations between specific and nonspecific afferents on the apical dendrites of pyramidal cells. (From Kety, 1970.)

via the corpus callosum (transcallosal), provides a matrix of structural interaction capable of altering the spatial organization of neural coherence during attention and learning.

VII. INVOLUNTARY AND VOLITIONAL ATTENTION

How might the matrix of interactions between specific and non-specific sensory systems select or determine the neural sensory patterns which will become the focus of attention? This very critical question will be discussed more completely in Chapter 9. However, at this point, we wish to point out that phenomena such as the orienting response and the subsequent habituation of this response would seem logically to require internal neural models of input from the past (neural representational systems, see Chapters 6 and 7), which are matched or mismatched with ongoing sensory input (Sokolov, 1960; 1963). According to Sokolov (1960, 1963) a *mismatch* between ongoing sensory input and information from the past (i.e., a "novel" event) results in an orienting response. Repeated presentations of the novel stimulus result in the creation of a representational system. Consequently, there is a *match* between the newly created internal representation and the sensory event. This match results in habituation (i.e., abolition) of the orienting response.

Sokolov (1960) and others (Groves and Thompson, 1971; Thompson & Spencer, 1966) indicate that the neuronal model is located largely within the midline core complex. This contention is supported by electrophysiological analyses demonstrating short latency responses of reticular formation units to novel stimuli (Groves & Thompson, 1971; Horn, 1969) and by developmental studies which demonstrate the capacity of newborn infants to exhibit nonvolitional short latency orienting responses (Luria, 1966, 1973; Piaget, 1971b). The fundamental role of the midline core systems in mediating responses to novel stimuli is further suggested by studies of children born without a cortex or thalamus (anencephalic monsters) who exhibit normal orienting responses (Gamper, 1926). Thus, it can be argued that experience establishes a neuronal representation, located within the midline system, which can match or mismatch with ongoing sensory input. The result of this interaction is a differential biasing of modality-specific cortical elements and a focusing of excitatory drives within afferent and efferent cortical domains. It is the interaction of the representational systems in the midline core regions with sensory input that determines the initial routing of corticopetal drives.

Luria (1973) distinguishes "reactive" attention (orienting response) from "volitional" attention, based upon a distinction between subcortical versus higher cortical control mechanisms (temporal and frontal). In particular, he emphasizes the role of language development and socialization in humans as an important factor in the transfer of attention control from a basic reflex-type action in which a child is continually an involuntary respondent to novel stimuli, to a higher order of mental activity in which a child exhibits volitional control over his attention. However, even in the case of higher order attention control, the reticular formation most likely plays an important organizational function. If one were to extend the Sokolov (1960) model to volitional attention, then the argument can be made that higher cortical systems are capable of creating an internal representational system which, through a diencephalic–reticular interaction, can shape or bias the direction of attentional activity in a manner analogous to novel stimuli (Thatcher, 1976b).

VIII. MECHANISMS OF ATTENTION SHIFT AND ATTENTION FIXATION

Green and Arduini (1954) were among the first to observe a reciprocal relation between the hippocampus and cortex during the orienting response. The presentation of a novel stimulus elicits electrocortical desynchronization and alpha block paralleled by a low-frequency, high-amplitude slow wave response in the hippocampus, called *theta rhythm.*

The hippocampus is considered to play a fundamental role in the mediation of the orienting response as well as attention shifts (Grastyán *et al.,* 1959). In animals, lesions of the hippocampus result in impaired orienting reactions (Henrickson *et al.,* 1969; Roberts *et al.,* 1962), behavioral perseveration (Ellen & Wilson, 1963; Kimble & Kimble, 1965), and impairment in the ability to inhibit ongoing responses or to reverse established response patterns (Kimble, 1968). In humans, hippocampal removal results in a severe memory deficit associated with the inability to store experienced events (Milner & Penfield, 1955; Penfield & Milner, 1958).

The close reciprocal linkage between the electrical behavior of the hippocampus and cortex suggests that there is an integrated corticolimbic system responsible for mediating attention shifts. This linkage is all the more relevant in light of the fact that the hippocampus is a junctional structure bridging neocortical function and diencephalic–reticular motivational systems which appear to be inextricably related during attention and learning.

Two separate but interrelated systems located in the reticular formation and diencephalon control hippocampal desynchronization and hippocampal theta rhythm. For example, high-frequency stimulation (100 Hz) of the central gray or the dorsolateral tegmental reticular formation elicits hippocampal *arousal,* manifested by an intense *hippocampal theta rhythm* (Anchel & Lindsley, 1972; Torii, 1961). Stimulation of the medioventral tegmentum, on the other hand, produces a different pattern, with *hippocampal desynchronization* (Anchel & Lindsley, 1972; Macadar & Lindsley, 1973).

Anchel and Lindsley (1972) conducted an extensive investigation of anatomic systems mediating these different hippocampal states: arousal (theta) and desynchronization. A combination of surgical lesions, electrical stimulation, and reversible cryogenic blocks allowed the following conclusions to be drawn: (1) There is an anatomically continuous system of cells involved in hippocampal theta activity, extending from the tegmental reticular formation (involving the nucleus reticularis pontis oralis, nucleus coeruleus, and the central gray; Macadar & Lindsley, 1973) to the dorsomedial region of the hypothalamus, to the medial septum and via the fornix to the hippocampus. (2) A second pathway exists, involved in hippocampal desynchronization, that overlaps somewhat with the theta system in the reticular formation (desynchronization can be elicited in isolation by stimulation of nucleus raphe and nucleus reticularis pontis caudalis; Macadar & Lindsley, 1973), but becomes completely separate along the course of the medial forebrain bundle and in the lateral hypothalamus.

These two parallel and reciprocally related hippocampal control systems extend a great distance along the rostrocaudal plane of the diencephalon and brain stem. The medial septum is believed to be the "pacemaker" which controls hippocampal theta rhythms (Petsche *et al.,* 1962; Stumpf, 1965). This septal pacemaker appears to be analogous to thalamic pacemakers that regulate cortical EEG. As discussed in the first chapter, desynchronization of cortical EEG is paralleled by desynchronization of the thalamic pacemaker system. Likewise, desynchronization of hippocampal theta appears to involve desynchronization of the septum (Anchel & Lindsley, 1972).

Thus, reciprocal and coordinated action of cortical and hippocampal systems appears to be under control of the reticular formation. This is indicated on several grounds. Stimulation of various reticular regions can either activate or deactivate hippocampal and cortical systems in a reciprocal fashion (Abeles, 1967; Kaada, 1960). Tegmental reticular fibers, passing via the dorsal longitudinal fasciculus of Shutz (Nauta, 1958), terminate in both the medial septum (the hippocampal pacemaker) and in the midline thalamus (the cortical pacemaker). As reviewed in Chapter 3, the ventral rather than the dorsal reticulothalamic pathway (see Fig. 3.14) exerts desynchronizing drives on the thalamus and cortex (Schlag & Chaillet, 1963) and simultaneously exerts synchronizing influences on the hippocampus via the medial septum (Anchel & Lindsley, 1972; Macadar & Lindsley, 1973; Torii, 1961).

Thus, some of the significant aspects of the anatomic substrate modulating and coordinating hippocampal and cortical activity have been elucidated. The question remains: How does this interrelated system function during attention?

A likely answer is derived from phylogenetic considerations as well as by anatomic and physiological fact. The brain stem–diencephalic structures, as discussed in detail in Chapter 5, constitute the primal brain of man, whence stem man's feelings and subjective sensations. Complex motivational and emotive systems are housed within the primitive diencephalic and limbic systems (termed the "centrencephalic" system by Penfield, 1969).

The evolutionary development of the cortex created a powerful analytic and discriminative system to elaborate and extend, in terms of goal-structured behavior, the drives within the primal diencephalic brain system. All our experiences are colored by feeling. Those sensations which are most significant to us as individuals command our attention, influence our decisions and become assimilated effortlessly as experiences. Clearly, there must be a coordinated interplay of the diencephalic–brain stem systems with the analytic and discriminative functions of the cortex.

The hippocampus, which although phylogenetically ancient is nonetheless a primitive cortex, occupies a pivotal or junctional position between the lower diencephalon and the neocortex. Through interactions with cortical and diencephalic structures, certain brain stem nuclei are inhibited while others are facilitated as the response to novelty unfolds (orientation).

The fact that the hippocampus is vital for the mediation of orienting is well established. There is some controversy regarding the role of the hippocampus in information processing and memory consolidation during learning (Adey et al., 1960; Bennett, 1970; Bennett & Gottfried, 1970; Elazar & Adey, 1967; Grastyán et al., 1959). However, given the junctional nature of the hippocampus and its abundant opportunities to mediate reciprocal diencephalic–cortical inter-action, it would indeed be surprising if the role of the hippocampus were confined only to orienting. Studies by Anchel and Lindsley (1972) and Stumpf (1965) suggest a much broader range of functions for this structure. For instance, stimulation of the medial hypothalamus or cryogenic blockade of the lateral hypothalamus produces similar behavioral effects consisting of head turning, and searching and orienting movements. These effects are associated with hippocampal theta waves and cortical desynchronization or activation. These manipulations reveal systems operating during the orienting response or during attention shifting.

On the other hand, stimulation of the lateral hypothalamus and cryogenic blockade of the medial hypothalamus causes an arrest of ongoing activity and "attentive fixation of gaze." This behavioral state is accompanied by hippo-campal desynchronization and represents attention fixation. Thus, hippocampal-cortical-diencephalic interactions appear to underlie both attention fixation and attention change.

In addition to the above, another set of forebrain structures, in conjunction with the reticulohippocampal circuit, appears to mediate attention change. As discussed earlier, lesions of the hippocampus result in an impaired ability to inhibit ongoing responses or to reverse established response patterns (Kimble, 1968). A very similar behavioral deficit follows lesions of the septal region (Butters & Rosvold, 1968; Donovick, 1968; Fried, 1971; King, 1958; McCleary, 1961; Zucker & McCleary, 1964). For this reason, Kimble (1968) and Altman and his co-workers (1973) conclude that the septum and hippocampus represent one level of a *braking* system which becomes active when it is necessary for animals to withhold responses such as in the presence of a threat, or when there is no reward or attention must be shifted. In contrast, a very different and perhaps opposite role of response facilitation is believed to be mediated by the cingulate and anterior and medial thalamus (Douglas, 1967; Altman et al., 1973; Vanderwolf, 1969, 1971). The first suggestion of this was provided by the work of Kaada (1951, 1960) who demonstrated that in cats stimulation of the septal–subcallosal region produces inhibition of ongoing motor activities while stimulation of the cingulate facilitates motor movements. Lesions of the cingu-late and the thalamic projection pathways to the cingulate result in an increase in freezing behavior (Lubar, 1964; Thomas & Slotnick, 1962; Ursin et al., 1969) and a decrease in the sensory threshold necessary to trigger motor action (Thomas & Slotnick, 1963). These contrasting roles of the cingulate and the septum–hippocampal system were demonstrated most clearly by Ursin and his colleagues (1969) in an experiment in which three groups of rats (one with septal lesions, one with cingulate lesions, and one control group) were placed in the middle of an alleyway. At one end of the alleyway was a food cup from which the animals had previously received a painful shock while at the other end

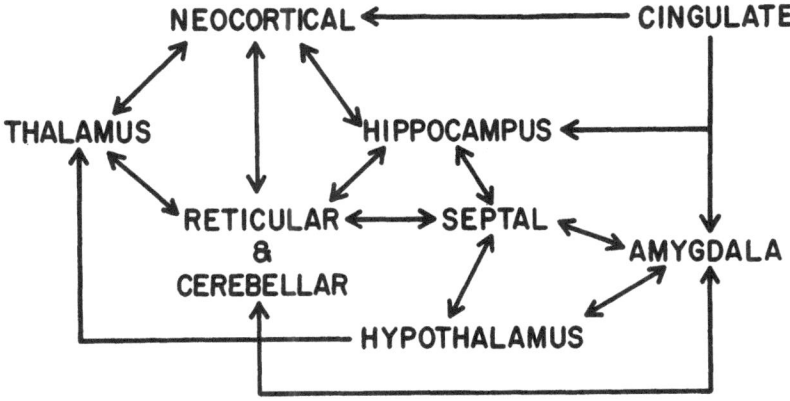

FIG. 4.20 A diagrammatic representation of some of the anatomic components mediating attention. See text for discussion of interactions.

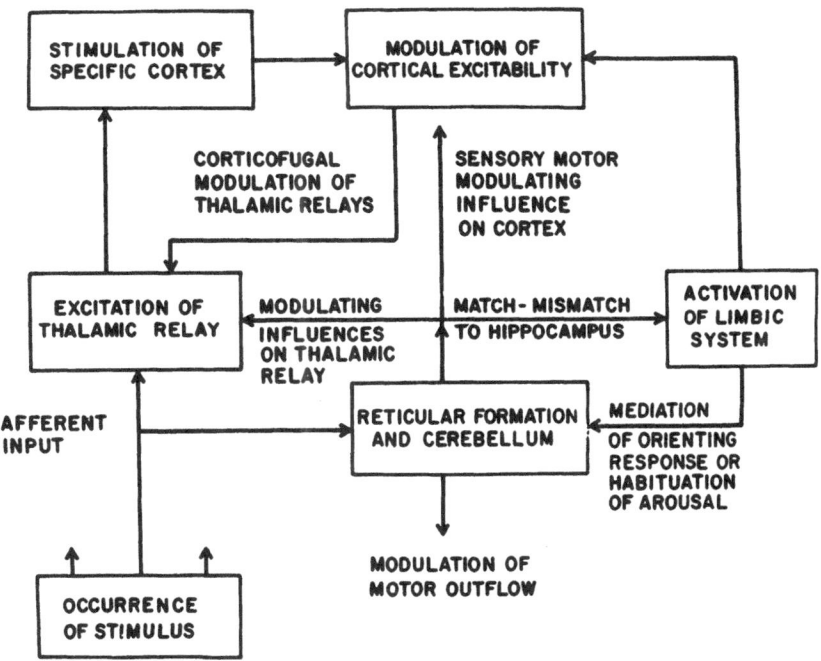

FIG. 4.21 A diagrammatic representation of some of the salient features of fixation of attention mediated by anatomic systems in Fig. 4.16. See text for explanation.

was a safe compartment. It was shown that the septal lesion animals, being unable to inhibit responding, ran to the food cup, the normal controls ran to the safe compartment, while the cingulate lesion animals, being unable to initiate responding, froze in the middle of the alleyway. Data indicate that cingulate and medial thalamic lesioned rats are not deficient in the execution of motor behavior, but rather they are deficient in the *initiation* of movement (Lubar, 1964; McNew & Thompson, 1966; Vanderwolf, 1969, 1971). As reviewed by Altman and his co-workers (1973), considerable data show that the medial thalamic–cingulate system stands in dynamic balance to the entorhinal–hippocampal–septal system; one is involved in initiation of movement while the other stops or inhibits movement. Attention shift requires both the inhibition of ongoing behaviors and, frequently, the initiation of an active behavior such as the orienting response. Thus, a very complex but interrelated system involving reticular, diencephalic, limbic, and cortical structures mediates attention shift and attention fixation.

The last structure to be mentioned in the attention system is the frontal cortex. As discussed earlier, this structure holds a very important position in the hierarchy of neural control systems (see Chapter 1). The frontal cortex exhibits extensive two-way connections with the medial dorsal nucleus of the thalamus and with the hypothalamus. Cortical areas 9 and 10 are connected via the cingulum with the temporal lobes and entorhinal cortex. Thus, the frontal lobes possess extensive connections with limbic system structures as well as with sensory and motor neocortex. In humans, lesions of the frontal lobes result in perserveration of motor acts (Luria, 1966, 1973) as well as in a deterioration of mental concentration and a reduced capacity for abstraction and sustained intellectual performance (Denny-Brown, 1951; Freeman & Watts, 1949; Partridge, 1950). Frontal lesion patients show impairments in the ability to constructively direct or control their attention (Luria, 1966). Thus, the frontal lobes, via an interaction with subcortical structures, particularly the hippocampal–cingulate system, clearly play a vital role in the mediation of attention.

The foregoing discussion of brain mechanisms of fixation of attention have revealed that this process is extremely complex, involving many transactions between different functional systems and a variety of dynamic processes, probably including the activation of memories related to the estimation of novelty or familiarity of the stimulus. The final figures (Figs. 4.20 and 4.21) in this chapter summarize the salient functional and anatomic features of this process.

5
Neurophysiology of Emotion

What I Have Lived For

Three passions, simple but overwhelmingly strong, have governed my life: the longing for love, the search for knowledge, and unbearable pity for the suffering of mankind. These passions, like great winds, have blown me hither and thither, in a wayward course, over a deep ocean of anguish, reaching to the very verge of despair.

I have sought love, because it brings ecstasy—ecstasy so great that I would often have sacrificed all the rest of life for a few hours of this joy. I have sought it, next, because it relieves loneliness—that terrible loneliness in which one shivering consciousness looks over the rim of the world into the cold unfathomable lifeless abyss. I have sought it, finally, because in the union of love I have seen, in a mystic miniature, the prefiguring vision of the heaven that saints and poets have imagined. This is what I sought, and though it might seem too good for human life, this is what—at last—I have found.

With equal passion I have sought knowledge. I have wished to understand the hearts of men. I have wished to know why the stars shine. And I have tried to apprehend the Pythagorean power by which number holds sway above the flux. A little of this, but not much, I have achieved.

Love and knowledge, so far as they were possible, led me upward towards the heavens. But always pity brought me back to earth. Echoes of cries of pain reverberate in my heart. Children in famine, victims tortured by oppressors, helpless old people, a hated burden to their sons, and the whole world of loneliness, poverty and pain make a mockery of what human life should be. I long to alleviate the evil, but I cannot, and I too suffer.

This has been my life. I have found it worth living, and would gladly live it again if the chance were offered me [from *The Autobiography of Bertrand Russell,* Vol. I, McClelland and Stewart Limited, Toronto, 1967, p. 13].

117

I. INTRODUCTION

As R. B. Livingston (1967) has emphasized, feelings stem from the most primordial areas of the central nervous system, from the tissues lining the central canal and periventricular passages. This matrix of cells comprising the central core regions is made of small, finely myelinated neurons, each with a plentitude of interconnections providing a multidimensional substrate for interaction. Activity in this matrix is primary to our subjective experience, dominating the subjective "I." Everything else is adventitious to it, since it is the source of our appetites and satisfactions. Feeling states, generated within this matrix, were the original content of our consciousness and they will, most probably, be the last.

Some workers believe that what we call "mind" arises from internal experiences of feeling. More specifically, H. Cantril and W. K. Livingston (1963) conceived of a phylogenetic origin of what we consider "mind" from states of feeling. They perceived the emergence of full consciousness, knowledge, and strategic goal planning as built up from an initially crude but basic awareness of feeling. In the newborn, crude and poorly differentiated feeling states (e.g., general excitement, distress, and delight; Bridges, 1932) are present while there is little sensory–motor coordination. During development, there is a gradual emergence of a body image, an awareness of self, of other individuals and objects, a voluntary governing of sensory–motor skills, and finally a conscious directed control over thought and the creation of predictions about the future. Studies of anencephalic monsters (children that lack a neocortex or thalamus) demonstrate that the primal diencephalic core structures mediate primitive feeling states and are responsible for feelings present during the earliest stages of maturation.

This primordial but integrative property, called *feeling* or *emotion,* is made up of sensory, motor, and motivational aspects. In other words, emotion can manifest itself as behavior or an internal experience.

II. ASPECTS OF EMOTIONAL BEHAVIOR

There are a great many types of behavior that are summed under the term "emotional." Some of these involve the somatic musculature. For instance, in man, emotional responses involve crying, laughing, smiling, screaming, running in flight, rage, startle responses, and a multitude of facial expressions capable of showing combinations of emotion. Animals show some of these responses.

In both man and animals, *autonomic* responses are an important aspect of emotional behavior. There is the pallor of fear and fainting, caused by circulatory changes. There are heart rate and blood pressure changes, as well as glandular secretions, that ready the muscles for flight or attack (see Morgan, 1965). By "autonomic" is meant responses of the autonomic nervous system.

This is a system of bundles of nerve fibers and neurons which extends throughout the body and provides innervation to smooth muscles and glands. Although many of the soma of nerve cells of the autonomic nervous system are organized into a chain of ganglia, or clusters of nerve cells, lying along the length of the body external to the spinal cord, this system is, nonetheless, closely coordinated with the central nervous system. Not only does the brain send and receive messages from the autonomic nervous system via the hypothalamus and mesencephalon, through some of the cranial nerves and the spinal cord, but many of the peripheral actions resulting from functions of the autonomic nervous system feed back onto the brain through the blood supply. Two reciprocally antagonistic categories of autonomic activity can be discerned peripherally, those which decrease the concentration of a transmitter substance called adrenalin (referred to as *parasympathetic* actions) and which increase adrenalin concentration or oppose its effects (called *sympathetic* actions). Much of contemporary pharmacology relates to the construction of chemical compounds capable of altering the balance between these two classes of action at the level of some particular bodily organ or function (Goodman & Gilman, 1965).

Just as two reciprocal classes of autonomic activity exist peripherally, so also can two intimately but reciprocally interrelated anatomic systems be identified within the central nervous system. These classes exert opposite effects on the peripheral autonomic system and upon different neuronal systems within the brain itself. Because of the major role of these systems in determining behavioral valence and feeling tone, they will be discussed in this chapter in some detail. No further attention, however, will be paid to the peripheral autonomic system. The reader wishing further information about peripheral autonomic activity, as well as about autonomic aspects of the activity of the central nervous system, is referred to Gellhorn (1957).

A. Autonomic Activity and Emotions

Some experimenters have speculated that the differences between "pleasantness" and "unpleasantness" arise from sensations produced by the two general categories of autonomic activity, parasympathetic and sympathetic (Morgan, 1965). The arguments in favor of such a view are that feelings of warmth are generally pleasant and related to the parasympathetic dilation of blood vessels. Various functions concerned with eating, which is a pleasurable activity, are parasympathetically governed, for instance, the secretion of saliva and gastric juices in anticipation of food. Also, sexual behavior involves vasodilation and certain muscular responses which are also parasympathetic in origin.

However, there are clear exceptions to any rule that pleasure is parasympathetic (Gellhorn, 1961; Morgan, 1965). For instance, crying is parasympathetic. Bad odors, as well as good food, elicit parasympathetic responses. Vomiting is parasympathetic and the motor activity of the bladder and rectum, which may be greatly augmented in fear or joy, is also parasympathetic. Thus, there is no simple relationship between the activity of the autonomic system and emotions. In general, the autonomic system, in terms of hypothalamic regulation or homeostasis, is adaptive in its function, lending service to a great many preparatory behaviors.

B. Emotional Experience

People and animals not only act emotionally but they feel emotional. Verbal reports confirm this in people although it can only be inferred from the behavior of animals. MacLean (1966, 1969) divides emotional experience into three categories: basic, specific, and general as shown in Fig. 5.1. Basic effects relate to bodily needs such as hunger, thirst, the urges to breathe, to defecate, to urinate, to have sexual outlet, and so forth. Specific effects are those activated by the specific sensory systems, such as the feeling of disgust in smelling a foul odor, pleasure in being rubbed or tickled, or the feeling of pain after a noxious stimulus. General effects are more abstract, involving the feelings of love, terror, fear, sadness, depression, foreboding, familiarity or strangeness, feelings of reality or unreality, anguish, the desire to be alone, feelings of certainty and conviction, as well as paranoid feelings and anger.

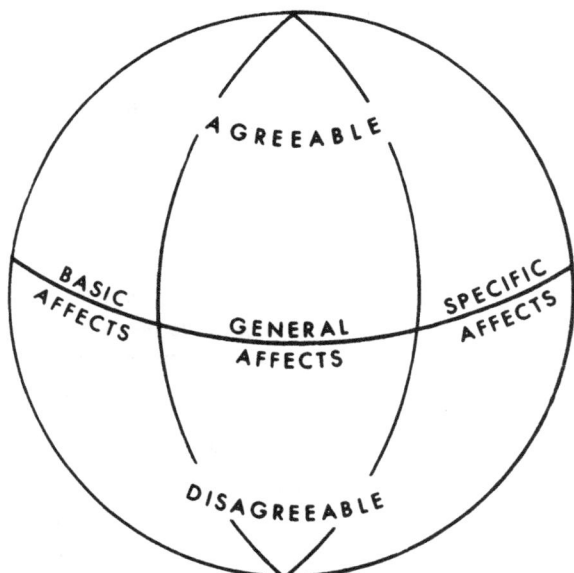

FIG. 5.1 Hypothetical schema for viewing emotional affect. There are two general categories of feeling (agreeable and disagreeable) that are divided into basic, general, and specific grouping. (From MacLean, 1970.)

In some epileptics, with foci located in the limbic lobes, feelings can arise in vacuo, that is, in isolation independent of any evoking condition. For instance, as mentioned earlier (Chapter 1), epileptic auras which precede a seizure may involve feelings of conviction, absolute truth, deja vu, strangeness, or other subtle feelings such as, in the case of Dostoyevsky, "a sensation of existence in the most intense degree (MacLean, 1970)." These reports indicate that the most subtle aspects of feeling, indeed man's most guarded and precious internal states, have a physical basis within the limbic lobes.

As reported by MacLean (1970), it is significant that some epileptics experience an alternation of opposite feeling, suggesting that there is a reciprocal innervation or balance of opposites within the anatomic substrates of emotion. The idea of balanced opposites playing a fundamental role in the production of affective states is supported on both physiological and anatomic grounds and will be discussed in the section to follow.

An important perspective of the integrative aspects of emotion is provided by phylogenetic considerations of the development of the mammalian brain. This approach is particularly valuable since modern man possesses certain evolu-

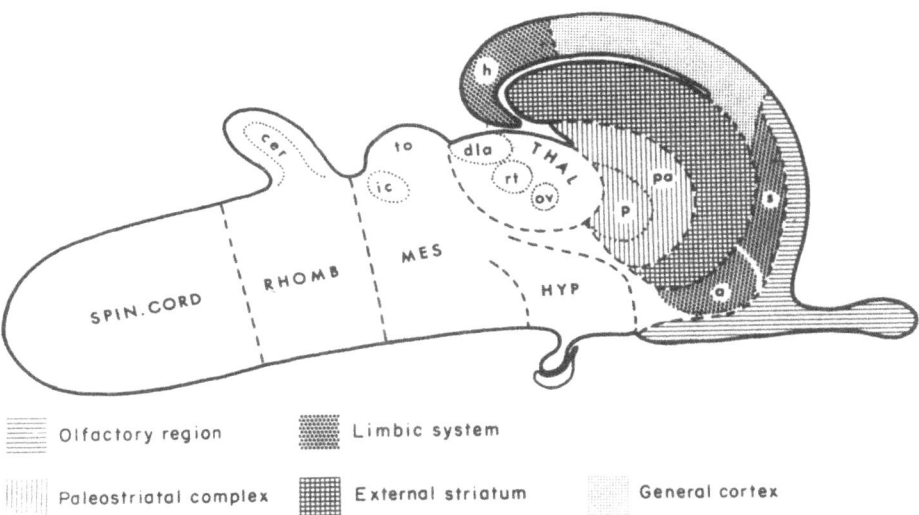

FIG. 5.2 Diagrammatic representation of the central nervous system of a nonmammalian vertebrae in longitudinal view. The diagram is based on recent findings with respect to connectional patterns in avian and reptilian brains in particular, showing the principal divisions of the neuraxis and, in various line and stipple patterns, the major territories of the cerebral hemisphere. Abbreviations: a, amygdala; cer, cerebellum; dla, dorsolateralis anterior of thalamus; h, hippocampus; HYP, hypothalamus; ic, inferior colliculus; MES, mesencephalon; ov, nucleus ovoidalis thalami; p, paleostriatum primitivum; pa, paleostriatum augmentatum; RHOMB, rhombencephalon; rt, nucleus rotundus thalami; s, septum; SPIN. CORD, spinal cord; THAL, thalamus; to, optic tectum. (From Nauta & Karden, 1970.)

tionary relics left from his ancient past. Compare, for instance, the organization of the premammalian brain (reptilian), shown in Fig. 5.2, with the mammalian brain represented in Fig. 5.3. Note that the spatial relationships between spinal cord, rhombencephalon, mesencephalon, hypothalamus, and limbic system (a, b, and h) remained basically constant throughout evolution. The major differences are in terms of the striatum, thalamus, and the development of the neocortex.

Figure 5.4 illustrates the relative stability of the limbic system, in comparison to the neocortex, in the development of the mammalian brain. It can be seen that there has been a considerable proliferation or mushrooming of the neocortex (and related thalamic and striatal structures), while the limbic system has remained constant.

MacLean (1970) discussed the comparative studies of the psychology of behavior which indicate that the intrinsic mechanisms of the rhombencephalon

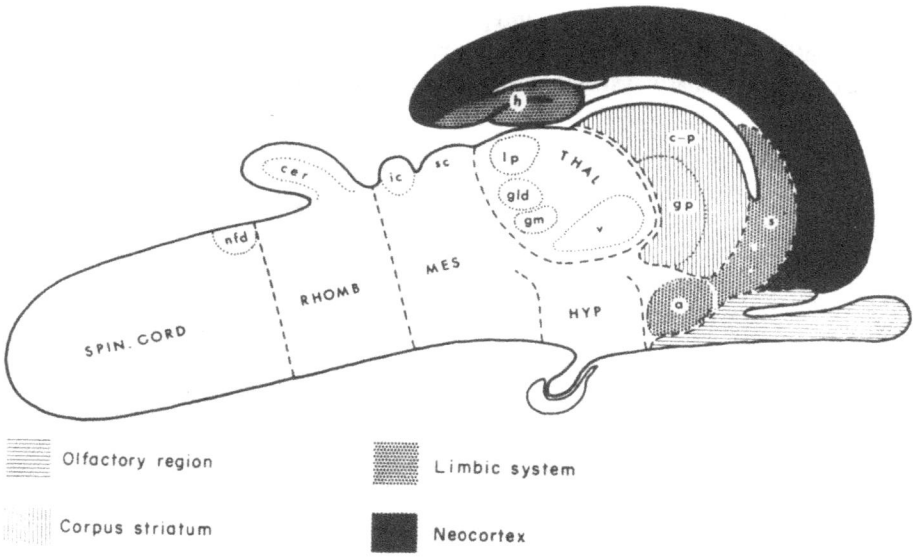

FIG. 5.3 Schematic drawings representing the mammalian brain. In comparisons with the diagrams of the nonmammalian brain shown in Fig. 5.2, the most pronounced differences appear in the composition of the pallial mantle in which the general cortex has become replaced by neocortex; the apparent absence of the nonmammalian external striatum; and the appearance of a circumscript somatic sensory nucleus (v) in the thalamus receiving, among other somatic sensory lemnisci, part of the spinothalamic tract and most of the medial lemniscus originating in the nuclei of the dorsal funiculus (nfd). Major conduction pathways afferent and efferent to the neocortex have been indicated in slightly bolder line. Abbreviations: a, amygdala; cer, cerebellum; c-p, caudoputamen; gld, lateral geniculate body; gm, medial geniculate body; gp, globus pallidus; h, hippocampus; HYP, hypothalamus; ic, inferior colliculus; lp, nucleus lateralis posterior of thalamus; MES, mesencephalon; nfd, nuclei of the dorsal funiculus; RHOMB, rhombencephalon; s, septum; sc, superior colliculus; SPIN. CORD, spinal cord; THAL, thalamus. (From Nauta & Karden, 1970.)

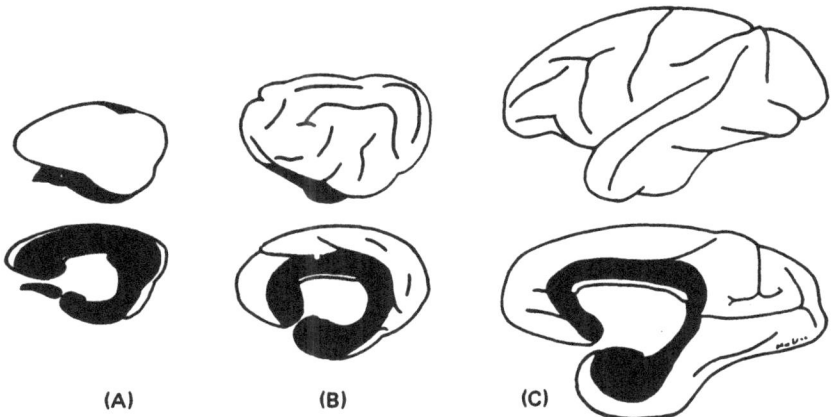

FIG. 5.4   Lateral and medial views of brains of rabbit (A), cat (B), and monkey (C) drawn roughly to scale. This figure illustrates that the limbic lobe (dark shading in medial view) is found as a common denominator of the cerebrum throughout the mammalian series. Surrounding the brain stem, the limbic lobe contains most of the cortex corresponding to that of the paleomammalian brain. The greater part of the neocortex, which mushrooms late in evolution, occupies the lateral surface. (From MacLean, 1954.)

and mesencephalon mediate the maintenance of primordial stabilities, such as homeostasis and somatic posture. For instance, in reptiles complex stereotyped behaviors are present which are based on ancestral learning and memory in the evolutionary sense, that is, *instincts.* This primitive brain is devoted largely to instinctual functions, which have been divided into two general classes: (1) preservation of the *species,* which involves functions such as establishing territory, homing, mating, and breeding; and (2) preservation of the *individual,* which involves such things as hunting, finding shelter, flight, attack, imprinting, and the formation of social hierarchies (MacLean, 1970). Reptilian behavior indicates a strong inherited devotion to behavior patterns, with only limited modifications occurring by learning.

In the individual, it is instructive to consider the behavior of an anencephalic human (Gamper, 1926) who lacks a neocortex and basal ganglia. Such a child exhibits a number of complex, integrative behaviors; for instance, those related to self-preservation such as rooting, suckling, swallowing, as well as cardiovascular, respiratory, and thermal regulation. The child also exhibits waking, sleeping, laughing, smiling, crying, general distress, and pleasurable excitement. Thus, we can conclude that the rudimentary aspects of basic, specific, and general emotions are present in the primordial regions of the human brain.

The most unique function of the rostral parts of the brain, particularly the forebrain and neocortex, appears to lie in the perception of goals and the assignment of goal priorities, as well as in the patterning of behavioral strategies devised for the achievement of such goals. The forebrain, in conjunction with the lower diencephalon and brain stem, serves to elaborate and extend performance beyond the simple capabilities of the premammalian brain. With the development of the neocortex came increased analytic and discriminative powers. As MacLean (1970) jokingly points out, the evolutionary development of the mammalian paleocortex was nature's way of providing a "thinking cap" to emancipate it from stereotyped behavior.

The mammalian forebrain has access to, and possesses mechanisms for, the integration of all available sensory information regarding both internal and external environments. The neocortex can be likened to an analytic and discriminative amplifier that gives the animal a better picture of its internal and external environment, thus providing greater adaptive capabilities (MacLean, 1958). Indeed, in mammals memories of current or recent experiences often take precedence over ancestral memories in guiding behavior. The fact that in mammals neocortical connectiveness can be modified by experience further emphasizes this point (Blakemore & Cooper, 1970; Hirsch & Spinelli, 1970).

III. ATAVISTIC BEHAVIOR

It is extremely important to bear in mind that humans, although the endpoint of current phylogeny, have a primitive system serving as the core of their brains. It is believed by some that the limbic system serves as an interactive buffer between the primordial feeling states and the analytic properties of the neocortex. The massive human neocortex allows for goal projections, the creation of strategies, and for discriminative functions. However, below this modifiable analytic forebrain, inhibited and modulated by the results of civilizing experience, lies the powerful machinery of the primitive brain, constantly generating pressures for atavistic behavior. Undoubtedly, many of the most tragic aspects of interpersonal behavior, including aggressive violence and war, arise from our failure to analyze, to recognize, or to anticipate the stereotyped products of this literally subhuman machinery. Perhaps one of the most important services neuroscience can render will be to help human beings learn how to anticipate, identify, inhibit, or consciously deal with the feelings and impulses which it generates. Such explicit, applied neuroscience might well be included in the elementary school curriculum.

IV. HIERARCHICAL ORGANIZATION
OF EMOTIONAL SYSTEMS

In this section the treatment of emotional behavior will revolve around specific neural structures. The task will be to proceed systematically from the brain stem to diencephalon to cortex in order to present the integrative and hierarchical aspects of affective behavior.

A. The Spinal Cord and Brain Stem

Clear somatic emotional patterns and visceral responses are present in animals whose brains were transected at the bulbar and midbrain levels (Kelly *et al.*, 1946; Macht & Bard, 1942). Such animals may growl, hiss, or spit, lash their tails, thrash with their forelegs, increase breathing, and show all the signs of rage in normal animals. However, these responses are not integrated, and usually occur in isolated nondirected fragments. Midbrain and bulbar animals never show the integrated "rage reaction" that one can see in normal animals.

A significant aspect of emotional behavior in midbrain and bulbar animals is that emotional reactions become more fragmentary the lower the brain stem section is made (Kelly *et al.*, 1946). Thus, it may be concluded that various items of rage behavior have their "seats" in the brain stem, but that integration of these various components of behavior must occur in a higher center.

B. The Hypothalamus

This structure has been called the "seat of emotion" (Morgan, 1965). Although such a term can be misleading, it points to the fact that the hypothalamus is the principal system in which the various components of emotional behavior are integrated into patterns. As will be discussed later, the hypothalamus receives powerful excitation and inhibitory influences from other limbic structures as well as from reticular and cortical systems.

1. Decorticate behavior. The classic studies showing the role of the hypothalamus in the mediation of emotion were performed by Bard and his associates (Bard, 1928). In some studies, only the neocortex of cats was removed. In others, the neocortex and thalamus were removed, leaving intact the hypothalamus and brain stem. In yet other cats, the midbrain was sectioned below the hypothalamus.

Emotional reactions were studied before and after these various operations. One striking result was a lowered threshold for rage following complete decortication, including the cingulate gyrus and ventral cortical areas such as the amygdala, hippocampus, and pyriform lobe. This suggested that cortical structures exert a tonic or constant inhibitory influence on subcortical systems. Some clarification of this point was provided by more discrete lesions that involved only removal of the neocortex, sparing the cingulate and ventral cortical structures. This operation produced a completely tame or placid animal. If one subsequently removes the cingulate or ventral cortical regions, then the animal becomes ferocious (Bard & Mountcastle, 1947).

2. Shortcomings of lesion studies. However, one cannot conclude that inhibitory influences on subcortical rage systems arise within the cingulate and ventral cortex. Unfortunately, subsequent studies (see Goddard, 1964) show that surgery, such as that performed by Bard and Mountcastle, can produce irritative foci at the periphery of the lesions. Such foci can activate subcortical rage centers. Studies by Ursin and his co-workers (1969), Vanderwolf (1967), and Lubar (1964) demonstrate that removal of the cingulate by itself has little effect on emotion. The animals appear to be more cautious, with less emission of spontaneous behavior. Lesions of the amygdala cause placidity and not rage (unless there is an irritative focus; Goddard, 1964). Thus, the results regarding cortical inhibitory control over subcortical structures are still somewhat ambiguous. It should be noted, however, that stimulation studies demonstrate the presence of both excitatory and inhibitory corticofugal drives on hypothalamic and limbic structures (Brady, 1960; Gloor, 1960; Goddard, 1964).

An important finding from the Bard and Mountcastle studies was that normal integrated rage behavior occurred only if the lower diencephalon and upper mesencephalon were intact. The conclusion, therefore, was that there is a focal mechanism in the hypothalamus, extending to the upper mesencephalon, that integrates emotional responses (Morgan, 1965).

3. Stimulation studies. Further understanding of the role of the hypothalamus in affective control is revealed by electrical stimulation studies. For instance, stimulation of lateral hypothalamic regions at high intensities produces well-organized attack behavior (Egger & Flynn, 1962; Nakao & Maki, 1958). Yasukochi (1960) reported specific differential effects of stimulation in the anterior, medial, and posterior hypothalamus. In anterior hypothalamus fear responses are most commonly elicited, in the medial and lateral hypothalamus aggressive behaviors are observed, while in the posterior hypothalamus general orienting responses and arousal as well as exploratory behavior are elicited (Hess, 1949; Ranson & Magoun, 1939).

Stimulation studies frequently yield ambiguous results. This is because of current spread from the site of stimulation and also because fibers in passage, originating and terminating in distant brain regions, are activated by the electrical currents. Nonetheless, such studies do indicate that there is an extensive and rather diffuse matrix of cells involved in fear and flight behaviors (Zanchetti, 1967). While the hypothalamus is heavily implicated as an efferent organizing center, coordinated fear and flight can be elicited by stimulation of areas of the limbic system, extending from the temporal cortex and amygdala (Ursin, 1960) to the caudal diencephalon and rostral midbrain, including the periaqueductal gray and reticular tegmentum (Zanchetti, 1967). Further elaboration of this rather extensive fear and flight system will be considered later.

4. Emotional expression. Lesion experiments have also aided in the understanding of the role of the hypothalamus in emotional expression. Posterior hypothalamic lesions usually produce stolid, somewhat somnolent animals (Ranson, 1939). Lesions of the more medial and ventromedial aspects of the hypothalamus cause generally ferocious animals (Wheatley, 1944). The well-directed rage behavior following ventromedial hypothalamic lesions indicates that some system has been released from inhibition. Such systems are probably located in the lateral and posterior hypothalamic regions, since lesions of these areas result in placidity or tameness (Ranson, 1939), while stimulation produces rage (Nakao & Maki, 1958). Thus, there is evidence for complex reciprocal interactions within the hypothalamus involved in affective expression. Such reciprocity, suggesting a homeostatic balance of opposites, is in evidence throughout the hypothalamus.

5. Regulation of consummatory behavior. Lesions of the ventromedial hypothalamus in rats give rise to a condition of hyperphagia in which animals eat nearly continuously and become obese (Brobeck *et al.*, 1943; Brooks, 1946; Heatherington & Ranson, 1942). Lesions of the lateral hypothalamus frequently inhibit appetite in rats (Anand & Brobeck, 1951a, b; Teitelbaum & Stellar, 1954). Thus, reciprocal interactions also appear to be involved in hypothalamic regulation of appetite and food intake. Stimulation studies elucidate further the nature of this reciprocal relation. For instance, stimulation of the ventromedial hypothalamus can arrest eating and decrease appetite in rats (Anand & Dua, 1955a, b) whereas stimulation of lateral hypothalamic regions can induce eating in satiated animals (Miller, 1957, 1960). Similar results are produced by selective deposition of catecholamines and acetycholine in the hypothalamus and reveal that there exists a "satiety" or "stop eating" system with a "center" which is adrenergic and located in the lateral hypothalamus (Grossman, 1960, 1962a, b, 1964). Eating and drinking systems are also distinguishable on neurochemical as well as anatomic grounds. Although there is some anatomic overlap in the distribution of hypothalamic eating and drinking systems, it has nonetheless been shown that cholinergic stimulation (e.g., carbachol or acetylcholine) produces vigorous and prolonged drinking (Grossman, 1962a, b, 1964) while adrenergic stimulation induces eating (Grossman, 1966). Cholinergic mechanisms in the lateral hypothalamus have also been linked with aggressive bheavior. For instance, Smith and his colleagues (1970) produced aggressive behavior and killing in rats which could be reversed or arrested with cholinergic blocking agents. Similar findings have also been noted for deposition of cholinergic agents in the septum and amygdala (Igic *et al.*, 1970).

6. A comment about "centers." Although the results of localized lesions, stimulation, or injections into hypothalamic regions are often interpreted as indicating the presence of a localized center mediating some vegetative function, these interpretations can be oversimplifications. More careful analysis (e.g., Teitelbaum, 1962) reveals that these various functions involve reciprocal relationships between anatomic systems distributed vertically along many different levels of the brain, up to and including the cortex. Such vertical distribution of function, in contrast to exclusive localization in structure at a single level, is a general principle of functional organization, particularly in the human brain (Luria, 1966, 1973). Failure to recognize this principle has caused a great deal of energy to be dissipated in fruitless controversy about narrow localization of function. This issue is discussed in greater detail in Chapter 9.

7. Ergotropic–trophotropic balance. Hess (1954) originated the terms *ergotropic system* and *trophotropic system* to distinguish two poles of a proposed hypothalamic integrative mechanism. The ergotropic system was an activating system, preparing the body for interaction with the environment and believed to be located primarily in the anterior hypothalamic regions. The trophotropic system was a passive system associated with relaxation, drowsiness, and internal recuperation and thought to be located more posteriorly. The use of methods to alter serotonin and noradrenalin concentrations in the hypothalamus has provided recent support for this ergotrophic and trophotropic distinction. For instance, Brodie and his co-workers (1959) demonstrated a heavy concentration of both serotonin and noradrenalin in the hypothalamus and linked these chemicals with system activation and deactivation. Brodie and his colleagues pointed out that dopa, the amino acid precursor of both dopamine and norepinephrine, produced EEG activation and a behavioral syndrome like that for ergotropic predominance. A link between serotonin and the trophotropic system was demonstrated by the slowing of the EEG which was produced by deposition of serotonin in the hypothalamus, particularly in posterior regions.

8. Self-stimulation. Self-stimulation behavior is readily produced through electrodes located in the lateral and anterior hypothalamic regions (Olds, 1956). Such behavior consists of performing some responses in order to turn on stimulation of a particular set of brain regions (defined as *positively reinforcing* or "rewarding" regions) or to terminate stimulation of a different set of brain regions (defined as *negatively reinforcing* or "aversive" regions). Punishing or aversive regions are located more centrally and caudal to reward or pleasure-producing sites. Thus, a dichotomous, pleasure–punishment system exists within the hypothalamus and upper mesencephalon. The involvement of biogenic amines in the control of self-stimulation in rats has been established. Wise and Stein (1969) have shown that self-stimulation is enhanced by amphetamines and retarded by either reserpine or alpha-methyl-*p*-tyrosine (an inhibitor of catecholamine synthesis).

9. Pharmacology of goal-directed behavior. Further work by Stein and associates (Stein, 1964, 1968; Stein & Seifter, 1961; Wise & Stein, 1969) has led to the suggestion that rewarded or goal-directed behavior is controlled by a specific system of noradrenergic neurons in the brain. The noradrenergic cell bodies are located in the lower brain stem (the locus coeruleus is a primary location; Moore *et al.*, 1973) and send axons via the medial forebrain bundle to form noradrenergic synapses in the hypothalamus, limbic system, and the temporal and frontal cortex. Electrical stimulation of the medial forebrain bundle produces a powerful rewarding effect and also elicits certain consummatory behaviors, such as feeding and copulation. Lesions of the medial forebrain bundle, or pharmacologic blockade of noradrenergic systems, cause severe deficits in goal-directed behavior and the loss of consummatory reactions (Stein & Wise, 1971; Teitelbaum & Epstein, 1962).

10. Possible etiology of schizophrenia. These findings have led Stein and Wise (1971) to suggest a chemical etiology of schizophrenia involving an impairment of noradrenergic reward systems by 6-hydroxydopamine (6-hydroxydopamine is an autoxidation product and metabolite of dopamine, capable of inducing degeneration of certain nerve endings and consequent depletion of norepinephrine; Porter *et al.*, 1963). Their proposal was supported by animal experiments in which catatonic-like behavior (waxy flexibility) and significantly lower self-stimulation rates followed intraventricular injections of 6-hydroxydopamine.

Hypothalamic self-stimulation has been used as an animal model to evaluate the relative effectiveness and mode of operation of antidepressants (Stein, 1968; Wise & Stein, 1969). Annau *et al.* (1973) recently evaluated the effects of the antidepressant tranylcypromine in rats, with self-stimulating electrodes located in the septum, anterior hypothalamus, and posterior hypothalamus. Tranylcypromine potentiated self-stimulation in the posterior hypothalamus, and inhibited all self-stimulation in the septum. These results suggest the modus operandi of the antidepressant drug tranylcypromine is via a reciprocal interaction that involves simultaneous excitation of posterior hypothalamus and inhibition of the septum; again, a balance of opposites.

C. The Thalamus

It was suggested around the turn of the century (Cannon, 1929; Head, 1920) that the thalamus plays a vital role in the mediation of emotional states. Although there has been controversy surrounding this claim, it appears that the thalamus is only secondarily involved in emotional expression. For instance, in cats stimulation of the dorsal medial nucleus of the thalamus evokes a fearlike crouching response which is aversive, since cats will learn to avoid stimulation (Roberts, 1962). Removal of the dorsal medial nucleus in humans appears to reduce anxiety and aggressive or assaultive behavior in some psychiatric patients (Spiegel *et al.*, 1951). Stimulation of the posteroventral nu. of the thalamus in monkeys (Delgado, 1955) and cats (Brady, 1960; Roberts, 1962) elicits anxiety, defensive and offensive movements, and vocalizations that appear to be "painlike" in nature (Roberts, 1962).

Stimulation of the posteroventral thalamus, however, may elicit emotional responses which are probably secondary to "pain," since this is a somesthetic relay center and part of the pain system. Thus, the posteroventral thalamus is probably not a part of the emotional system.

The anterior thalamic regions, including the dorsal medial nucleus, are positively reinforcing in self-stimulation experiments (Olds, 1956). However, the anterior thalamus is an integral part of the limbic system (receiving inputs from the hypothalamus and projecting to the cingulate), while the dorsal medial nucleus contains a large number of amygdaloid axons, some of which terminate but many of which pass through to the frontal cortex (Krettek & Price, 1973). The dorsal medial nucleus also receives a large afferent contribution from hypothalamic nuclei and apparently serves as a relay station between the frontal cortex and hypothalamus (Truex & Carpenter, 1964). Thus, emotional reactions produced by stimulation of the thalamus are largely limbic in nature. Stimulation of specific sensory regions, as well as the diffuse sensory thalamus, usually results in an arrest reaction, that is, attention or arousal, but little emotional behavior (Grastyán & Angyan, 1967).

D. The Limbic System

Figure 1.17 shows the rather detailed anatomic organization of the limbic system. Prior to 1937, this set of structures, particularly the septum, hippocampus, and amygdala, were thought to be part of the "olfactory" brain, since they carried connections with the olfactory tract and tubercle. As mentioned earlier, the limbic lobe (limbic means "forming a border around") is a common denominator in reptilian and mammalian brains. This fact, plus the extensive anatomic connections of the limbic lobe to the hypothalamus, suggested to J. W. Papez in 1937 "that the hypothalamus, the anterior thalamic nuclei, the gyrus cinguli, the hippocampus, and their connections constitute a harmonious mechanism which may elaborate the functions of central emotion, as well as participate in emotional expression." In more general terms, Papez emphasized the transmission of the "central emotive" processes that flow in the hippocampal formation (hippocampal gyrus, dendate gyrus, and amygdala) to the mammillary bodies via the fornix. Efferents from the hypothalamus then coursed downward to the brain stem and tegmentum and also upward through the anterior thalamus to the unsulate gyrus and then back to the hippocampus, thus forming a complex but integrative loop system (see Fig. 5.5; Section IV.D.2).

1. Kluver–Bucy syndrome. Almost simultaneously with the publication of Papez' theoretical effort, Kluver and Bucy (1939) reported striking effects on emotional behavior in monkeys following bilateral removal of the temporal lobes, including the hippocampus, presubiculum, pyriform lobe, and amygdaloid complex. The preoperatively wild and intractable *rhesus macaques* were made docile and tame by the operation, showing signs of neither fear nor anger while exhibiting bizarre alterations in affective state. The five major effects of bilateral lobectomy, identified by Kluver and Bucy, were: (1) "psychic blindness" or *visual agnosia*; (2) oral tendencies, particularly a strong tendency to examine all objects by mouth, as well as hyperphagia and other dietary changes; (3) "hypermetamorphosis," an apparently irresistible compulsion to contact every object as quickly as possible; (4) hypersexuality, not only in amount of sexual activity, but in the range of stimuli to which sexual responses were initiated (stuffed animals, experimenter's gloves, immature monkeys, pingpong balls, and so on); (5) taming effects, in which the previously "wild" feral monkeys could be handled with little or no danger to the experimenter. Of particular significance was the fact that unilateral lesions or bilateral lesions that were restricted to the neocortex and spared the limbic system failed to produce dramatic changes in behavior.

Following Kluver and Bucy's (1939) initial findings, a concerted effort was made to determine the specific structures responsible for the various aspects of the syndrome. Compulsive oral behavior and decreased emotional responsiveness to aversive stimuli were observed following extensive frontotemporal ablations (Fulton, 1951). However, there were no alterations in sexual behavior and much of the observed effects could be accounted for by trauma and nonspecific loss of neural tissue, although visual discriminations, independent of emotionally motivated effects, appeared to be impaired by such lesions (Blum et al., 1950; Pribram & Mishkin, 1956).

The portion of the ablated tissue responsible for the Kluver–Bucy syndrome appears to be in the limbic lobe, but more specifically in the amygdala. For example, in the monkey, lesions carefully restricted to the pyriform lobe, amygdaloid complex, and hippocampus (sparing neocortical regions) result in loss of fear and anger responses, docility, and compulsive oral behavior without gross motor or sensory deficits (Smith, 1950). Lesions restricted to the amygdala, sparing the hippocampus and pyriform lobe, produce docility and reduced fear and aggressive responses in monkeys (Thompson & Walker, 1951; Walker et al., 1953). Rage and fear could still be elicited, but only at significantly raised thresholds. Similar findings, implicating the amygdala specifically in the phenomenon of docility, psychic blindness, and altered sexual behavior, have been reported in rats (Robinson, 1963; Yamada & Greer, 1960), cats, agoutis and lynxes (Green et al., 1957; Schreiner & Kling, 1953, 1956), in baboons and dogs (Fuller et al., 1957; Mishkin, 1954; Mishkin & Pribram, 1954; Pribram & Bagshaw, 1953) and in man (Sawa et al., 1954; Terzian & Ore, 1955).

The very interesting studies by Schreiner and Kling (1953, 1956), involving amygdalectomies in wild and intractable agoutis and lynxes, point out the pivotal role of the amygdala in the mediation of fear and aggressive behavior. It is of some interest that the state of hypersexuality produced by amygdalectomies in Schreiner and Kling's cats (1954) can be abolished by castration and hormone therapy. This was one of the earliest indicators that the amygdala is involved in the control and regulation of neuroendocrine function. Subsequent lesion and stimulation studies have demonstrated a close linkage between the amygdala and neuroendocrine release, particularly in regard to gonadotrophic hormone (Koikegama et al., 1958; Sawyer, 1955) and ACTH (Mason, 1959).

It is significant that lesions or stimulation of the hippocampus have little if no effect on emotional behavior (Pribram & Kruger, 1954), although clear deficits in attention and learning following hippocampal removal have been noted in animals and man (Kimble, 1963; Milner & Penfield, 1955; Teitelbaum, 1964). On the other hand, stimulation of the amygdala gives rise to a wide spectrum of affect, such as rage, flight reactions, and sexual responses in animals (Brady, 1960; Gloor, 1960; Goddard, 1964). In addition to affective emotional responses, amygdaloid stimulation in man can modify levels of awareness, introduce confusion of consciousness, and interfere with memory recording mechanisms (Chapman *et al.*, 1954; Feindel & Penfield, 1954; Penfield & Jasper, 1954).

Lesions of cingulate gyrus and anterior thalamus have generally unspectacular effects on emotionality, although there have been reports of reduced emotionality (Bard & Mountcastle, 1947; Brady, 1960). Stimulation of the anterior thalamus in cats, which is the intermediate nucleus between the mammillary bodies and the cingulate, causes arousal and alerting responses at low intensities and fearlike crouching at higher intensities (Hunter & Jasper, 1949).

Stimulation of the amygdala can elicit fear or flight behavior, depending upon the location of the stimulating electrodes. Ursin and Kaada (1960) demonstrated a topographic organization (although there was some overlap) of different functions within the amygdaloid complex. Fear (flight) and aggressive reactions are produced from two rather separate zones within the basolateral nucleus.

2. Septal–amygdaloid relations. Thus, the data clearly indicate that the amygdala is in a pivotal position within the limbic complex in regard to the production and regulation of certain aspects of affective state. However, the function of the amygdala is very complex, since it is only one side of a dipole in opposition with the second side which is the septum. This appears to be the case, since ablation of the septal nuclei (particularly the lateral septum) results in ferocious and savage behavior (Brady & Nauta, 1953).

From the fact that septal lesions cause ferocity and savageness, it has been concluded that the septum normally exercises a restraining influence on the hypothalamus. Since lesions of the amygdala usually cause placidity and docility, it is assumed that the two structures are opposed to each other, the amygdala being excitatory and the septum inhibitory, with their major influence directed toward the hypothalamus. This view is warranted on both anatomic and physiological grounds. Figure 5.5 shows a diagrammatic perspective of the limbic "ring," with the amygdala and septum in dynamic balance of the two poles of the ring. Both the amygdala and septum possess strong connections with the hypothalamus, via the medial forebrain bundle (MFB), as well as abundant reciprocal connections. This anatomic picture suggests that a balance of opposites occurs within the limbic system and, in fact, is a fundamental feature of this system.

As mentioned earlier, some epileptic patients frequently shift state from one emotional extreme to the opposite in a fashion suggestive of reciprocal interaction. The series of reciprocal relations within the hypothalamus described

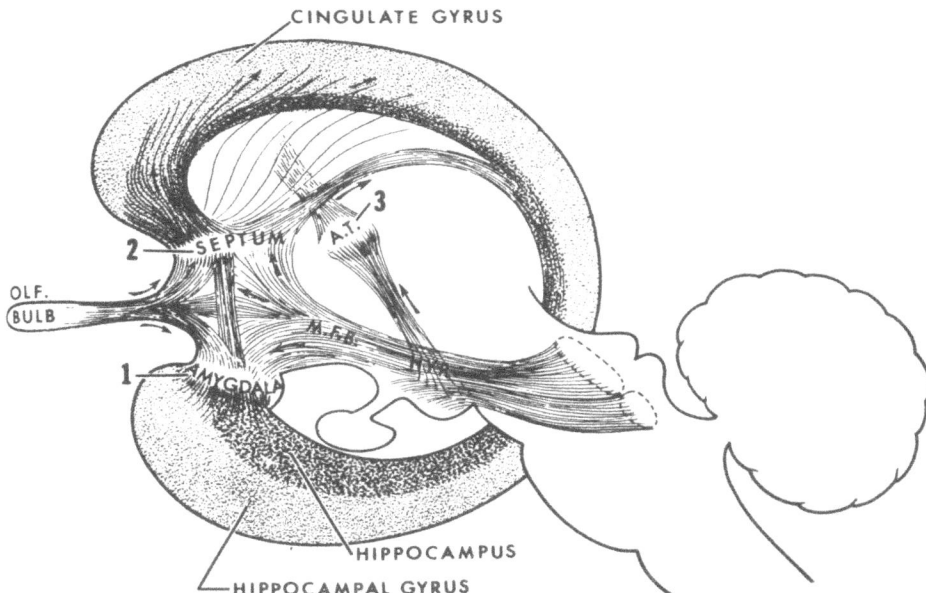

FIG. 5.5 The limbic system comprises the limbic cortex and structures of the brain stem with which it has primary connections. This diagram shows the ring of limbic cortex in light and dark stipple and focuses on three pathways (1, 2, and 3) that link three main subdivisions of the limbic system. Abbreviations: A.T., anterior thalamic nuclei; HYP, hypothalamus; M.F.B., medial forebrain bundle; OLF, olfactory. (From MacLean, 1958, 1970.)

earlier appear to extend into the limbic system and characterize the operation of the polar septal–amygdala interaction.

Additional support for the idea of balanced opposites is provided by a series of experiments conducted by King and Meyer (1958). These workers removed the septum and amygdala in combination, in an attempt to "titrate" the emotive state. The operations were performed in two sequences: the septum followed by the amygdala, and the amygdala followed by the septum (three days between operations). In the septum–amygdala order, the amygdalectomy completely wiped out the hyperemotionality and the rage responses caused by the septal lesion. When the order was reversed, the initial amygdalectomy caused very little behavioral change. This appears to be because the rats used in the study were docile and tame preoperatively. The subsequent septal lesion (three days later) produced an increase in emotionality, but this hyperemotionality was considerably less than that caused in a septal animal without amygdalectomy. It is not understood why amygdalectomy following septal lesions completely eliminates the septal effect, while only partially altering the septal effect if the lesions occur in the reverse order. However, it does seem clear that the two structures work in opposition to each other such that the removal of one tends to cancel out the effects of removal of the other.

E. The Neocortex

The classic studies of Bard and Mountcastle (1947) demonstrate a possible inhibitory influence of neocortical and allocortical origin upon subcortical, particularly hypothalamic, systems. Electrical stimulation of the neocortex generally has very little effect on emotionality. An exception to this was reported by Ursin (1960) in which rage responses were produced in cats following stimulation at high intensities of the temporal lobes. However, this is not surprising since the temporal lobes and the underlying entorhinal cortex have extensive classical and neurochemical relations with the amygdala, particularly the basolateral complex.

The one cortical region most frequently implicated in emotional or affective state is the frontal cortex. Although stimulation of the frontal cortex has little effect, the consequences of lesions to this area are well known (see Brady, 1960; Luria, 1966, 1973). The presumed therapeutic emotional changes observed to follow prefrontal resection have frequently been rationalized in terms of the intimate anatomic and functional relationship of the frontal cortex with the affective integrative mechanisms of the diencephalon, principally the dorsal medial nucleus. This seems appropriate, since extensive interconnections exist between the frontal cortex and dorsal medial thalamus, which in turn interacts with the hypothalamus via the rostral periventricular fiber system. It seems clear that extensive limbic system influences exert important mediating effects on diencephalic–cortical interactions. However, evaluations of the effects of frontal or prefrontal resections on behavior (see Fulton, 1948, 1951) clearly show that any improvement in emotional behavior is outweighed by other adverse consequences of frontal lesions, such as control over impulsive behavior and general intellect deterioration.

6
Information Representation

I. INTRODUCTION

Throughout this chapter, emphasis will be focused on space–time patterns of neural behavior. Orderly neural behavior (information) underlies perception, thinking, memory, and movement, and is common to all functional neural activities. Many forms of such ordered neural behavior or information exist. There is information about the external world (external sensory information) and information about internal sensations (internal sensory information); there is information about information (memory, consciousness, and awareness) and there are complex mental operations one can perform on information (thinking, planning, analyzing, creating, and so forth).

These forms of information all involve *internal representations*. For instance, the tree swaying gently outside the window has a representation inside the head. Obviously, the internal representation does not resemble a tree. It is in some other form, most likely involving distributed but coordinated neural activity shifting and changing in an orderly fashion as the tree sways. A formal definition of a representation is *any structure of which the features symbolize or correspond in some sense to some other structure. For any given structure, there may be several equivalent representations* (MacKay, 1969a). The primary goal of this chapter is to elucidate some of the neural features of representational systems.

II. INFORMATION AND REPRESENTATIONAL SYSTEMS: A DEFINITION

Information theory is primarily concerned with three aims: (a) to isolate and extract the invariant features of a representation; that is, to isolate the "kernel" features of the representation which do not change following reformulation. An entire branch of mathematics is concerned with this problem, for example, representation theory or abstract group theory; (b) to treat quantitatively the abstract features of the processes that make a representation; and (c) to give quantitative meaning to the several senses in which the notion "amount of information" can be used. In this chapter we are concerned mostly with how representational systems are created in the brain, what might be their invariant features, and most importantly, what type of descriptors must be used to describe neural representational systems.

Because neural representational systems are dynamic structural features of the brain (in space and time), they are amenable to quantitative analysis. Much of present day quantitative analysis of information has arisen from communication theory, which is concerned, very simply, with making a representation in one space A of a representation already present in another space B. By space is meant a coordinate mathematical framework, or manifold, in which the elements of a representation are ordered (MacKay, 1969a).

The foregoing discussion of representational systems provides the necessary background for a working or operational definition of information. A definition which is broad in scope and applicable to neurophysiology (MacKay, 1969a) is that information constitutes a change in a representational system. According to this definition information, in its most broad sense, is that which creates, adds to, or alters an existing representation.

Given the working definition of information just provided, it is important to ask: How is information about the external world represented in the brain? More specifically, how are patterns of sensory excitation transformed into representational systems? What do representational systems look like in space and time? What are some of the possible mathematical descriptions of representational systems? In recent years, a tremendous amount of knowledge has accumulated on the subject of sensory neurophysiology. Classical sensory neurophysiology is concerned with the detailed anatomy and physiology of the various classical modalities of sensation (auditory, visual, olfactory, kinesthetic, vestibular, and somaesthetic). It is, however, beyond the scope of this volume to discuss in detail the subject of sensory neurophysiology. The main purpose of this section is to acquaint the reader with basic mechanisms by which sensory representational systems are formed. The visual system will be emphasized, although generalities can clearly be made to the other sense modalities. Detailed information on classical sensory neurophysiology can be obtained in the following references: Adrian (1928), Aidley (1971), Barlow (1961, 1972), and Polyak (1957).

III. ORGANIZATION OF LOCAL
REPRESENTATIONAL SYSTEMS

In recent years, extensive experimentation has been directed toward elucidating the neural basis of form or feature perception in vision. Microelectrode analyses of retinal, lateral geniculate, and visual cortical systems have revealed basic mechanisms by which elementary visual features are coded. Extensive reviews of these analyses are available (Barlow, 1961, 1972; Blakemore, 1974). In general, it has been shown that single neurons of the primary visual cortex (area 17) respond optimally to lines or slits exhibiting a particular orientation in space (Hubel & Weisel, 1962). Furthermore, a single visual cortical cell responds to stimuli that impinge on particular sets of retinal receptors. The area of the retina to which a cell responds when stimulated or the region in visual space subtended by that retinal area is called the receptive field of that neuron.

There are a number of important differences between the sensitivity to different stimuli displayed by cells located in the different levels of the sensory system. Cells in the primary sensory cortex (area 17) respond to simple stimuli such as lines or edges. Cells in secondary sensory cortex (areas 18 and 19) respond to angles, rectangles, and other complex stimuli (Hubel & Wiesel, 1959, 1962, 1963, 1965). In general, the receptive field of area 17 neurons is smaller than the receptive fields of area 18 or 19 neurons (Hubel & Wiesel, 1962). This latter finding is consistent with the anatomic fact that the number of different sources of afferent input converging upon a cell progressively increases as one moves from peripheral (i.e., retinal) to higher cortical levels (Barlow, 1969, 1972; Hubel & Wiesel, 1962). There is also an increased proportion of cells that exhibit multimodal responses as one moves from primary visual areas to secondary and tertiary visual regions (Morrell, 1967; Thompson *et al.*, 1963). Thus, the dimensionality of neural response, in terms of the size of the receptive field, the complexity of stimuli, and the number of sense modalities to which cells respond increases from primary to secondary systems.

The accumulated evidence from stimulation, lesion, and electrophysiological studies supports the following set of summary conclusions regarding the neural basis of visual perception: (1) no matter which level is examined, it is clear that populations of neurons are involved in the coding of sensory information; (2) the primary visual system is involved in the coding of elementary visual features such as edges or lines; (3) the secondary system is involved in *synthesizing* or combining elementary sensory features into wholes; (4) the various levels of the sensory systems are hierarchically organized; and (5) multilevel interactions occur in which secondary systems influence primary systems and vice versa. These conclusions will be discussed in detail in the following pages, while considering three rather broad topics of interest: (a) hierarchical organizations of representational systems; (b) multidimensionality of sensory unit responses; and (c) evoked potential correlates of sensory representation.

A. Hierarchical Organization of Representational Systems

The fact that sensory systems are hierarchically organized is well established. The evidence showing an integrative hierarchy of systems is derived from stimulation, lesion, electrophysiological, and anatomic considerations. It is generally agreed that cortical sensory systems (visual, auditory, and somato-sensory) are organized in terms of a primary sensory system in which specific sensory afferents from specialized receptors synapse in specific thalamic "relay" nuclei, which in turn send their outputs to the cortex where they terminate in particular cell layers. Columns of vertically organized neurons (Hubel & Wiesel, 1963) that exhibit comparatively little intracortical interactions are found in this primary projection region (area 17, see Fig. 1.14). Surrounding the primary region is a secondary cortical system that receives afferents from the primary regions and contains neurons that respond to many modalities of sensory stimulation. The secondary areas are made up of a large number of small neurons that exhibit extensive intracortical interconnections.

Some of the cytoarchitectural differences between the primary visual cortex and adjacent secondary cortex are shown in Table 6.1. The most distinct differences between the two areas are in terms of the architecture of layers II, III, and IV. Layer IV, containing large pyramidal cells and receiving geniculo-cortical afferents, is more developed in area 17 than in area 18 (see Truex & Carpenter, 1964).

Table 6.1 shows the relative proportions of cell types in the various layers of the primary and secondary systems and demonstrates the cytoarchitectural differences between these cortical regions. There are a larger number of small cells that have extensive intracortical connections and there is a greater or more complex interaction between neurons in area 18 than in area 17.

Studies using strychnine neuronography (McCulloch, 1943; Thompson et al., 1963) show that associational interactions are greater in the secondary cortex compared to the primary. For instance, excitatory spread is confined within the primary sensory systems. In comparison, strychnine neuronographic methods show that considerable excitatory spread occurs in secondary areas, and can even result in the invasion of contralateral sensory regions (McCulloch, 1943).

TABLE 6.1

Number of Cells in Layers of the Primary and Secondary
Visual Cortical Systems in Man and Their Percentages of the
Total Number of Cells in These Areas[a]

Area	Layers			Total number
	II–III	IV	V	
17	143.4 (25.9%)	166.4 (33.1%)	49.3 (8.9%)	538.2
18	392.1 (51.9%)	152.9 (20.2%)	75.9 (10.2%)	756.8

[a]From Luria (1973).

There is considerable evidence showing a progressive increase in the complexity of representational systems within sensory organizations. Single unit analyses of visual cortical regions show that area 18 and 19 neurons respond to more complex stimuli than area 17 neurons (Hubel & Wiesel, 1962, 1963). Area 17 cells, responding to slits or edges of specific orientation, are involved in the creation of the elementary building blocks from which complex representations become synthesized. The synthesis of complex wholes from elementary parts appears to involve areas 18 and 19. For example, lesions of area 17 in humans result in specific *scotomas* or "blind spots" in the visual field. Lesions of area 18, on the other hand, interfere with the ability to combine elementary sensations into complete patterns. For instance, patients with lesions of the secondary visual zones are not blind and can see individual features. However, they are unable to synthesize basic elements into complete forms and require methods of deduction to determine what they are looking at.

The following quote from A. R. Luria (1973) illustrates this point: "This patient carefully examines the picture of a pair of spectacles shown to him. He is confused and does not know what the picture represents. He starts to guess. 'There is a circle . . . and another circle . . . and a stick . . . a cross bar . . . why, it must be a bicycle?' Or a patient looks at a picture of a peacock with different colored feathers in its tail, but not recognizing it as a whole, he surmises 'This is a fire, here are the flames' [p. 116]." In general, lesions of the secondary visual cortex (area 18) result in what is termed *visual agnosia,* an inability to perceive form.

A similar deficit in the ability to synthesize elementary sensations occurs following damage to secondary zones in other modalities (somatosensory or auditory). For example, lesions of the secondary somatosensory system (Brodmann's areas 3, 2 & 1) result in *parietal tactile agnosia* (Nielsen, 1946), or, as Denny-Brown and his associates (1952) described it, "a disturbance of the synthesis of tactile sensation causing defects of tactile perception of shape (*amorphosynthesis*)." Lesions of secondary auditory regions (Brodmann's area 22, see Fig. 1.14) do not affect the perception of simple sounds, but do impair the ability to differentiate groups of simultaneously presented sounds or patterns of sound (Butler *et al.,* 1957; Goldberg *et al.,* 1957; Luria, 1966, 1973).

Comparably, electrical stimulation of primary sensory areas in people elicits elementary sensations. For instance, in the visual area (17) this may be a flash of light or colored spots (*photopsia*) occurring in a specific area of the visual field. On the other hand, stimulation of secondary visual systems (area 18) frequently produces complex hallucinations involving whole forms, such as images of flowers, or animals or familiar people (Forester, 1936; Penfield, 1958, 1969; Penfield & Perot, 1963).

It is clear, therefore, that there is a hierarchy of progressively more elaborate representational systems. This conclusion is further supported by investigations of lesions of tertiary cortical zones, such as the areas that form the boundary between the secondary temporal, parietal, and occipital zones (e.g., Brodmann's areas 29 and 40 or areas 37 and 21; see Fig. 1.14 which shows the nomenclature devised by Brodmann for different cortical regions). These are areas of overlap of the various sensory modalities. In humans, lesions of these zones result in deficits in higher order functions, such as the formation of symbolic relationships or the synthesis of logical grammatical structures. For instance, such patients demonstrate a good understanding of individual words but they cannot grasp the meaning of a sentence as a whole (Luria, 1959, 1966, 1973; Luria & Tsvetkova, 1968).

Thus, there are multiple sensory representational systems that are hierarchically organized and exhibit dynamic spatial–temporal relationships. Some of the dynamics of the operation of complex representational systems are revealed by a closer analysis of deficits caused by lesions. For example, many studies show that the degree of deficit is related to the size of the lesion (Lashley, 1929, 1942, 1950; Luria, 1966, 1973).

Methods of tachistoscopically presenting visual stimuli to patients with localized occipital cortical lesions support this conclusion (Luria, 1973). The more severe the cortical damage the longer the exposure duration of visual stimuli required before synthesis can occur. Thus, the efficiency of secondary regions to synthesize representational systems is affected, depending on the amount of cortical tissue removed. This suggests that secondary systems perform a "holistic" or gestalt type of function, that is, the function of secondary systems is to coordinate sets of elementary patterns. Damage to the system diminishes this organizational capacity.

Consistent with this conclusion is evidence indicating that damage to secondary systems *reduces the number of elements represented in synthesized representational systems.* As Luria (1973) states, the Hungarian neurologist Bálint (1909) was the first to report that patients with lesions of the anterior occipital cortex (near the boundary of the inferior parietal region) exhibit a definite and distinctive *decrease in their range of visual perception.* This deficit differed from that caused by lesions of the optic tract or primary visual cortex by the fact the deficit was measured in units of meaning rather than in units of space. That is, "the patient could see only one object at a time regardless of its size (whether it is a needle or a horse) and was completely unable to perceive two or more objects simultaneously [p. 121]."

Similar observations in patients with lesions of secondary sensory zones have subsequently been made (Denny-Brown & Chambers, 1958; Denny-Brown *et al.,* 1952). As Luria (1973) reports: "They cannot place a dot in the centre of a circle or a cross, because they can perceive only the circle (or the cross) *or* the pencil point at any one time; they cannot trace the outline of an object or join the strokes together during writing; if they see the pencil point they lose the line, or if they see the line, they can no longer see the pencil point [p. 121]." This deficit is called *simultaneous agnosia.*

A similar phenomenon has been noted in animal studies. For instance, studies by Pribram (1960) demonstrate that monkeys with bilateral damage to the inferior temporal gyri respond to significantly fewer simultaneously presented stimuli. A narrowing of the range of simultaneous tactile perception has also been noted in patients with secondary parietal damage (Denny-Brown & Chambers, 1958; Denny-Brown et al., 1952). These findings are significant, since measurements in normal humans show a limit of simultaneous perceptual capacity which is approximately seven items (Miller, 1956).

These data indicate that the degree of perceptual complexity, in terms of the number of elements that can be synthesized to form a representation or the number of simultaneous representations, is reduced following secondary sensory damage.

Some of the mechanisms that control the perceptual capacity of sensory systems have been elucidated in animal experiments. Studies by Spinelli and Pribram (1966, 1967) and Gerbrandt et al. (1970) indicate that attention mechanisms control the complexity and perceptual capacity of the sensory systems. Spinelli and Pribram (1967) have shown that *the size of the receptive field of primary visual neurons increases or decreases following stimulation of secondary (associational) cortex.* Some of the results of the Spinelli and Pribram experiment are shown in Fig. 4.12.

A study by Luria (1973) has shown that the problem of simultaneous agnosia can be ameliorated by injections of caffeine. For instance, an injection of .95 ml of a 1% caffeine solution enabled a patient with anterior occipital damage to perceive two and in some cases three objects simultaneously. The increase in perceptual capacity lasted 30 to 40 min, the duration of action of the caffeine. These studies indicate that excitatory cortical influences can alter the integrative capacity of primary visual systems and that the perceptual deficits following secondary sensory damage arise because of diminished cortical action upon the primary systems, as well as diminished integrative capacity of the secondary systems themselves. This conclusion is supported by studies showing that changes in the "complexity" of interaction of neurons in the primary visual system can occur following stimulation of secondary sensory cortex (Spinelli & Pribram, 1966, 1967; see Chapter 4, Section V.C).

These studies suggest that the *amount* of "synchronization" or "desynchronization" within populations of primary visual cortical cells can be modulated by the action of the secondary sensory systems. These studies further emphasize the role of attention mechanisms in the control of perceptual phenomena. As discussed in Chapter 4, shifts in attention involve changes in the integrative capacity of primary systems, which is reflected by the dynamics of neural cooperative behavior within these systems. The problem of attention control over sensory input and the role of association and frontal cortical regions in sensory integration was discussed previously (Chapter 4).

Poststimulus histograms. If the activity of a single neuron is affected by presentation of some stimulus or occurrence of some event, whether inside or outside the nervous system, the firing pattern of that neuron will be altered. Such alterations may be manifested over a long period of time relative to the time required for the afferent information to reach and activate the cell. This occurs because the consequences of occurrence of the event continue to reverberate in different circuits or loops of neurons, with persistent influences impinging upon the neuron under observation (see Chapter 2). These effects may be difficult to discern clearly, either because they cause subtle changes in the activity of the unit or because of variability in the unit behavior due to other influences.

A powerful method for obtaining a clear picture of the effect of a stimulus upon a neuron is to record the firing patterns displayed by the cell during a period of a few hundred milliseconds after occurrence of the event, and to superimpose a large number of firing patterns elicited by repetitive presentations of the same stimulus. The period throughout which the response to a single stimulus is monitored is called the *analysis epoch*. The average firing pattern obtained by superimposing or summating a set of firing patterns, with the analysis epoch phase-locked to the time of stimulus presentation, is an "average poststimulus spike histogram." Usually, this term is shortened to poststimulus histogram, or abbreviated as PSH.

PSHs are usually constructed by using special purpose computers called "average response computers," or small general purpose minicomputers. In either case, the principle is the same: occurrence of the specified event or stimulus triggers the computer at time t_0. All spikes emitted by the cell under observation during a brief interval, Δt_1, after t_0 are counted and stored in one register, or bin, in the computer. The computer then indexes to an adjacent register and stores there the number of spikes emitted by the cell during the next time interval, Δt_2.

This process is continued until the sequence of time intervals sampled equals the analysis epoch. Thus, the analysis epoch is divided into a sequence of N equal time intervals t in duration. This duration, Δt, is a parameter which can be set by the observer to provide any desired resolution of the firing pattern. Analysis of the firing pattern caused during the analysis epoch following a single stimulus thus produces a sequence of N numbers, each number N_i corresponding to the number of spikes which occurred during the corresponding time interval, Δt_i. The sample of firing pattern occurring during a single epoch is often referred to as a "sweep."

In order to compute the PSH, the process just described is repeated a number of times. Each time, the number of spikes produced during each of the specified time interval, Δt_i, is *added* to the number already stored in the corresponding bin, accumulated from previous sweeps. As the number of sweeps is increased, that is, as the number of observations of firing patterns elicited by repeated stimuli is increased, the sequence of N numbers stored in the bins corresponding to the sequence of time intervals Δt_i describes *the probability of cell firing as a function of time after the stimulus.* If the stimulus has no reproducible effect on the cell, all N numbers will be approximately equal. In that case, a bar graph, or histogram, displaying the value of each of the N numbers successively as a function of time, would be approximately flat across the duration of the analysis epoch. Increased discharge or excitation caused by the stimulus would cause a peak in the contour of the histogram, while inhibition would cause a trough. The time at which the peak or trough occurred would indicate the latency of the effect. Usually, a PSH displays a complex sequence of peaks and troughs showing that the response of a cell to a given stimulus is a complex sequence of facilitation and inhibition of its activity. Usually, the temporal pattern of modulation of firing observed in the PSH is quite typical for a particular stimulus and will alter when the stimulus is changed.

B. Multidimensionality of Sensory Unit Responses

These computer calculations of the averaged responses of single units to sensory stimulation, called "poststimulus time histograms," typically show a wide range of stimulus features to which a cell will respond (Pettigrew *et al.,* 1968; Spinelli, 1967; Spinelli & Barrett, 1969; Phelps, 1973). This is not to say that sensory cells do not respond optimally to particular stimulus features. However, sensory neurons exhibit a "tuning" type of behavior which involves diminishing responsivity to nonoptimal stimuli. Figure 6.1 shows an example of a visual area 17 cell that exhibits tuning features to a particular orientation of line. Note that while the cell shows a distinctive behavior in response to a "preferred" stimulus, it nonetheless responds to other orientations. Computer averaging PSH methods help to quantify the range of responsivity of particular cells (Blakemore, 1974; Morrell, 1967; Spinelli, 1967; Spinelli & Pribram, 1967). The results of studies using computer averaging indicate that the building blocks of visual perception are created by the response of a constellation of cells in which individual neurons respond at different rates to different stimuli; that is, *a given cell may participate in the creation of several elementary representations, but the contribution of the cell to any particular representation diminishes as the stimulus feature is "detuned."* For instance, the cell in Fig. 6.1 is shown to respond optimally to a particular orientation. The same cell also responds, as measured statistically, to the representation of other orientations. Similar studies, using computer statistical analyses, have emphasized the multidimensional character of sensory neurons (Ben-Ari & LaSalle, 1972; Fox & Norman, 1968; Freeman, 1972a, 1975; John & Morgades, 1969; Verzeano, 1972).

Recent psychophysical and evoked potential analyses have shown that neural populations exhibit specific responses to particular spatial frequencies (Blake-

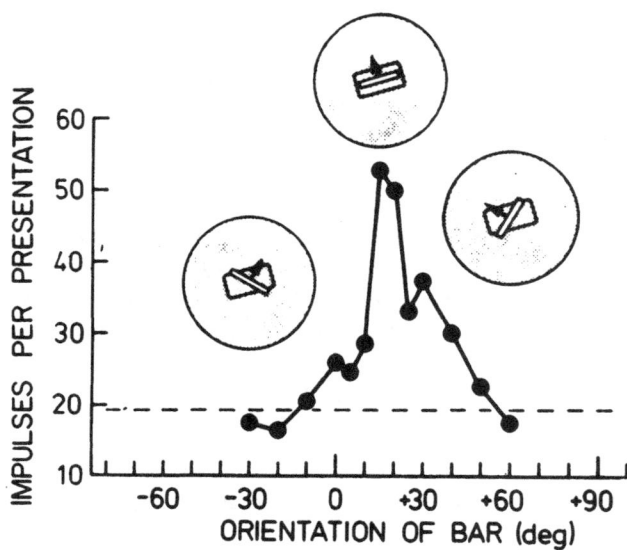

FIG. 6.1 The orientational tuning curve for a complex cortical cell. The stimulus was a bright bar, generated on a display oscilloscope, as shown in the inset diagrams where the large rectangle is the receptive field. Each point is the mean number of impulses produced during six successive sweeps at that orientation. The dashed line is the mean spontaneous discharge in the same period of time, in the absence of any stimulus. (From Blakemore, 1974.)

more & Campbell, 1969). A spatial frequency is defined by the number of lines per unit retinal area which comprise a visual stimulus. The more dense the lines, the higher the spatial frequency.

It has been proposed that sensory neurons create mosaics of *spatial Fourier terms* and that the statistical integration of the Fourier terms constitutes a representation of a visual image. The Fourier analysis involves the decomposition of complex periodic functions by a set of elementary sine waves. Similarly, Pollen and his co-workers (1971) and Pollen and Taylor (1974) suggest that visual representational systems are synthesized from the combination of elementary neural responses, using the Fourier series. Although this idea has been criticized on the grounds that it requires a completely linear neural system with a high degree of tuning (Mittenthal *et al.*, 1972), it nonetheless provides a simple schema by which "whole" neural representational systems can be synthesized. Similar integrative ideas have been offered by Pribram (1969) and others (Longuet-Higgins, 1968; Van Heerden, 1968) in holographic models. The holographic models, which rely extensively on Fourier analysis, provide a very simple mechanism to explain how complex representational systems may be delocalized in space.

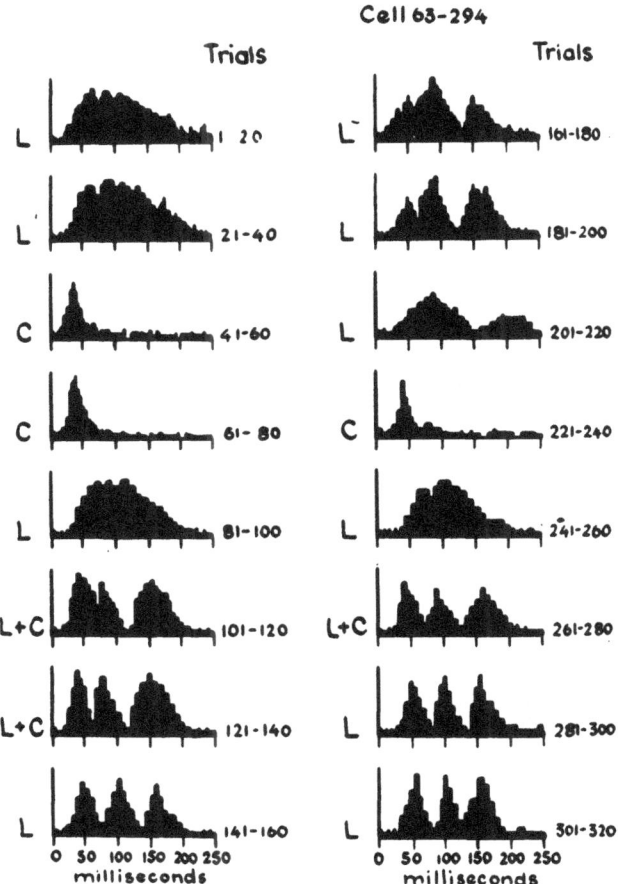

FIG. 6.2 Response modification in a polymodal cell from area 18 of the cat. "Preferred" visual stimulus (L) for this cell was a light line at 2 o'clock in its receptive field. Click stimulus (C), 30 dB above human auditory threshold, was also effective in producing a response, although with a different pattern. L + C indicates preferred visual and acoustic stimuli combined. The kind of stimulus is indicated on the left and trial number on the right. It can be seen at bottom left that after 40 pairings of L + C, L alone elicits a modified response. Further explanation in text. (From Morrell, 1967.)

Another example of the multidimensional character of single unit responses is shown in Fig. 6.2. In this example, Morrell (1967) recorded from a single cell in area 18 of the cat visual cortex. The top two histograms show the response of the cell to a light stimulus. The third and fourth histograms show the response of the cell to a click stimulus. When the light stimulus is paired with the click stimulus, the cell exhibits a complex modified response (histograms 6 and 7), which is *not* a simple sum of the light and click histograms. This modified

CAT: N8

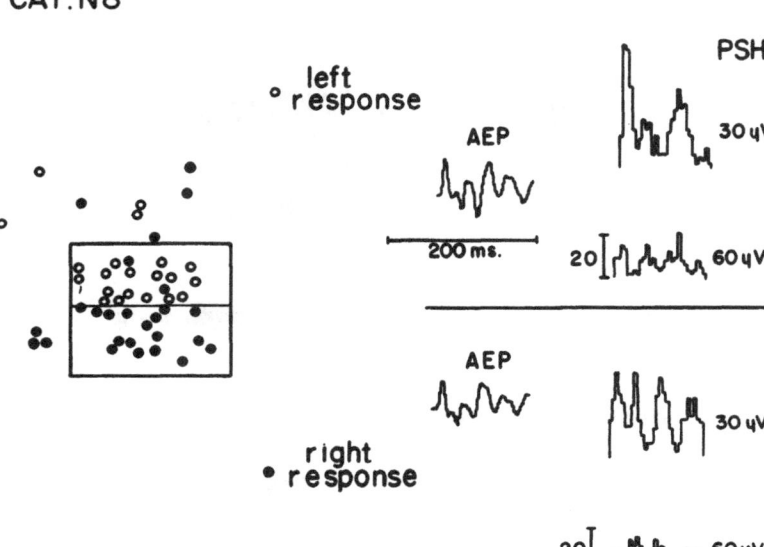

FIG. 6.3 Averaged evoked potentials (AEP) and poststimulus histograms of unit discharge (PSH) recorded from the same microelectrode (in the lateral geniculate) when the cat interprets the generalization stimulus as "go left" (top of figure) or "go right" (bottom of figure). Two distinctly different PSH contours are elicited depending on the interpretation the animal gives the stimulus. PSH differentiation between left and right responses occurs primarily to lower amplitude units in the $\leqslant 30\ \mu$V window. PSHs for higher amplitude units (in $> 60\ \mu$V window) do not show a clear differentiation.

response pattern persists following the light and click pairing (histogram 8). The behavior of the cell exhibits properties of learning in that the modified pattern can be extinguished and reestablished. This figure shows that secondary visual cells are not only polymodal in character but can exhibit modified response patterns through an association linkage. Similar learning type responses from cells in the primary sensory system, including retinal ganglion cells, have also been noted (Ben-Ari & LaSalle, 1972; Chow *et al.*, 1968; Lindsley *et al.*, 1967).

More specific examples of a population of cells participating in multiple representational systems have been provided by the work of John and Morgades (1969b) and in a current study by Ramos *et al.*, (1974, 1976). These workers recorded responses of groups of cells in cats trained to respond differentially to two different frequencies of a flickering light. After the differential response was well established, a generalization stimulus was presented at a frequency inter-mediate to the two conditioned stimuli. Half of the time, the animal interpreted the intermediate stimulus as stimulus frequency 1 and performed the corres-

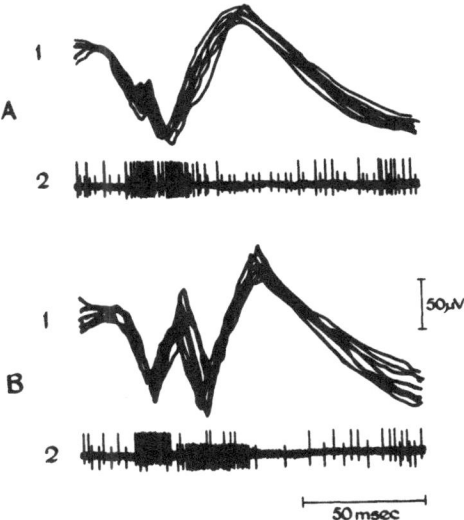

FIG. 6.4 The superimposed tracings (A1) show the response of the visual cortex to light flashes, in a curized cat preparation. The same microelectrode from which (A1) was recorded detected two cortical units which show different spike amplitudes, seen in (A2). The neuron producing the larger spike fire vigorously during the early component of the evoked potential, while the neuron producing the smaller spike fires sporadically throughout the interval and is apparently unresponsive to the light. The cat was then exposed to a classic conditioning procedure in which the flash was paired with foot shock. Following the procedure a late component appeared in the evoked potential as shown in (B1). Note that in (B2) the large spikes occur primarily during the early component, while the small spikes are now strongly timelocked to the late component. These data provide suggestive evidence that, in some cases, the readout or endogenous component is mediated by a separate class of cells. (Reprinted from John, 1969.)

ponding response. On other trials, the animal treated the intermediate stimulus as stimulus frequency 2 and performed a response consistent with that frequency. Poststimulus time histograms showed that different unit response patterns were displayed to the identical intermediate stimulus, depending on how the animal interpreted the stimulus. Figure 6.3 shows the poststimulus histograms of cells, as well as averaged evoked responses, elicited by identical stimuli which were interpreted in different ways. The poststimulus histograms at the right of Fig. 6.3 establish that the same population of cells can exhibit differential response patterns to identical stimuli. These data indicate that a single cell or a specific population of cells can participate in *multiple* representational systems. While these data indicate that a given cell may participate in more than one memory or representation (see Chapters 8 and 9) other data indicate that specific cells may also be more restricted, participating in specific representations. For example, Fig. 6.4 shows that the behavior of a population of cells producing spikes of small amplitudes was correlated with one evoked potential waveform, but not another. Thus, it would appear that both situations are true, that is, some cells are members of multiple representations, while other cells are more unidimensional. Conceivably, there are gradations in the degree to which cells participate in multiple representational systems. It is likely that the degree of dimensionality will be shown to increase as the loci of cells studied moves from primary to secondary sensory organizations.

C. Evoked Potential Correlates of Sensory Representation

It was argued in previous sections that neural representational systems involve the cooperative behavior of large populations of cells. In Chapter 9 the concept of a function is defined in terms of the cooperative behavior of populations of neurons. Rather than proposing that individual functions are localized to some isolated cell group or narrow region of cortex, it is most consistent with the data to assert that functions arise from the cooperative activity of the constituents of an extended system. Individual components or members of the system are frequently distributed in wide areas of the brain. It is therefore most appropriate to speak in terms of functional systems composed of distributed yet coordinated populations of cells that operate as a whole.

Since representational systems are distributed in space, one must bear in mind that evoked potentials recorded from one or a few gross electrodes (especially scalp electrodes) only reflect a small aspect of any functional system. This is particularly true if we realize that the neurons whose activity contributes to the voltage gradients monitored by a recording electrode may be members of a variety of functional systems.

As discussed in Chapters 9 and 10, individual neurons interact statistically on a multidimensional basis. In these chapters data are presented demonstrating that considerable freedom exists in the behavior of the individual neurons which comprise an ensemble. Such individual variability notwithstanding, degrees of order and replicability of response are manifested within a population of cooperatively acting neurons. In this regard it is important that the orderly response of a large population of synaptic generators contributes significantly to the EEG and the averaged evoked potential (Creutzfeldt *et al.*, 1966a; Humphrey, 1968; see also Chapters 12 and 15).

Although the evoked potential reflects the activity of only one portion of the volume of a distributed system, the averaging process makes visible those features of the patterns of synaptic activity that are reproducible across time (see Chapter 10). An ensemble of neurons participating in a certain function exhibits a particular *spatiotemporal* organization, that is, patterns of coherent or correlated synaptic events will occur in different portions of the system in a temporal sequence determined by anatomic connections, transmission times, and similar parameters of the ensemble. If this spatiotemporal pattern is initiated by some stimulus and if it possesses stable invariant features, then it will contribute significantly to the waveshape obtained when a number of potentials evoked by successive presentations of that stimulus are averaged, with appropriate synchronization or "timelocking" to the stimuli. (Average evoked response computation is discussed in Chapter 9).

It is most probable that sensory representational systems involve specific spatiotemporal organizations, with groups of cells distributed in space but cooperative in action. It also seems to be true that a functional system of cells will exhibit a characteristic spatiotemporal pattern corresponding to the representation of a particular sensory stimulus. Thus, it would be expected that successive presentations of a stimulus will cause successive activations of the corresponding spatiotemporal organization. The results of evoked potential analyses indicate that indeed this is the case.

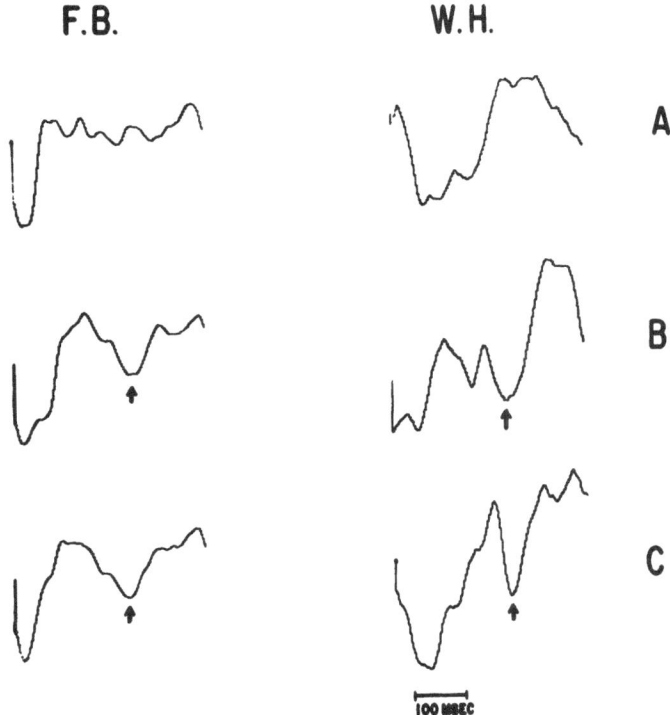

FIG. 6.5 Averaged evoked potentials elicited by different information displays (the letter A or B or C) in two subjects (F. B. and W. H.). Recordings are from the left parietal lobe (P_3). The arrows show the presence of a common late process produced by Bs and Cs. The waves differ significantly in terms of both early and late components. (From Thatcher & John, 1975.)

TABLE 6.2
EP Factor Loadings for the Letters A, B, and C[a]

		Factors[b]		
		1	2	3
Subject W. H.	A	27	0	0
		83	0	0
		87	9	0
		89	10	0
		93	4	0
		86	1	5
		41	15	2
	B	25	50	6
		26	26	5
		1	86	3
		2	93	0
		4	95	0
		0	94	0
		0	83	5
	C	19	58	13
		19	56	19
		29	51	12
		23	32	40
		17	16	64
		48	6	13
Subject F.B.	A	77	17	0
		86	11	0
		88	11	0
		76	17	0
	B	21	75	0
		16	81	0
		15	83	0
		16	80	0
		16	81	0
		17	69	0
		19	56	0
		17	60	0
	C	24	53	21
		29	46	24
		27	55	16
		49	33	15
		66	23	8
		66	21	5
		59	30	1
		49	45	1
		39	50	4
		35	49	3

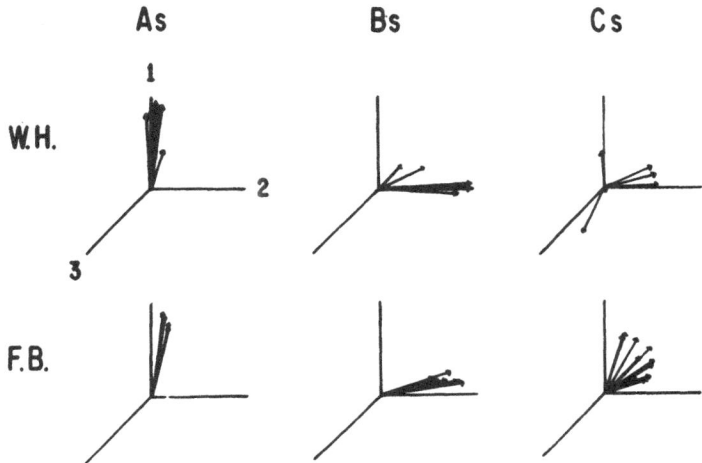

FIG. 6.6 Vectograms illustrating factor loadings for sliding averaged evoked responses (shown in Table 6.2) elicited by As, Bs, and Cs in two subjects (W.H. and F.B.). Factors 1, 2, and 3 from Table 6.2 form the axes of the vectograms. The length of a vector (arrow) defines the total energy of the evoked response in subspace (see Volume 2, Chapter 3, for factor analysis explanation). This shows that the morphology of the evoked potential, determined by the information content of a stimulus, can be succinctly quantified.

For example, Fig. 6.5 shows the evoked response recorded from occipital scalp regions of two human subjects who viewed a brief presentation of three different letters: A, B, or C (Thatcher, 1974a, b, 1976b; Thatcher & John, 1975). It can be seen that the average response waveform for the two subjects is distinctly different for each of the letters. The letters were presented in a counterbalanced manner and equated for luminescence, so that momentary excitability changes and arousal effects were distributed uniformly over time. These data indicate that the neural representation of each of the three letters involves a distinct and differentiated spatiotemporal organization. The stability of the representational systems as revealed by the evoked potential is demonstrated by factor analyses of sliding averages (see Table 6.2 and Fig. 6.6). These analyses demonstrate that specific wave processes are consistently associated with different letters. (The application of factor analysis is discussed in Chapter 3 of Volume 2.)

Other studies have provided similar results. For instance, White (1969) demonstrated unique and distinct evoked potential waveshapes elicited by circles,

[a]From Thatcher and John (1975).

[b]Sliding averages of evoked responses produced by As, Bs, and Cs were made by averaging 5 evoked responses in sequential order; that is, A ERs 1 + 2 + 3 + 4 + 5, then A ERs 2 + 3 + 4 + 5 + 6, then A ERs 3 + 4 + 5 + 6 + 7, etc., for each letter. This moderate amount of averaging was sufficient to create factor loadings specific to the letter displayed. Factor 1 loads heaviest for letter A, factor 2 for letter B, and all three factors, but particularly factor 3, for letter C.

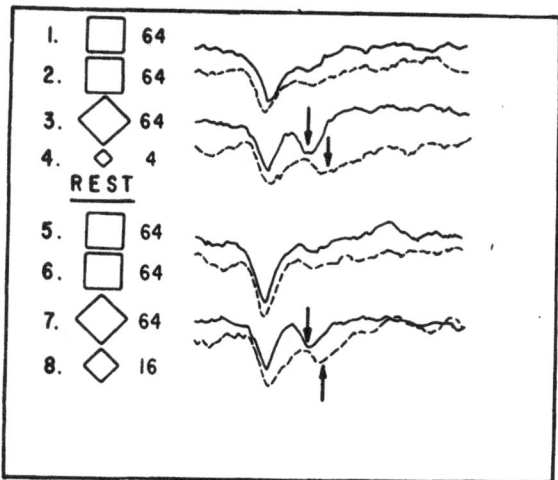

FIG. 6.7 Averaged responses from two sessions with the same subject, separated by 30 minutes. All averages based upon 100 repetitions of the stimulus, and a 500-msec analysis epoch. Negative upward. Responses 1, 2, 5, and 6 were to squares with an area of 64 square inches, responses 3 and 7 to diamonds 64 square inches in area, response 4 to a diamond of 4 square inches, and response 8 to diamond of 16 square inches. Note the new components, marked by the arrows, which appeared in 3 and 7 when the square was rotated 45°. (Data from John *et al.*, 1967. Copyright © 1967 by The American Association for the Advancement of Science.)

horizontal lines, checkered patterns, and radial lines. John and his associates (1967) demonstrated different and replicable evoked potential waveshapes elicited by squares, diamonds, circles, and different words printed with letters of equal area. Similar results have been reported in other studies (Clynes *et al.*, 1967; Fields, 1969; Herrington & Schneidau, 1968; Pribram *et al.*, 1967). Experiments have also shown that different evoked potential waveforms can be elicited by different colors (Burkhardt & Riggs, 1967; Clynes, 1965; Riggs & Sternheim, 1969).

The relevance of the analysis of evoked potentials for our understanding of information processing in the brain is demonstrated with particular clarity in Fig. 6.7. This figure shows that the relevant variable determining the evoked potential waveform is the "meaning" of the stimulus, independent of stimulus size (John *et al.*, 1967); that is, the presentation of diamond and square stimuli elicited invariant wave processes specific to the concept "square" or "diamond." As can be seen in Fig. 6.7, the area of the retina activated by the stimulus was unimportant compared to the "identity" of the stimulus itself. As mentioned earlier, anterior occipital lesions produce deficits measured in units of meaning and not in units of size. Correspondingly, Fig. 6.7 provides an example of an evoked potential correlate of a representational system which reflects the meaning of a stimulus and not its size.

These data support the conclusion that representational systems are constituted by stable spatiotemporal organizations within cooperative ensembles of cells. The evoked potential, although it reflects only relatively local processes in a portion of a distributed system, nonetheless shows us important features of this organization. The clinical significance of these conclusions is discussed in later chapters of this volume.

Studies have shown that evoked potentials also reflect spatiotemporal organizations involved in higher order processes. For instance, a long latency (250–320 msec) positive evoked potential component (called the P300) is correlated with the level of uncertainty when a sequence of unpredictable stimuli is presented to human subjects (Donchin & Cohen, 1967; Sutton et al., 1967). This component is markedly diminished or absent when the stimulus sequence can be predicted by the subject (Sutton et al., 1967; Weinberg et al., 1970). These studies suggest that the brain compares neural representations of expected with existing sensory events and that the certainty of the occurrence of the sensory event determines, in part, the magnitude of the comparison process. The P300 is nonspecific in that it (1) reflects a comparator operation rather than a particular stimulus feature, and (2) occurs with different modalities of stimulation at similar latencies over widespread anatomic regions.

However, identifiable processes related to specific memories may also be reflected by evoked potential waveshape. As discussed in Chapters 9 and 10, John and co-workers (John, 1963, 1967a, 1972; John et al., 1973) have demonstrated that evoked potential waveforms in cats can be elicited by neutral or irrelevant stimuli which are good facsimiles of waveforms produced by meaningful stimuli. John (1967a, 1972) refers to such endogenously produced waveforms as "readout" potentials. Others have termed these processes "emitted potentials" (Begleiter et al., 1973; Weinberg et al., 1970, 1974). Thus far, such potentials (emitted or readout) have been observed in humans in a brightness discrimination paradigm (Begleiter et al., 1973), in expectancy studies (Weinberg et al., 1970, 1974), and in studies requiring a subject to imagine a previously presented display (Herrington & Schneidau, 1968; John, 1967a).

IV. ANATOMIC AND TEMPORAL CHANGES IN THE ORGANIZATION OF GLOBAL REPRESENTATIONAL SYSTEMS

In the previous section, evidence was presented for the participation of rather circumscribed local populations of cells in representational systems. Although such phenomena reveal local participation, nonetheless representational systems are distributed in space. A single recording channel, whether monitoring multiple units or evoked potentials, must necessarily reflect only a small part of the total functional system.

What are the global characteristics of representational systems? In particular, do such systems possess a fixed anatomic structure or can representational systems expand or contract? Do representational systems exhibit a particular shape in space and time on a global level? Do representational systems have a characteristic morphology or geometry and, if so, does the geometry itself possess informational value?

The answers to these questions necessarily rely on studies that employ simultaneous recordings from many areas of the brain.

Most of the studies that will be reviewed will be from behaving subjects. Although acute preparations, involving animals immobilized in a stereotaxic apparatus and artificially respirated, have yielded information about representational systems, for the most part such studies are extremely limited. Active interaction with the environment seems to be a prerequisite for the manifestation of complex representational systems.

A. Expansion of Global Representational Systems

It is important to distinguish between activity related to arousal or nonspecific excitability changes (related to the onset of a stimulus) and activity characteristic of global representational systems. As discussed in Chapters 3 and 4, the onset of a novel stimulus usually results in diffuse or widespread desynchronization of neural systems. In simple classical conditioning paradigms, this initial widespread activation gradually abates and becomes localized to the sensory and motor regions subserving the conditioned response (Beck et al., 1958; Gastaut et al., 1957; Lindsley et al., 1950; Morrell & Jasper, 1956). This phenomenon has been interpreted (Voronin & Sokolov, 1960) as a "tuning" of the parts of the nervous system that are most concerned with the performance of the response. The representational systems are reflected by the persistent and stable activation pattern that remains after the diffuse changes have receded. The diffuse (habituation-prone) pattern represents an initial process of increased alertness and readiness to respond, involving a general increase in cortical tone or excitability (see Chapter 4). This appears to be the requisite condition for the gradual development of complex representational systems.

The distinction between the development of dynamic global representational systems and the arousal reaction is illustrated by the work of John and Killam (1960). These workers studied assimilated or labeled rhythms in the EEG of performing cats, in a paradigm involving a progressive increase in the "relevance" or "significance" of a conditioned stimulus. (Labeled or assimilated rhythms, which reflect the retrieval of temporal information, are discussed in detail in Chapters 7, 9, and 10.) The first stage of the study involved presenting a background flickering light while cats pressed a bar for food. The flickering light was a noncontingent event. The initial presentation of the flicker resulted in a widespread distribution of labeled rhythms which in a few trials diminished and almost completely disappeared. When the behavior of the animals was brought under the control of the flicker, so that bar pressing occurred only in the presence of the flicker, then the anatomic distribution of labeled rhythms increased. Next, after discrimination training in which lever pressing was rewarded in the presence of one frequency and not another (go–no go discrimination), the anatomic distribution of labeled rhythms increased further. Finally, a differential approach–avoidance response was established to two different frequencies of flicker. This resulted in a further increase in the amplitude and distribution of the assimilated rhythms. Thus, as John and Killam (1960) concluded, the increased signal value of the CS resulted in an increase in the number of structures or in the amount of total brain tissue involved in the evaluation of the signal and the subsequent performance of the animal. The results of the John and Killam study also emphasized the intimate interplay between expansion of representational systems and attention mechanisms (see Chapter 4).

In an earlier study John and Killam (1959) showed that a complex anatomic reorganization of neural processes in which labeled rhythms increase in amplitude in some structures and decrease in others parallels the establishment of an instrumental response. This observation was subsequently confirmed by John and Killam (1960). Thus, it would appear that in addition to a general expansion of neural systems during learning, there may also be a "shaping" or refinement process in which certain neural populations are turned on while others are turned off. This particular and very important aspect of representational systems will be discussed in a later section.

Other evidence showing anatomic expansion of local representational systems is offered by the work of Woody and Engel (1972). These workers, using microstimulation of the pericruciate cortex of the cat, first determined the response thresholds of cells that project to different facial muscles (e.g., the nose and eyelid). The cats were then trained to either twitch their nose or blink their eyes to a tone CS. If a nose twitch was conditioned, there was a decrease in the threshold of "nose" cells that project to the nose and a significant increase in the number of cells projecting to nose muscles. If, on the other hand, eye blink was conditioned, there was a significant decrease of threshold as well as an increase in the number of cells that project to eye muscles. A similar anatomic expansion of systems mediating a behavioral response has been demonstrated by Olds and co-workers (Disterhoft & Olds, 1972; Olds et al., 1972; Segal & Olds, 1972). These workers recorded the evoked response of single units from a wide variety of brain regions during pseudoconditioning in which two different tones and a food pellet were presented in a random order. These responses were subsequently compared to responses elicited during a conditioning procedure in

TABLE 6.3
Total Numbers of Units Showing Sensory and Learned Responses with Different Latencies in Different Brain Areas[ab]

	N	Sensory					Learning					%	%
		20	40	60	80	X	20	40	60	80	X	20	80
Frontal cortex	24	0	2	1	1	20*	1	3	1	4	15	4	37
Sensorimotor cortex	32	0	3	0	0	29	1	0	1	5	25*	3	22
Auditory and visual cortex	13	0	2	0	0	11	2**	2	1*	2	6	15	54
Subiculum	18	0	0	0	0	18	1	2	2	2	11**	6	39
Dentate	24	0	1	1	0	22	1	1	2	2	18	4	25
CA3 field of hippocampus	63	0	0	3	1	59	3*	6	3	8	43*	5	32
CA1 field of hippocampus	36	0	1	2	1	32	0	2	2	3	29	0	20
AML thal	71	4	8	7	8	44	5*	6*	11*	10	39	7	44
Vent N and lat gen	23	1	3	1	0	18	0	0	5	4	14	0	39

Region															N
Medial geniculate	14	6	3	0	0	5	2	2	4	2	0	0	6	14	57
Post thal	17	2	6	2	0	7	9*	1	5	1	5	5	2	53	88
AD retic	32	2	2	2	0	26	0	5	5	0	3	0	19*	0	41
V tegm	14	2	0	2	0	10	3*	2	1	1	1	3	7**	21	50
Pontine reticular formation	26	7	4	3	0	12	3	6*	3	1	1	0	13*	12	50
Tectum	14	9	1	0	0	4	0	1	1	0	0	0	11	0	21
Msc corp call	4	0	0	0	0	4	0	0	0	0	0	2	4	0	0
Msc limbic	7	0	0	0	0	7	1	0	0	2	4	0	0	0	43
Msc expyr	11	1	0	1	2	7	0	3	0	0	0	8	0	0	27

[a] From Olds et al., (1972).

[b] The latency intervals (0–20, 20–40, 40–60, and 60–80 msec) are marked according to the high end of the interval. N = total number of cases tested; X = number of cases which did not show responses with latencies of 80 msec or less. Abbreviations: AML thal, anterior, medial, and lateral groups of thalamic nuclei; vent N and lat gen, ventral nucleus of thalamus and lateral geniculate; post thal, posterior nucleus of thalamus; AD retic, anterodorsal part of the midbrain reticular formation; V tegm, ventral tegmentum and posterior hypothalamus and adjacent zone incerta; msc corp call, miscellaneous points in or near the corpus callosum. Each asterisk indicates that one individual from the adjacent number of cases had a "possible" latency of 10 msec.

which one of the tone stimuli CS⁺ was paired with food. In the latter procedure the CS^+ was presented for 1 sec and then followed by the presentation of a food pellet. Unit responses to the various stimuli during the pseudoconditioning control and during conditioning are shown in Table 6.3. In this table the number of units recorded from each brain region and the latency of response from the onset of the tone are shown for the pseudoconditioning (sensory stimulation) and conditioning (learning) procedures. It can be seen that there is a widespread increase in the distribution of responses (in both space and time) as a function of conditioning. Some structures, particularly thalamic and cortical, exhibited stronger and more persistent responses than in other regions (e.g., pontine reticular formation and the tectum, which showed decreases).

In another study in which rats were classically conditioned to different CS–UCS intervals, an expansion of representational systems was demonstrated to occur as a function of the CS–UCS interval (Thatcher, 1970, 1975). In the latter study (see Fig. 6.15), EEG rhythms at frequencies specific to the frequency of a past CS appeared in certain structures at particular times and exhibited an expansion of anatomic structure markedly similar to that observed by Olds and co-workers.

B. Changes in Complexity of Global Representational Systems

Yoshii *et al.* (1957) were one of the first to observe that evoked response waveshapes to a conditioned stimulus, recorded from a variety of brain regions, tend to become similar as conditioning proceeds. John and his colleagues (1963, 1964) quantified this phenomenon by using cross correlations and factor analysis. Further, John (1967a) has shown that, in general, the similarity between evoked responses in different brain regions increases as significance is attached to a stimulus. This phenomenon is shown most clearly in Fig. 6.8. In this figure it can be seen that evoked response waveshapes recorded during a control period exhibit a variety of differences across structures, but that these differences disappear as the animals are conditioned.

Such findings led John to conclude that information is represented by the temporal pattern of coherence within distributed populations of cells. A large amount of data from diverse paradigms has been accumulated by John in support of this conclusion. Most of these data are from animals performing a frequency discrimination (Bartlett & John, 1973; John, 1967a, 1972; John & Killam, 1959, 1960; John & Morgades, 1969b; John *et al.*, 1973). The basic paradigm involves establishing a given behavioral response to one frequency of stimulus repetition and a different response to another frequency. No pattern discriminations have been involved, and behavioral decisions have generally been binary (i.e., go—no go or left versus right). In the initial stages of training, such a discrimination is somewhat difficult. However, in well-trained or overtrained animals the frequency discrimination is between stimuli possessing low informa-

EVOLUTION OF VISUAL EVOKED RESPONSE

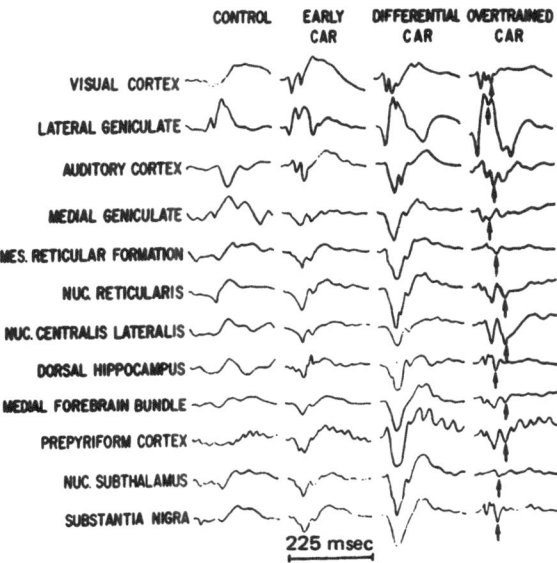

FIG. 6.8 Evolution of visual evoked responses. Control, average responses evoked in different brain regions of a naive cat by presentation of a novel flicker stimulus. Several regions show little or no response, and different regions display differing types of response. Early conditioned avoidance responses (CAR) to the same stimulus shortly after elaboration of a simple CAR. A definite response with similar features can now be discerned in most regions. Differential CAR, changes in the response evoked by the flicker CS shortly after establishment of differential approach–avoidance responses to flicker at two different frequencies. As usual, discrimination training has greatly enhanced the response amplitude, and the similarity between responses in different structures has become more marked. Overtrained CAR, after many months of overtraining on the differentiation task, the waveshapes undergo further changes. The arrows point to a component usually absent or markedly smaller in behavioral trials on which this animal failed to perform. (From John, 1972. Copyright © 1972 by The American Association for the Advancement of Science.)

tional value, that is, the stimuli are unidimensional since the significant parameter is the relative length of the time interval between identical events. Although such information seems to be represented by the temporal pattern of coherent discharge in extensive populations of cells, the level of complexity of representational systems mediating such frequency discrimination is probably of a low order.

Recent data indicate that, in addition to information represented in terms of coherent discharge patterns in populations of neurons, *increased complexity of a representational system involves an increased differentiation, in space and time, of groups of coherent cell populations.* Support for this idea is provided by John and Morgades (1969b), who mapped evoked potentials and multiple unit responses in widespread brain regions during a frequency discrimination. Considerable homogeneity between evoked response waveforms and poststimulus histograms in different brain regions was noted when animals made correct responses. However, when animals hesitated, acted confused, and performed incorrectly, the anatomic homogeneity of response deteriorated. This phenomenon of decreased homogeneity related to incorrect responses has been observed in other experiments (John, 1967a; John & Killam, 1960). These findings indicate that complex interactions occur when there is difficulty in the information resolving process. Conceivably, the phenomenon observed in these studies indicates that there was a mismatch between activity representing the peripheral flicker signal and the internal representational systems which mediate the behavioral responses.

Clear examples of complex global representational systems are offered by Thatcher (1970, 1975) and by Olds and co-workers (Disterhoft & Olds, 1972; Olds *et al.*, 1972; Segal & Olds, 1972). Thatcher (1970) conducted four separate experiments, each involving a different CS duration (40 sec, 60 sec, 80 sec, and 120 sec) followed by foot shock. In each of the experiments, *half* of a group of rats were classically conditioned to a 3.3-Hz flicker and the other half were classically conditioned to an 8.2-Hz flicker. Each rat was given one flicker—foot shock (CS–UCS) pairing an evening for three consecutive evenings, using the appropriate CS frequency. On the day following the last conditioning trial, an intermediate 5.8-Hz test flicker was presented while the animals bar pressed for food. Depending on CS duration, this resulted in the appearances of a conditioned emotional response (CER). Power spectral analyses of the EEG, recorded while the animals were immobilized (i.e., showing a CER), demonstrated statistically significant increases in the proportion of power ($\mu V^2 \, sec^{-1} \, cycle^{-1}$) at 3.3 Hz, in animals conditioned with 3.3-Hz flicker. On the other hand, animals conditioned at 8.2 Hz exhibited significantly greater power at 8.2 Hz and 4.1 Hz than animals conditioned at 3.3 Hz. These "labeled rhythms" were reliably present, and occurred in particular structures at particular times following the onset of the indifferent 5.8-Hz flicker. These data revealed *local* representational systems by the appearance of labeled rhythms in a particular structure. Existence of the *global* representational system was revealed by the fact that these phenomena appeared in many local systems, operating in an ordered and coordinated fashion. Figure 6.9 shows a series of group difference waves, which demonstrate these findings. The group difference waves were constructed by subtracting the proportion of power at 3.3 Hz in animals conditioned at 3.3 Hz from the proportion of power at 3.3 Hz in animals conditioned at 8.2 Hz. All points above the zero line (see MRF, Fig. 6.9) represent greater power at 3.3 Hz in animals conditioned at 3.3 Hz, while points below the zero line represent greater power at 3.3 Hz in animals conditioned at 8.2 Hz.

Three observations, shown in Fig. 6.9, should be emphasized: (1) The number of structures exhibiting statistically significant differences at the labeled fre-

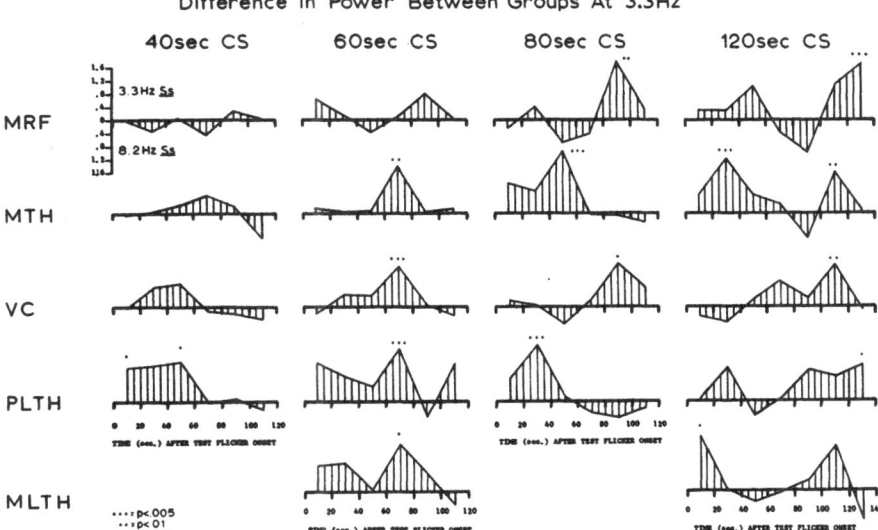

FIG. 6.9 Difference in mean power between groups at 3.3 ± .25 Hz. Difference waves were constructed by determining difference in power at 3.3 Hz (±.25) for each 20-sec data epoch. All peaks above the zero line represent greater power at 3.3 Hz (±.25) in Ss conditioned at 3.3 Hz than in Ss conditioned at 8.2 Hz. Points below the zero line show that there was greater power at 3.3 Hz (±.25) in Ss conditioned at 8.2 Hz. Mesencephalic reticular formation, MRF; visual cortex, VC; midline thalamus, MTh; posterior lateral thalamus, PLTh and medial lateral thalamus, MLTh. In this figure, the anatomic and temporal distribution of 3.3-Hz labeled activity is shown to change as a function of CS duration. Note that peak power consistently occurred during intervals associated with anticipated foot-shock and that anatomic expansion and increased complexity occurred as CS duration was lengthened. (From Thatcher, 1975.)

quency (asterisks) increased as a function of CS duration. Thus, there was a clear anatomic expansion in the distribution of labeled rhythms as a function of CS duration. (2) The CS–UCS interval was consistently reconstructed. Peak significance occurred at a time after the onset of the 5.8 Hz test flicker which corresponded to the duration of the CS in training. (3) Labeled rhythms appeared in particular structures at particular times. This was seen most clearly with CS durations greater than 60 sec, in which early peaks and early–late dual peaks appeared in thalamic structures while late peaks occurred in the visual cortex (VC) and ventral mesencephalic reticular formation (MRF). This example shows that reconstruction of an interval of time (the CS–UCS interval) involves the sequential action of a set of different local anatomic systems (Thatcher, 1975). Furthermore, reconstruction of the CS–UCS interval involved the reconstruction of past events that made up the interval (see Chapter 7).

The point most relevant to the topic of this chapter is that there is an increase in the complexity of the global representational system as a function of the

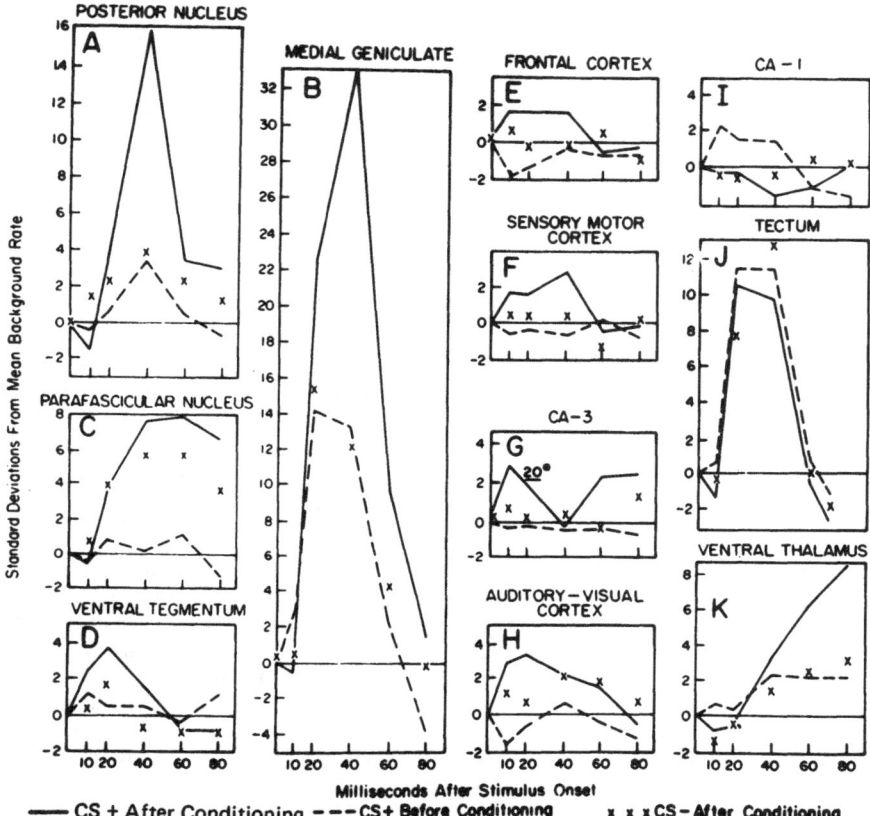

FIG. 6.10 Samples of statistically portrayed responses from different brain areas. Dashed lines: responses to the CS⁺ prior to conditioning; solid lines: responses to CS⁺ after conditioning; "x": responses to CS after conditioning. In these poststimulus histograms, the background activity (first 100 ordinates) was used to establish a mean and a standard deviation. Then the eight ordinates (80 msec) after stimulus onset were taken in groups of two (1 and 2 = 20, 3 and 4 = 40, and so on) and the average of each two was stated as a number of standard deviations from the mean background rate. Also the first ordinate considered separately (1 = 10 msec) was stated as a number of standard deviations from the mean background rate. Thus the 10-msec point was based on only half as much data as the other four points and its data were also included in those of the 20-msec point. (From Olds *et al.*, 1972.)

duration of CS. The behavioral data from the study just cited showed that significantly fewer animals exhibited a CER as the CS duration was lengthened; that is, increased complexity of the global representational system was correlated with increased difficulty in relating the CS with the UCS.

These data suggest that the shape or topology of representational systems is related to information; that is, *information is represented by temporal coher-*

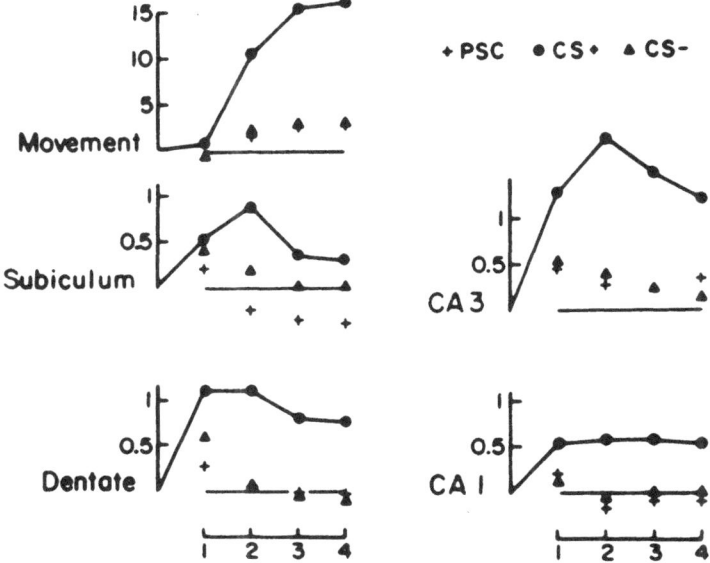

FIG. 6.11   Averaged Q scores for all units in the various areas (subiculum, dentate, and hippocampal CA3, CA1) expressed as deviation from mean background rate during pseudo-conditioning (PSC) in response to CS⁺ and to CS⁻; abscissa: value of Qs in standardized scores; ordinate: four periods of the CS–US interval, each 250 msec in duration. (From Segal & Olds, 1972.)

ences in populations of cells as well as by the constraints, in anatomy and time, which are placed on the development of the system.

Data by Olds and co-workers appear to support this conclusion. Figure 6.10 shows the deviation of discharge rate in cells recorded from a variety of brain regions in the work of Olds and his colleagues, discussed earlier (Olds *et al.*, 1972). The solid lines represent the response of cells to the CS⁺ (1000-Hz tone) after conditioning, while the dashed lines represent the response of the same cells to the same stimulus before conditioning. Note that particular structures exhibit significant increases in discharge rate at particular times following the onset of the stimulus. As in the Thatcher (1975) study, Olds and his co-workers (1972) and Disterhoft and Olds (1972) found that the posterior thalamus exhibited the most consistent and earliest responses (see Fig. 6.9). These data again suggest that complex representational systems are comprised of particular spatiotemporal configurations of cooperative cell activity. Both Figs. 6.9 and 6.10 illustrate that many structures share time intervals (see Table 6.3). This finding further indicates that the global representational system is made up of smaller but distributed systems that are sequenced in time. An additional example of an anatomic–temporally ordered representation is shown in Fig. 6.11. In this figure Q scores are standard scores of the deviation from background discharge due to the CS^+. Note that the different regions of the hippocampal system are differentiated in time. This suggests that changes in unit discharge rate are not by themselves significant, but rather information is contained within the anatomic order of change.

As will be described in Chapter 10, information is represented in terms of patterns of discharge within populations of cells. Since the population of cells exist in a three-dimensional space it is likely that information within local representational systems is constituted by a particular configuration of activity in space and time. These considerations suggest that there is a hierarchy of "geometries" of cellular activity constituting representational systems; that is, there are shapes or topologies of local neuron populations, there are shapes or topologies of distributed systems, and in the case of complex representations, the topology extends across widespread populations of neurons.

7
The Neural Representation of Time

I. INTRODUCTION

We first considered titling this chapter simply "Time." However, it is clear that time is merely an abstraction and, in a concrete sense, does not exist. To avoid the confusion which stems from the concept of time, when it is erroneously used as a concrete entity, the word "change" should be used instead since this word is fundamental to time. Change reflects a particular category of relationships between events. In this sense time is an abstraction based upon a succession of events. Event A is succeeded by event B. The regular occurrence of such relationships provides the basis by which objective time can be understood. It is believed that the subjective perception of time (e.g., duration, time passing, temporal perspective, simultaneity, and so forth) can best be understood by first examining the means by which objective or external time is measured. The last statement may at first seem naive. However, an understanding of subjective time must ultimately stem from the fact that a living organism operates on a world that exhibits succession and change. How is succession and change represented in the brain? What are the mechanisms by which the brain manipulates and makes use of time? What are some of the basic processes that underlie time estimation?

This chapter will not attempt to answer all these questions in detail. Rather, basic experimental data will be reviewed which deal with the elementary aspects of neural time representation. The general question that will be asked is: what is the nature of the internal references used for the subjective estimation of time? It is felt that the answer to this question will provide a basis for understanding many of the other aspects of time. However, in order to determine the internal references for subjective time it is necessary to first consider the role of a reference in the measurement of objective or external time.

II. EXTERNAL OR OBJECTIVE TIME

For better or worse, man is in a universe that exhibits ordered change. "Time's arrow" is an inexorable movement of events which follows the laws of cause and effect and exhibits a specific direction. Can time go backwards? Modern physicists seriously explore this possibility, although with little success (Overseth, 1969). It is indeed remarkable to consider the evolution of the intelligence of man which can reflect on the order of events, anticipate their future occurrence, and manipulate relationships with predictable consequence. Modern technological society puts a high premium on the precise accounting of the relationship between events. But how is this accomplished in today's sophisticated societies? Very simply, by establishing a reference (i.e., a clock) so that the events that constitute time's arrow can be anticipated and quantified. Today, the cesium crystal which oscillates between two energy levels, precisely 9, 192, 631, 770 Hz, is the clock reference. Thus, the elementary unit of time, the second, is defined by a highly regular or ordered event change. Of course not all events are regular. Some events occur only once in an infinity of time, while others are frequent but random with respect to the clock reference. The important point is that objective time depends on a reference; that is, measurements are made by comparing one set of events with another set.

The idea of a reference system is of fundamental importance in physics. The early laws of motion, as posited by Newton, assumed that all bodies were either at rest or were moving with respect to three-dimensional space, which was an unchanging and absolute reference. Gottfried Leibnitz, Newton's rival, rejected the theory of absolute space and argued that space was merely a system of relationships between the bodies in it and had no existence independent of these bodies. The statement "this body is in motion" does not imply some kind of intrinsic motion but only a change of position of the body relative to another body. The latter constitutes a reference system which is conveniently thought of as a set of three mutually perpendicular coordinate axes.

Albert Einstein generalized the above notion of relativity to pertain to all natural phenomena. Einstein focused his theoretical discussions on the concept of simultaneity of two events taking place at different locations. He was able to demonstrate that the simultaneity of two events occurring at different places would be interpreted differently by observers in two coordinate systems moving relative to each other. Measurements of simultaneity depended entirely on the relative motions of the reference systems. Einstein stressed that there was no absolute space, but rather a coordinate reference system of space–time. Also, he abandoned the idea of instantaneous transmission of effects and replaced it by the constant of the speed of light.

In summary, relativity theory leaves us with an environment constituted by events whose relations are changing. Concepts of space and time are understood in terms of specific frames of reference. Space is understood in terms of the relations between objects and time is understood in terms of the relative change of events.

III. THE CONSTRUCTION OF SUBJECTIVE TIME

Studies of the developing child show that subjective time, in parallel with the development of the awareness of space, is constructed little by little and involves the elaboration of a system of relations (Piaget, 1971a). More specifically, time sense develops ontogenetically and involves learning and the construction of a set of operations by which one makes comparisons between events (Piaget, 1971b). The title of this section was chosen to emphasize that time sense is an adaptive creative process that enables an individual to manipulate and extend his control over the environment.

Obviously, cultural factors play an important role in determining the way in which an individual conceives and uses time. But again this fact helps emphasize the adaptive aspect of subjective time.

Now, what is meant by the "construction" of subjective time? Very briefly, subjective time is constructed through the action of an internal reference or signal. This internal signal develops ontogenetically and parallels the development of memory and the "self" (Piaget, 1971b). A clearer perspective of the nature of this process can be obtained by considering the various aspects of subjective time operations. For convenience, subjective time can be divided into at least four (nonmutually exclusive) categories (Ornstein, 1969): (1) succession (change) and simultaneity; (2) awareness of time passing (the specious present); (3) estimates of duration; and (4) time ordering (temporal perspective).

The first category is actually fundamental to the other categories. At the root of succession is the concept of change and, as discussed earlier, this involves the representation of a relationship between events. Event A can be discriminated from succeeding event A' when a minimum interval of time elapses (in the case of the human observer this is 20 to 60 msec). The relationship between simultaneity and succession is clear since two events will be perceived as one if they occur successively within a short interval of time. It is important to emphasize that both succession and simultaneity involve a comparison, namely, comparisons between the neural representations of events. In 1962 Halliday and Mingay (see Efron, 1963a) measured the latency of cortical evoked potentials following stimulation of the toe and index finger. They found that the evoked potentials elicited by the toe stimulus occurred approximately 20 msec later than the potentials evoked by the finger stimulus. Psychophysical experiments showed that the stimuli delivered to the toe and finger were not perceived as simultaneous unless the toe stimulus preceded the finger stimulus by 9 to 17 msec. Halliday and Mingay (Efron, 1963a) concluded that "the central nervous system does not correct for the time error produced by conduction differences and that the perception of simultaneity occurs when two sensory messages reach some point in the nervous system at the same time [p. 261]." Efron (1963a) extended these findings by demonstrating a consistent time error for the awareness of simultaneity of 2 to 6 msec when stimuli are delivered to left and right symmetrical parts of the body. For left-handed individuals the delay was in stimuli delivered to the right half of the body, while for right-handed individuals the delay was in stimuli delivered to the left half of the body. This led Efron (1963a) to conclude that the sensory messages received by the nondominant hemisphere are transferred by a pathway (probably the corpus callosum) to the dominant hemisphere for language where the conscious comparison of the time of occurrence of two sensory stimuli occurs.

Awareness of time passing also appears to involve an active comparison. The comparison in this case is between short or long term memory and information processing of immediate events. It is closely linked to the awareness of succession but, as William James (1890) suggests, time passing involves an interplay between the present and the immediate past. Awareness of time passing seems to involve consciousness of the occurrence of events. Using Halle and Steven's (1962) model of speech recognition, it is possible to argue that consciousness is a creative process involving the synthesis of an internally generated signal that is continually being matched and mismatched with a constellation of sensory input, memories, feelings, and other constitutive factors occurring in the present. Consciousness would thus be a process whereby a reference signal, which is part of the constructed self (developing gradually over years), is continually compared, on a multidimensional basis, with sensory, affective, and endogenous factors. If such occurs, then there should be a time delay for awareness, which is the time required for the process to complete itself. This view is one where consciousness and the awareness of time passing are a continual creative process that ceases during slow wave sleep and starts again in waking and REM sleep.

Estimates of duration can be considered in terms of both short durations and long durations. Estimates of durations seem to involve memory and other processes similar to those operating during awareness of time passing. Memories of durations from the past may be compared with durations occurring in the present. An important and distinctive feature of duration is that this process involves the use of time markers or tags to label the beginning and end of an interval. It is important to ask: how is the beginning and end of an interval represented in the brain? Or more precisely, what is the nature of the time markers? Examples of the neural representation of duration will be discussed in detail in a later section. However, suffice it to say that change in a stimulus event invariably results in the excitation of a set of neurons and in the inhibition of others. One consequence of this action is synchronization of neural ensembles. The synchronization process appears to be of fundamental importance in the creation of neural time markers. As discussed in Chapter 3 synchronization, which has neuroanatomic correlates, may be involved in initiating construction of representational systems. Neurophysiological data show that synaptic inhibition plays a basic role in the synchronization process (see Chapter 2). For example, intracellular analyses of visual cortical neurons show that synaptic inhibition disappears, while excitation is left largely unaltered near the interstimulus interval for flicker fusion (Kuhnt & Creutzfeldt, 1971). This finding supports the notion that inhibition is important in the segregation of neural representational systems in time (see Chapter 3).

Recurrent and lateral inhibition appear to be integral parts of the dynamics of nearly all brain activity. Recurrent inhibition has been shown to be of vital importance in thalamic synchronization processes (Thatcher & Purpura, 1972, 1973) and is believed to underlie the production of alpha rhythms (Andersen & Andersson, 1968) and cortical EEG (Creutzfeldt, et al., 1966a). Both lateral and recurrent inhibition are most likely of fundamental importance in the isolation and protection of neural representational systems in space and time.

A recent study, involving measurements of the discrimination of recency, has shown that damage to the frontal cortex results in a deficit in the ability to discriminate temporal order (Milner, 1971). It is interesting that the latter study showed a deficit in time ordering of verbal material in left frontal lobe lesioned patients but not in patients with right frontal lesions. On the other hand, significant deficits in time ordering of nonverbal material occur in patients with right frontal lesions while only slight deficits occur in patients with left frontal damage. This indicates that the frontal regions are also involved, either in the creation of time markers or tags, or in the retrieval and organization of material previously tagged.

In any case, if memories are retrieved through the use of tags or markers created by the onset of the event, then estimates of duration may involve an active comparison of memory segments from the past and stimulus input segments occurring in the present (consciousness referent). A change of brain state, such as by drugs or natural arousal, may affect the consciousness referent (created in the present) in a manner that could cause underestimation or overestimation of a standard interval.

Temporal perspective is determined to a large extent by social factors. "Temporal perspective" means the ordering of time based on a particular view of events that a given individual may choose. Such perspective, however, also involves references. The references may be past experiences or a specific plan or an idea that occurs at the moment. "Perspective" also means the manipulation of past events or the reorganization of ideas and relations to enable an individual to internally order either his own actions or events in time. The adaptive and constructive manipulation of events in terms of time as well as the general use of knowledge to predict the future are all consumed under the term "time perspective."

The ability to have time perspective is not given at birth. Rather, a gradual emergence of intellectual capacities parallels the development of time perspective (Piaget, 1971b). For instance, the development of concepts of space and of the relation of the self to others and objects develops hand in hand with time sense (Piaget, 1971b).

It is beyond the scope of this chapter to discuss in detail the nature of the various referents used in time sense. However, arguments can be put forth that match–mismatch processes involving internal references underlie many aspects of time sense. These match–mismatch processes all appear to involve the interaction and manipulation of neural representational systems (Thatcher, 1974a, 1976a, b). In the remainder of this chapter experimental data will be discussed in light of the previous discussion of time duration. Experimental data will be presented showing the existence of a neural representational system inextricably bound to the reconstruction of an interval of time. The purpose of this presentation is to derive some understanding of the basic neural elements involved in time representation. What is the basic element of neural time representation? Is it a movement through neural space, or the pulse character of a synchronizing process or a set of such processes? Based on the previous discussions a good starting point on the journey to answering these questions would be to first inquire into the nature of neural representational systems.

IV. NEURAL REPRESENTATIONAL SYSTEMS

As discussed in detail in Chapters 6 and 9, information from the external world is represented in terms of specific spatiotemporal organizations of neural activity. These internal organizations, which reflect information from the external world, are called "representational systems." A representational system is defined as any structure of which the features symbolize or correspond in some sense to some other structure, and for any given structure there may be several equivalent representations (MacKay, 1969a). In Chapter 6 it was argued that information, in its most broad sense, is that which creates, adds to, or changes a representation.

The precise mechanisms by which neural representational systems are created are currently unknown. However, the fact that neural representational systems exist and can be described in (at least) four dimensions has been clearly established.

Recent behavioral experiments have demonstrated functional relationships between representational systems and time estimation. These experiments have led Robert Ornstein (1969) to postulate what he calls the "storage size" hypothesis of time estimation. This hypothesis argues that the size of a representational system (which is related to the amount of information it contains) is directly related to the estimation of time. Neurophysiological experiments conducted in the late 1960s (Thatcher, 1970; Thatcher & Cadell, 1969) tend to support Ornstein's hypothesis by showing that a neural representational system can expand or contract (spatially) depending on the amount of information the system represents. The latter studies are important since they provide the only empirical evidence to date showing a relationship between the amount of neural space occupied by a complex integrated experience (which is related to the amount of information) and the reconstruction of an interval of time.

V. LABELED RHYTHMS AND TIME RECONSTRUCTION

As discussed in Chapter 6, the experiments by Thatcher and Cadell (1969) and Thatcher (1970, 1975) involved classically conditioning two groups of rats to two different frequencies of a flickering light-conditioned stimulus (a painful foot shock paired with either 3.3-Hz or 8.2-Hz CS). Each group was given several classical conditioning experiences with only one frequency of flicker. The rats were subsequently tested for their memory of the experience by presenting an intermediate frequency of flicker (5.8 Hz) while the animals bar pressed for food. When the rats saw the intermediate flicker they immediately froze (and exhibited fear responses), demonstrating recognition of the past association between the flicker and a painful foot shock (UCS). The electrical activity recorded while the animals were freezing was subjected to a power spectral analysis. This analysis allowed quantification of the amount of energy in each of several areas of the brain at specific frequencies. The study involved four separate experiments (and over 190 animals) using four different CS–UCS intervals (40 sec, 60 sec, 80 sec, and 120 sec). The amount of power (in μV^2/cycle/sec) in the EEG at 3.3 ± .25 Hz and 8.2 ± .25 Hz (and at harmonic frequencies, 4.1 Hz, 6.6 Hz, 16.4 Hz) was calculated for each animal for each of several brain structures. The relative amount of power (percent of total) at these labeled frequencies were then compared for the two groups of animals. It was found that animals conditioned at 3.3 Hz exhibited significantly greater percent power at 3.3h ± .25 Hz than did animals conditioned at 8.2 Hz. On the other hand, animals conditioned at 8.2 Hz exhibited significantly greater power at 8.2 ± .25 Hz and at 4.1 ± .25 Hz than animals conditioned at 3.3 Hz. Furthermore, it was shown that these representational systems (labeled rhythms) occurred in particular structures at particular times and reflected the CS–UCS interval. The results of this study are shown in Table 7.1. This table shows the intervals (following the 5.8-Hz test flicker onset) at which significant differences between groups occurred at the various labeled frequencies. The largest number of significant differences occurred in the 120-sec CS experiment and the smallest number occurred in the 40-sec CS experiment. Note that no significant differences occurred beyond 40 to 60 sec with a 40-sec CS, or beyond 60 to 80 sec with a 60-sec CS or beyond 80 to 100 sec with an 80-sec CS. This shows that an interval of time specific to the CS–UCS interval was consistently reconstructed.

TABLE 7.1

Anatomic–Temporal Distribution of Labeled Rhythms as a Function of CS Duration

CS Duration (sec)	Intervals during which statistical significance occurred ($p < .025$)						
	0–20	20–40	40–60	60–80	80–100	100–120	120–140
120	VC,[a] 8.2 MLTh, 3.3	MTh, 3.3 PLTh, 4.1	—	—	—	VC, 3.3 MTh, 3.3 PLTh, 3.3	MRF, 3.3 MRF, 4.1 PLTh, 3.3 PLTh, 6.6
80	VC, 8.2	PLTh, 3.3	MTh, 3.3	—	VC, 3.3 VC, 4.1 MRF, 3.3	—	—
60	—	MRF, 8.2	—	VC, 3.3 MTh, 3.3 PLTh, 3.3 MLTh, 3.3	—	—	—
40	PLTh, 3.3 PLTh, 4.1	—	PLTh, 3.3 PLTH, 6.6	—	—	—	—

[a]VC = Visual cortex; MRF = mesencephalic reticular formation; MLTh = medial lateral thalamus; MTH = midline thalamus; PLTh = posterior lateral thalamus.

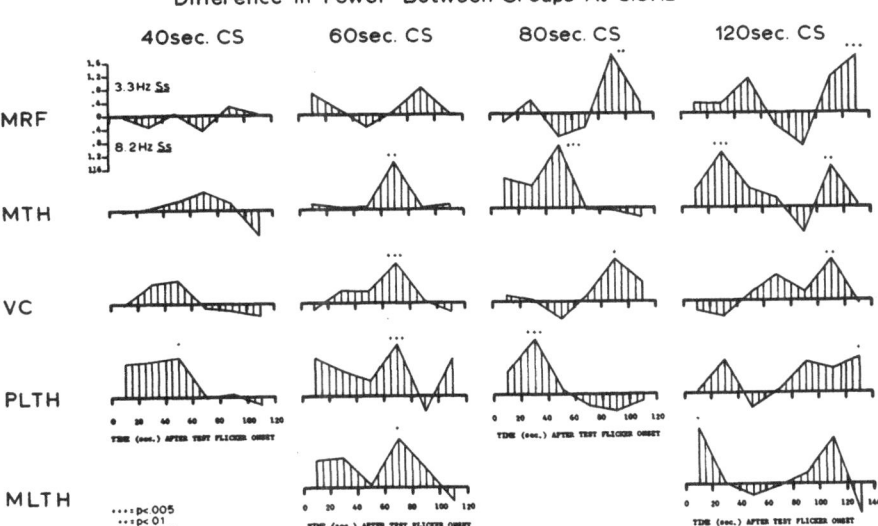

FIG. 7.1 Difference in mean power between groups at 3.3 ± 25 Hz. Difference waves were constructed by determining difference in power at 3.3 Hz (±.25) for each 20-sec data epoch. All peaks above the zero line represent greater power at 3.3 Hz (±.25) in subjects conditioned at 3.3 Hz than in subjects conditioned at 8.2 Hz. Points below the zero line show that there was greater power at 3.3 Hz (±.25) in subjects conditioned at 8.2 Hz. Mesencephalic reticular formation, MRF; visual cortex, VC; midline thalamus, MTh; posterior lateral thalamus, PLTh; medial lateral thalamus, MLTh. In this figure, the anatomic and temporal distribution of 3.3-Hz labeled activity is shown to change as a function of CS duration. Note that peak power consistently occurred during intervals associated with anticipated footshock and that anatomic expansion and increased complexity occurred as CS duration was lengthened. (From Thatcher, 1975.)

A clear picture of the anatomic and temporal evolution of the 3.3-Hz phenomenon (which was the strongest of the labeled rhythms) is shown in Fig. 7.1. This figure shows the difference in power between groups of 3.3 ± .25 Hz during successive 20-sec epochs beginning with the onset of the intermediate (5.8 Hz) flicker. All points above the zero line (see MRF 40-sec CS) represent greater power at 3.3 Hz in animals conditioned at 3.3 Hz than in animals conditioned at 8.2 Hz. All points below the zero line represent greater power at 3.3 Hz in animals conditioned with an 8.2-Hz flicker. Note that most of the points are above the zero line.

Several important features can be seen: (1) Progressively more structures exhibited significant differences as CS duration was lengthened. (2) The rhythms consistently represented the reconstruction of the CS–UCS interval. These two features provide the first evidence that the size of a representational system can expand as a function of CS duration. (3) There is a specific space–time organization of the rhythms; that is, particular structures exhibit labeled rhythms at specific times following the onset of the test flicker.[1]

The only variable manipulated in this study was CS duration. As CS duration was lengthened the flicker representational system became anatomically more widespread and temporally more complex. What is the significance of this phenomenon? Why is the CS–UCS interval reconstructed, and what might these evolving anatomic–temporal organizations mean? One answer is that these phenomena represent the action of a system involved in predicting or anticipating the time of expected foot shock. It has been mathematically proved that the best predictor of the future is based on an optimally succinct description of the past (Van Heerden, 1968). It is reasonable to argue, therefore, that the best way to predict foot shock is to reconstruct those events that, in the past, led to foot shock; that is, the results reflect the fact that the flicker CS was the most consistent event preceding foot shock. This interpretation is consistent with the fact that foot shock is the most significant past event determining the animals' behavior.

The results of this series of experiments are very complex. Differences are noted between 8.2 and 3.3 and the subharmonic 4.1. Each of these rhythms may reflect a local representational system which, in combination, constitute a global or very large representational system.

Is it possible to reduce this complexity to the operation of a set of very elementary operations? To do so would be desirable since, generally, simple models are the most comprehensive and provide the most readily testable predictions. In the text to follow a very simple loop model of time representation will be presented which is capable of explaining most of the major findings of the Thatcher (1970) study.

A. Neural Loop Model of Time Representation

How is an interval of time represented in the brain? A clue to the answer is provided by the well-documented fact that the brain is composed of a large number of neural loops (Lorente de Nó, 1938; Scheibel & Scheibel, 1958, 1965, 1966; see Chapter 2). These loops seem to be of various sizes nested within and between each other (Lorente de Nó, 1938; Verzeano, 1972). Evidence is available indicating that individual neurons may be members of many loops, that is, one neuron may be shared by more than one loop (John & Morgades, 1969a; Lorente de Nó, 1938; MacGregor *et al.*, 1973; Scheibel & Scheibel, 1965) and in this way serve as nodal points within oscillatory systems. Thus, it would be consistent with anatomy and physiology to argue that time is represented by the

---

[1] These differences only occurred in animals that suppressed responding and exhibited recognition of the past conditioning experiences. Animals that failed to suppress responding (55% with a 2-min CS) did not show statistically significant differences.

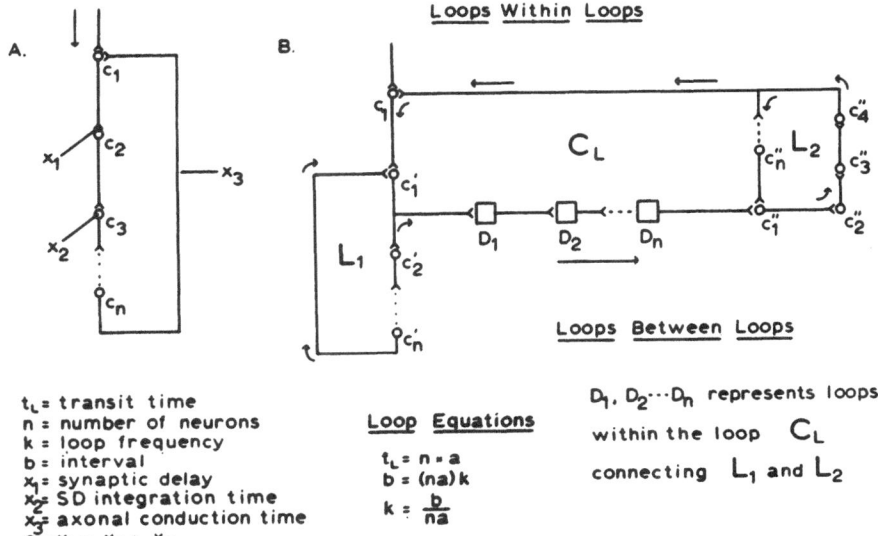

Loops Within Loops

t$_L$ = transit time
n = number of neurons
k = loop frequency
b = interval
x$_1$ = synaptic delay
x$_2$ = SD integration time
x$_3$ = axonal conduction time
a = x$_1$ + x$_2$ + x$_3$

Loop Equations

$t_L = n \cdot a$
$b = (na)k$
$k = \dfrac{b}{na}$

D$_1$, D$_2$···D$_n$ represents loops within the loop C$_L$ connecting L$_1$ and L$_2$

Loops Between Loops

FIG. 7.2 Neural loop model for interval representation. In this model time is represented by the circulation of activity within reinterant neural loops. The relationship between loop size, delay factors, and transit time is given in (A). The loop equation $K = b/na$ expresses a relationship between reiteration, loop size, and an interval of time (see text for details). The simple loop system in (A) can be expanded by coupling loops across distributed regions; see (B). In (B) a diagrammatic representation of loops nested within loops is shown. The purpose of this is to illustrate complexity possible with such a system. No representation is provided for the control of coupling or switching of loops.

circulation of activity within neural loops. In other words, the basic functional element of time estimation is a loop. The larger the loop or the greater the delay between elements in the loop, the longer the interval of time. Below is a formalized loop model of time representation.

Consider that time is represented by the circulation of activity in a loop of cells (see Fig. 7.2). A given interval of time can be represented by one circulation in a large loop or by a number of circulations in a smaller loop. In Fig. 7.2 assume that a signal initiated at the entering axon passes through the loop C_1, C_2, C_3, ..., C_n. Transit time (t_L) of the signal through the loop is the number of neurons (n) in the loop times the time for activity to pass through an individual neuron (a); that is, $t_L = nxa$. For simplicity consider that transmission time a is represented by factors such as (1) axonal conduction time, (2) synaptic delay, and (3) somadendritic integration time. For simplicity, assume that the neurons in the loop are similar, that is, they have the same transmission properties such that a is the same for all neurons and assumed equal to 1 msec. The time (t_L) required for transmission through a loop that consists of n neurons is given by $t_L = na = n$ (msec). If the CS interval b is 100 msec, then n is related to b, such that $b = t_L K$ for some integer K. In other words, b equals the time t_L for circulation within one loop times the number of circulations in that loop, K. Since $t_L = na = n$, then $n = b/K = 100/K$. Thus if $K = 1$ then $n = 100$; that is, there are 100 neurons in the loop. If $K = 10$, then $n = 10$; that is, there are 10 neurons in the loop. Since $b = t_L K = nK$ for some integer K, than K is the loop frequency; that is, the number of times the loop has to be traversed in order to match the interval b. The relationship of K to the interval b is given by $K = b/na$; that is, the number of circulations K that represent the interval b is inversely related to na.

The formula $K = b/na$ can apply to multiple loop systems of considerably greater complexity than that represented in Fig. 7.2. For instance, circulation of activity can occur in a system of loops arranged in such a way that completion of circulation in one loop activates a second anatomically different loop which then activates a third anatomically different loop, and so forth. In other words, K can represent the circulation of activity within a single loop or a series of loops, or both. The loop equation can also be applied to more abstract or generalized oscillatory systems such as those envisioned by Freeman (1972a, 1975) and others (Eigen, 1974).

B. Application of the Model to the Labeled Frequency Findings

Since the labeled frequency phenomenon was strongest and most reliable at 3.3 Hz, the loop equation $K = b/na$ will be applied to these data first (refer to Fig. 7.1). Consider the 40-sec and 60-sec CS experiments. In the 40-sec and 60-sec CS experiment 3.3-Hz labeled activity was confined (in the structures investigated) to the PLTh but extended through the 0 to 20 sec to the 40- to 60-sec interval. In the 60-sec CS experiment labeled activity was anatomically more extensive but the temporal extent within a structure was reduced; that is, the peak at 0 to 20 sec in the PLTh just reached the .05 level and there was a clear-cut dip in power at 40 to 60 sec. The finding that the within-structure temporal extent of the labeled activity was reduced when anatomic distribution was increased is explained by the loop model; that is, if time is represented by the circulation of activity in a loop system, then as the size of the system expands the number of circulations necessary to represent an interval decreases. This is represented by the equation $K = b/na$ where b = CS–UCS interval and na is a factor representing the size of the system. In the 40-sec CS experiment reconstruction of the CS–UCS interval involved the circulation of activity within a large number of anatomically confined loops. When the system expanded with a 60-sec CS, activity circulated across an anatomically distributed system (i.e., increased na) and the number of circulations (K) within a structure decreased. According to this model labeled frequency loops were established by experience with the CS, and the reconstruction of the CS–UCS interval involved reactivating the coupling relationships between the loops (or spirals).

The model can be extended to the 80-sec CS experiment (Fig. 7.1). With this duration of CS a further anatomic expansion of the system occurred. Correspondingly labeled frequency peaks appeared only during single 20-sec epochs. However, with this duration CS temporal relationships between structures appeared. According to the model such relationships represent circulation between structures. Thus, with an 80-sec CS, increased K is an increase in the number of circulations extending across structures. An important consequence of the model is that in order for K to increase, the loop system representing the interval b must be anatomically confined; that is, na is inversely related to K, thus if K increases na must be restricted.

The results of the 120-sec CS experiment are entirely consistent with this aspect of a loop model. As can be seen in Fig. 7.1 a 120-sec CS results in an increased temporal extent of labeled activity within a structure as well as increased temporal relations between structures. At this duration dual peaks (iteration) occurred within structures. These findings suggest an increased amount of circulation both within a structure and between structures (i.e., increased K). The labeled frequency phenomenon begins to resemble that seen in the anatomically confined 40-sec CS experiment (Fig. 7.1). This indicates that the loop system is restricted with CS durations that exceed 60 sec. The results of the 120-sec CS experiment are consistent with the notion that a loop system, involved in reconstructing the CS–UCS interval, begins to involute or reiterate when it is spatially restricted, and that this restriction occurs with CS durations greater than approximately 60 sec. A specific experimental prediction arises from this model: further increases in the CS–UCS interval should result in further increases in the temporal extent of activity within a structure until some maximum level is reached.

A parsimonious feature of a loop model is that it can explain differences between 3.3 Hz and 8.2 Hz. If $b = 303$ msec (for 3.3 Hz) and $K = 1$, then $n = 303$ neurons. If $b = 123$ msec (for 8.2 Hz), then $n = 123$ neurons. This means that the loop system representing 8.2 Hz is smaller than that for 3.3 Hz and thus for a given structure, there are a smaller number of neurons involved in the representation of 8.2 Hz. Given the signal-to-noise nature of this spectral technique one would therefore expect 8.2 Hz to be more difficult to detect than 3.3 Hz. This is what was observed in this study.

The loop model can also be applied to some features of the 8.2-Hz – 4.1-Hz phenomenon. A prominent 8.2-Hz – 4.1-Hz phenomenon occurred in the 40-sec and 120-sec CS experiments (see Table 7.1). This is expected, based on the loop model, because K is greatest with a 40-sec and 120-sec CS; that is, since there is a large number of circulations within anatomically confined systems in the 40-sec and 120-sec CS experiments 8.2 Hz and 4.1 Hz is easy to detect.

This model of the neural representation of time explains several aspects of phenomena observed in the Thatcher (1970) study and emphasizes, in particular, that long intervals of time may be represented by the circulation of neural activity within neural loop systems distributed across widespread regions of the brain. It emphasizes that much of the phenomenology of time can be reduced to coupled loop systems in the brain (see Chapter 3). The application of the loop model to the Thatcher (1970) results also emphasizes the possibility that time is directly related to spatial distribution. In other words, there is a space–time transformation. Anatomic order is translated into time. In fact, the data indicate that anatomy and time are indissolubly interrelated and that time reconstruction involves the reproduction of a geometric representation.

The loop model also has limitations. For instance, no account of inhibitory synaptic control was provided. This is a serious omission since inhibition is believed to be of fundamental importance in neural control processes (Purpura, 1970) and may, in fact "sculpture" the flow of excitation in space–time (see Section IV.C, Chapter 3). The model is also linear. A great many oscillations in the brain are nonlinear. This emphasizes another drawback of the model since it has been demonstrated that a dynamic time structure common to living systems can be described using nonlinear oscillators and, further, linearization of the same equations abolishes the time structure (Goodwin, 1963). The model, however, does give rise to several important and testable predictions. For example, the model suggests that a three minute CS would result in very powerful (but not in anatomically more distributed) labeled rhythms. Also, it suggests that rats with smaller brains would exhibit (e.g., in strains selected for small cranial size) between structure reiteration with shorter CS durations than in rats with larger brains. The latter prediction arises from the postulate that time intervals are represented by a finite system of neurons and that an eventual limit in the spatial extent of the representational system can be reached. When the limit is attained other processes, such as biochemical rate reactions or oscillator subharmonics, may operate to represent very long intervals of time. Finally, if time is indeed represented by the circulation of neural activity in loops, then cooling of the total brain or selected fiber systems should effect the space–time aspects of the labeled rhythms in a predictable manner.

The rather slow circulation times of very large loops postulated by the model are consistent with measured circulation velocities in hypothetical neural loops. For instance, Verzeano and co-workers (Verzeano, 1955, 1963, 1970, 1972; Verzeano & Negishi, 1960, 1961; Verzeano et al., 1965) have extensively examined circulation velocities in groups of cells. The velocities they observed ranged from .5 to 8 mm sec^{-1} Experiments recently reviewed by Verzeano (1972) indicate that neural activity moves within both spirals and closed loops. The loop model, both conceptually and in terms of velocity values, is consistent with Verzeano's results.

The results of the Thatcher (1970) and Thatcher and Cadell (1969) studies are also consistent with recent findings by Olds and co-workers (Disterhoft & Olds, 1972; Olds et al., 1972; Segal & Olds, 1972) demonstrating differentiated anatomic and temporal neural organizations in the rat (see Chapter 6). The latter studies also indicated that the posterior thalamus is an important structure in the possible initiation and organization of developing representational systems.

VI. CONCLUSIONS

There is considerable anatomic and physiological evidence demonstrating the existence of loops in the central nervous system. Thus, it is reasonable to argue that the basic timing element in the brain are loop systems. Such systems are both very large and very small and capable of forming a hierarchy of organizations. The evidence presented in this chapter argues that the representation of experience occupies space in the brain; and that, in fact, such representations possess a specific shape or geometry in space—time. The latter conclusion is also suggested by the results of others (Bartlett & John, 1973; Disterhoft & Olds, 1972; John & Killam, 1960; Olds et al., 1972; Segal & Olds, 1972). The results of the experiments presented in this chapter appear to demonstrate a representational system which exhibits a development in space as a function of time. The successful application of a loop model to these data suggests that space is translated into time through the sequential operation of neuroanatomically distributed loops. These loops appear to be part of a highly stable space—time structure.

The larger question of the subjective experience of time passing requires further analysis. However, data by Ornstein (1969) and others (Creelman, 1962; Fraisse, 1963; Frankenhaeuser, 1959; Michon, 1966) show that subjective estimates of time are related to information processing. Ornstein (1969) presents evidence indicating that the size of a stored representation of a past experience can directly influence time estimation. This suggests that the storage size hypothesis as well as the relationship between information and time may be understood in terms of expanding or contracting representational systems (Thatcher, in press).

But the question still remains: What gives rise to the awareness of time sense and time passing? Man most likely recalls experiences of time passing from his past and uses this as a reference to compare and estimate time passing in the present. In this sense a matching of information processing with an "internal reference signal" can serve as the basis for time estimation. In the experiment described earlier, it is possible that the labeled rhythms represent the actual synthesis of the reference signal. The loop model may reflect the basic physiological mechanism by which the internal signal is generated. In man, as contrasted to the rat, very different parameters may operate so that smaller loops or a simple cascading of time "chunks" are used to reconstruct an interval of time. Such a process would require more elaborate rules, but also less space in the brain. In any case, the main assertion of these arguments is that time and space are interrelated in the brain and that time sense is intimately related to information storage and processing.

8

The Chemical Basis of Memory

Since a memory can endure for a lifetime, although it be established by even a brief experience, men have long assumed that something is made in the brain when an event registers in consciousness. Over 2,000 years ago, Socrates said:

> There exists in the mind of man a block of wax, of different sizes and qualities in different men. This tablet is a gift of memory, the mother of the Muses, and when we wish to remember anything which we have seen or heard or thought, we hold the wax to it and in that material receive its impression as from the seal of a ring. We remember and know what is imprinted as long as the image lasts, but when it is effaced or cannot be taken, then we forget and do not know [Jowett, 1931, p. 254].

Speculations about the nature of memory, based upon descriptive anatomic studies, continued through the subsequent centuries. Experimental investigations began early in the nineteenth century. Such studies, consisting of correlations between anatomic damage and behavioral observations, were performed sporadically throughout the nineteenth century, and still continue. The use of electrical stimulation to study brain function began about 100 years ago, and this technique was almost immediately applied to the study of mechanisms of learning and memory.

With the advent of methods of classical and instrumental conditioning around the turn of the century and the development of more or less standardized procedures for establishing a specified conditioned response, it became possible to make quantitative statements about the extent to which an experimental manipulation affected the acquisition or retention of a new behavior. Consequently, a great proliferation of studies on the effects of brain lesions upon learning and memory occurred in the first half of the twentieth century.

After World War II, a new genre of studies of learning and memory appeared. The use of sensitive amplifiers and chronically implanted electrodes permitted observation of electrophysiological phenomena in the brains of animals as they acquired and performed conditioned responses. More recently, such techniques have become further refined so as to permit the continued observation of groups of neurons or even single neurons under such conditions. The development of special purpose computers and laboratory oriented general purpose minicomputers during the last 15 years has enabled high precision and sensitivity to be achieved in such studies and has provided a great impetus to investigations of the neurophysiology of learning and memory, resulting in a voluminous literature.

The biochemistry of memory has been the last area to come under systematic investigation. Almost all of the numerous studies into the biochemical reactions involved in learning and the formation of memory have occurred during the last 10 years. This field has grown in leaps and bounds and now occupies the attention of a large number of investigators.

Our discussion of the brain mechanisms of learning and memory will be organized functionally, according to the following questions: (1) What is the material basis of memory and how is it constructed? (2) Where is memory? (3) How do we remember?

I. WHAT IS THE BASIS OF MEMORY AND HOW IS IT MADE?

A. The Consolidation Phase

As Socrates argued, an experience must leave a residue in the brain. Whether that residue be a new capability or only the recollection of a pattern of sensations, something must be built which serves as the physical basis for storage of the details of a transient process. One of the most useful lines of research into the mechanisms of the storage process has arisen from the phenomenon of *consolidation*. Many experiments show that there is a labile period early in the registration of a memory during which the fixation of experience is susceptible to external interference. A variety of perturbations will cause erasure during this period although the same disturbances are ineffective some time later. The duration of this vulnerability of memory ranges from a few seconds to days or even weeks, depending upon the test situation, the experimental species, and the strength and type of interfering agent. This phenomenon was reviewed some years ago by Glickman (1961).

Consolidation theory was first formulated by Müller and Pilzecker in 1900 (see McGeoch & Irion, 1952) to account for deterioration in the recall of recently acquired verbal material as a function of the interpolation of other tasks. This retroactive inhibition was explained by postulating a perseverative neural process which could be disrupted by external stimulation. Observations of retrograde amnesia after cerebral trauma provided more direct physiological evidence for the existence of a consolidation phase. In a survey of over 1000 cases of head injury reported by Russell and Nathan (1946), amnesia for events up to one-half hour before the injury was found in more than 700 instances. The erasure occupied only a few moments in most cases.

Since those studies, a wide variety of agents has been demonstrated to cause a marked deterioration of memory with a severity which increases as the interval between the experience and the disrupting perturbation decreases. These agents include electroconvulsive shock (ECS), anesthesia, anoxia, convulsions induced by a variety of causes, hyperthermia, and cerebral ischemia. Interference with consolidation has also been accomplished by electrical stimulation or by spreading depression localized in a variety of brain regions. For recent reviews of the extensive literature on these topics, see Deutsch (1973a) on ECS and Schneider (1973) on spreading depression. The wide variety of perturbations which can block consolidation and the variety of anatomic regions in which interference has been effective suggest that no single brain region is responsible for consolidation.

B. The "Trace" Theory

Thus, there is a period following the occurrence of an event during which some process must be allowed to continue if that event is to be stored in permanent memory. The hypothesis most frequently offered to explain the phenomenon of consolidation is that it depends on the ability of specific neuronal circuits to sustain reverberatory activity, an idea originating from the anatomic studies of Lorente de Nó (1938). Hilgard and Marquis (1940), Hebb (1949), and numerous subsequent workers suggested that such reverberatory activity sustained the representation of an experience until permanent structural or chemical storage had been accomplished. This "trace" theory was reviewed by Gomulicki (1953). Evidence that stimulation causes a transient reverberation of neural activity has been provided by Burns (1954, 1958) and by Verzeano and Negishi (1960), who showed long-lasting recurrent patterns of unit discharge following stimulation, which was interpreted as evidence for continued circulation of the effects of the stimulus around a responsive neural network.

C. Facilitation of Consolidation

As various learning situations were used for the study of consolidation, it became evident that with appropriate procedural standardization, the amount of time required for consolidation of a particular learning experience by a certain species was quite constant. Numerous studies have reported that some drugs markedly increase and other drugs decrease consolidation time, presumably by facilitating or inhibiting the chemical reactions involved in the transition from the labile to the stable phase of memory. Such experiments have been thoroughly reviewed by McGaugh and Petrinovich (1965), and by Dawson and

McGaugh (1973). In general, anticholinergic substances, barbiturates, and compounds with depressant action tend to impair consolidation. Conversely, anticholinesterase drugs, stimulants, or convulsant drugs in subconvulsive doses tend to facilitate consolidation.

Many substances found to be facilitatory, such as strychnine, picrotoxin, nicotine, caffeine, amphetamine, pentylenetetrazol, or physostigmine, share an excitatory effect on the brain but are believed to possess different mechanisms of action. Thus, strychnine blocks postsynaptic inhibition, while picrotoxin blocks presynaptic inhibition (Eccles, 1962). The common effect of these two substances, in spite of their different locus and mode of action, suggests that the release of inhibition per se might be a crucial factor in facilitating consolidation. However, pentylenetetrazol has been reported to cause even more marked facilitation of consolidation than strychnine or picrotoxin (Irwin & Benuazizi, 1966). This substance blocks neither presynaptic nor postsynaptic inhibition, but seems to decrease the time required for neuronal recovery after discharge. Such an action might increase the maximum firing rate that a cell could sustain, and might accelerate neuronal transmission.

These findings might be interpreted to indicate that the drugs extended the period of reverberation, thus providing a prolongation of the time during which the memory might be stored. Exploring this possibility, Pearlman, Sharpless, and Jarvik (1961) showed that normal animals anesthetized ten minutes after a one-trial avoidance procedure suffered severe impairment of memory, while animals who received a strychnine injection immediately after the learning trial followed by anesthesia 10 min later showed excellent retention. Thus it appeared that the strychnine accelerated the process responsible for long-term storage, rather than merely extending the duration of the reverberatory process. It seems probable that such procedures either intensified the reverberatory activity or somehow facilitated synthesis of some substances involved in permanent storage of information.

In many of these studies, the drug effects might have been due to increased activity levels, attention, or sensory responsiveness during the subsequent behavioral trials. However, Breen and McGaugh (1961) showed that *posttrial* injections of picrotoxin enhanced the learning of mazes by rats. Westbrook and McGaugh (1964) studied the effects of *posttrial* injection of a stimulant on latent learning of rats that were permitted free exploration of a maze without reward. After exploration followed by the injection, reinforcement was introduced. The experimental group showed fewer errors in the maze than the controls. Apparently, rats injected with an analeptic drug stored more information about the floor plan of the maze discovered during the prior exploration than normal rats, yet no difference in behavior developed until reinforcement was subsequently introduced. Petrinovich, Bradford, and McGaugh (1965) showed that *posttrial* injection of analeptic drugs significantly extended the duration of the maximum period after which rats could make correct delayed alternation responses. These various data support the contention that such drug injections facilitate the memory storage process.

Thus, it appears that accelration of the reverberations of activity caused by a stimulus through a neuronal network, or prolongation of such reverberatory activity, somehow facilitates the permanent changes required for long-term storage of information about that stimulus.

D. Critical Substance and Critical Shift

This implies that reverberatory activity somehow contributes to some irreversible process in those cells that participate for the period necessary for consolidation to occur. Since all experiences do not seem to register in memory, it seems possible that a "storage" threshold exists. Elsewhere (John, 1967a, 1971), the hypothesis has been discussed in detail that the reaction which synthesizes the substance responsible for these long-term changes in neural reactivity is triggered by a greater change in the concentration of a critical cellular constituent than can be accomplished by a single neural discharge. Changes in the concentration of this *critical substance* might be opposed by homeostatic mechanisms operating at a rate adequate to compensate for the effects of low firing frequencies. However, sustained alteration in the firing pattern or rate might shift this equilibrium.

The rate at which homeostatic mechanisms within the cell can destroy or otherwise inactivate the critical substance may be proportional to its concentration or may be a constant. Let us assume that there is some maximum rate K_1 at which such dissipation of critical substance can occur, such that

$$\frac{-d[c]}{dt} = K_1$$

where $-d(c)/dt$ stands for decreases in concentration due to homeostatic processes.

Let us assume that the increase in the rate of production of critical substance for some time interval T after each neural discharge is K_2, such that

$$\frac{d[c]}{dt} = NK_2$$

where $d(c)/dt$ stands for increases in concentration due to N neural discharges within the time interval, T.

Within this time interval, T, the net change in concentration of the critical substance will be

$$\Delta C = \int_0^T (NK_2 - K_1)\, dt = (NK_2 - K_1)T.$$

Examination of this expression shows that there will be no increase in the concentration of the critical substance until $NK_2 > K_1$. Once the firing density of the cell alters sufficiently, the rate of synthesis of critical substance will exceed the rate of dissipation. Finally, let us assume that there exists some threshold change in concentration, *a critical shift*, required to initiate the permanent storage reaction in any particular cell, *i*. Let us denote the critical shift as ΔC_i.

Then those cells that can participate in the representation of an experience will be cells for which $(NK_2 - K_1)T > \Delta C_i$. T is equal to the consolidation time required by that cell, i. This formulation provides a basis for understanding why some events are stored in memory with sufficient intensity to influence subsequent behavior and conscious experience, while others seem to leave no effective residue:

1. In many cells responding to an event, NK_2 will fail to exceed K_1 and no change in critical substance will occur.

2. In many cells responding to the event, $(NK_2 - K_1)T$ will fail to exceed ΔC_i. Although a change in critical substance will occur, it will not be sufficiently intense and/or sustained to produce the minimum net change necessary for storage.

3. Factors such as connectivity and refractoriness from recent activity will prevent many cells from responding to the event.

4. Thus, only a selected subset of the cells upon which afferent input impinges will be capable of participating in the representation system for that impinging event.

The critical substance might not be a molecule common to all neurons, and the threshold for the critical shift might be different from state to state (since metabolic alterations accompany changes in state) as well as from cell to cell. Thus, the achievement of consolidation in a group of neurons sharing response to a particular stimulus might proceed at varying rates in different cells.

These assumptions create a picture of a network in which certain cells participate in reverberations after an afferent barrage. Each cellular discharge is considered to contribute a unitary increment toward a concentration change in a critical substance. Since discharge occurs in varying rates in various loops, increments accumulate at varying rates in different neurons. The rate of change of concentration must exceed some minimum in order to outstrip the homeostatic mechanisms of the cell. This adequate rate must be sustained for a sufficient time to achieve a critical shift in concentration necessary to trigger some storage reaction. Since only the net shift is considered crucial, this model depicts consolidation as a process occurring at different rates in circuits reverberating at different frequencies, all of which participate in the representation of the original event. Since these cells are the only neurons in the nervous system which have been altered by the event, the memory of that event must necessarily be somehow stored initially in this neural subset. Learning often takes numerous experiences although it can occur with a single trial. Certain events seem to register with reliable rapidity. If the necessary shift in concentration postulated above occurs in some or all the neurons of a particular mediating loop after a specific trial, consolidation for those cells can be considered accomplished as a consequence of that single experience. In many cells, the effects of that trial will fail to achieve the critical shift before neural activity returns to the usual level. Residual concentration changes would thereafter be expected to dissipate gradually because of metabolism and diffusion. Unless the cell is again set into sustained activity before normal concentration levels are restored, there will be no lasting effect of the experience in that neuron.

After any particular trial, some neurons are postulated to achieve an altered state (consolidation) in the nonincremental manner outlined. On subsequent trials, stimulus conditions and neural excitability will be somewhat different. Presumably additional sets of neurons will achieve consolidation in a similar way each time the event is repeated. The ability of these neural sets to alter behavioral performance, which is the usual operational criterion that learning has occurred, will depend on a variety of factors. These probably include the percentage of neurons in the population which have achieved consolidation by the relevant time, the variability of state of the system, and the complexity of the stimulus input and of the operationally defined response. The ability of these altered sets of neurons to mediate altered behavioral response in a reliable way might be expected to increase more or less gradually, that is, incrementally, depending upon the response criterion. In this view, information storage can occur without effecting overt behavior. Whether one chooses to call such storage "memory" becomes a matter of definition. This "homeostatic" model has been discussed and supporting data set forth by several workers (Dawson & McGaugh, 1973; McGaugh & Petrinovich, 1965).

E. Threshold of Consciousness

There are several reasons why such speculations are relevant to psychological and psychiatric considerations. First, it seems quite possible that experiences which fail to cause an overt change in performance, that is, which cannot be consciously recalled, may nonetheless cause a permanent alteration of a portion of the neural system affected by that experience. Such changes, which might be called "subconscious," might well have significant functional consequences. Second, particular states might alter the storage threshold of various brain regions in a differential way so that certain kinds of experience were far more likely to achieve permanent changes in some brain regions than in others. Similarly, particular states might alter the retrieval threshold, that is, the threshold for activation of storage systems, in a differential way so that certain kinds of experiences are more readily recalled. Thus, it seems likely that the ease of recall of an experience would depend both upon the nature of the experience and the state of the brain when the experience occurred or when it was recalled.

Finally, it is well known that various drugs create characteristic profiles of excitability and inhibition in different brain regions. Not only does the foregoing discussion imply that certain drugs might facilitate or impede particular types of learning, it also provides a clear rationale for the expectation that certain drugs might selectively facilitate or impede the retrieval of particular types of experience because only those cells in which a critical shift had occurred could subsequently participate in the representation of that specific event.

F. The Multiple Trace Theory

The previous discussion was presented from the relatively simplified viewpoint of what might be called "dual-trace" theory in that it postulates a labile, short-term memory trace which is probably reverberatory in nature, and a long-term memory trace which is stable and somehow constructed as a consequence of the reverberatory activity. Several recent experiments suggest that this picture probably has to be modified.

In mice and goldfish, consolidation is severely disrupted by postacquisition injection of puromycin, an inhibitor of protein synthesis (Agranoff & Klinger, 1964; Flexner, Flexner, & Stellar, 1963). Davis and Agranoff (1966) have shown that puromycin blocks long-term retention if injected immediately after the training session but not if injected one hour later. Yet, if the goldfish are left in the training tank for as long as three hours after completion of training, the period of puromycin susceptibility is correspondingly lengthened. This suggests the existence of an intermediate storage process capable of holding a memory for extending periods before permanent fixation. In addition, puromycin injection does not prevent acquisition of a learned response immediately afterward. However, the performance of such puromycin pretreated animals gradually deteriorates after several hours, and no long-term retention is displayed (Barondes & Cohen, 1966). Similar results have been obtained with pretrial injection of acetoxycycloheximide. Performance of learned responses is unimpaired three hours after training, but deteriorates severely by 6 hr. Similar phenomena were encountered in the work of Albert (1966a, b), who showed that consolidation of a one-trial learning task could be blocked by posttrial application of cortical spreading depression or cathodal polarization. Under certain circumstances, recall could be demonstrated for a few hours after this interference, but performance gradually deteriorated and no long-term retention was established. Yet, an intermediate holding process apparently existed because subsequent application of surface anodal polarization reversed these effects, allowing consolidation to proceed and long-term storage to occur. Geller and Jarvik (1967) have made a report on what seems to be a related phenomenon. Animals tested for retention of a one-trial learning task immediately after recovery from ECS show unimpaired performance that gradually deteriorates. This observation indicates that performance during this interval was mediated by a temporary holding mechanism. Additional evidence for the existence of an intermediate holding process comes from the work of McGaugh (1967), who has reported that the injection of strychnine up to three hours after ECS almost eliminates the disruption of consolidation of one-trial learning caused by electro-convulsive shock.

These various results suggest that there may be an intermediate holding mechanism among the processes of information storage in the brain. This intermediate trace is of too long duration to be plausibly attributed to reverbera-tion, and its ability to survive ECS, spreading depression, and cathodal polariza-tion further excludes this explanation. It consists possibly of a biochemical template that gradually decays. Although adequate for the mediation of recall for several hours, this mechanism is not responsible for long-term storage. It is possible that these multiple traces are organized serially, so that a short-term reverberatory trace induces the intermediate holding mechanism which in turn brings about the processes responsible for long-term storage. Alternatively, these multiple traces might be organized in parallel so that the intermediate memory

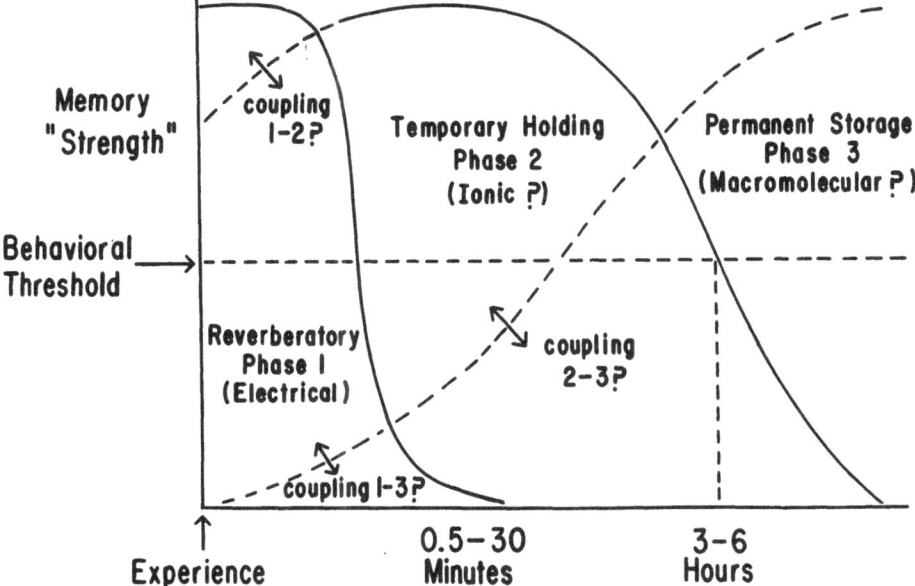

FIG. 8.1   The multiple trace theory of consolidation of memory envisages three phases in the storage of information: (1) a short-term process, probably consisting of reverberatory neuronal activity, representing the immediate effects of an experience upon the brain; (2) an intermediate holding process, perhaps mediated by the binding of extracellular potassium or calcium, responsible for temporary retrieval; and (3) a permanent storage process, probably consisting of RNA-mediated changes in synthesis of proteins or catecholamines, perhaps induced as a result of reactions initiated during the reverberatory activity (John, 1970).

process mediates temporary retrieval while consolidation is occurring independently. Obviously, several possibilities can be envisaged. Although further research will be necessary to clarify these questions, we should bear in mind that there are different kinds of memory with different decay times, and that each process may be based upon a different biochemical mechanism. A model of a multiple trace mechanism is shown in Fig. 8.1.

G. Does Critical Shift in Some Cells Contradict Statistical Theory?

It may seem at this point that the foregoing argument outlining the processes by which changes in specific cells would occur after an experience actually provides support for the connection formulation challenged in the following chapters (Chapters 9–11). If changes occur in specific cells due to reverberation so that a consolidation process can take place only in those cells, they therefore participate discretely in the storage of information. Clearly, no theory could hope to explain memory without invoking permanent alterations in a definite set of brain cells. Information cannot be stored in a vacuum. Certainly changes must take place in particular cells, and such changes might even consist of synaptic alterations. The critical question is how such changes represent the storage of information and whether activation of a memory or readout of the stored information requires the discharge of unique cells or "pathways" whose activity represents the past experience in a deterministic way.

The essential feature of this model is that information about an experience is stored as changes in specific brain cells but does not influence behavior deterministically. In the preceding discussion, a process has been described by which a large number of cells located in multiple brain regions is affected by the occurrence of an experience in such a way as to bring about some long-lasting consequence, the details of which remain to be discussed. These cells are presumed to have been selected fortuitously and to be distributed in parallel diffusely throughout the brain. The modification of future behavior of the organism as the result of this stored information, involving readout of the stored memory, need not require the participation of any specific cells so that activation of a particular pathway mediates the acquired response. The essential features of the proposed mechanisms are that readout may be accomplished probabilistically, that different cells may control the same behavior on different occasions, and that any given cell may contribute to the storage of multiple experiences and to the performance of a variety of learned responses. The effect of an experience is postulated to consist of alteration in the properties of many cells in various anatomic regions. The consequence of this alteration is suggested to be a change in the probability of coherent activity in neural populations when the specified stimulus is subsequently presented. The replacement of baseline or random activity by coherent activity in an ensemble of cells is proposed as the informationally relevant event.

In this view, the activity of any single cell is important from an informational viewpoint only insofar as it contributes to the time course of coherence in a neuronal ensemble, that is, to a mode of oscillation. It is suggested that the level of coherence or signal-to-noise ratio represents the significance or reliability of the information being processed, while the specific information content is reflected by the average activity of the ensemble through time, that is, by the mode of oscillation. These hypotheses remove the representation of experience from any particular cell and free the identification of familiarity from dependence on any pathway. At the same time, the proposed mechanism demands that definite changes occur in a discrete group of cells.

H. Mechanisms of Stable Information Storage

1. Rapid turnover of brain compounds. What mechanism might mediate the storage of information in the stable phase? Since memories are extremely resistant to erasure, persisting through sleep, unconsciousness, and excitement for a good part of a lifetime once consolidation has been completed, it seems reasonable to argue that during the consolidation phase something must be made. Although reverberatory neural activity may be the basis for short-term representation during the labile phase, long-term memory cannot be attributed to enduring reverberation. Whether information is stored by the actual growth of new connections between nerve cells or the synthesis of substances inside neurons or glial cells, whether the responsible mechanisms operate in a deterministic or a statistical manner, these processes must require changes to occur in the matter of which the brain is constructed—changes in structure or composition. What might be the chemical nature of the change which makes for stable storage? Since information stored in the nervous system can often be retrieved through an individual's lifetime, can one detect or demonstrate a stability in the matter of which brain is made that will provide a counterpart for the stability of memory? Although the gross morphology of brain seems to be fairly characteristic and stable, radioisotope turnover measurements and other methods provide insight about whether or not any of the molecular species composing the structure are in fact static and persist throughout a lifetime once they are laid down. Such isotope turnover data are available for most of the compounds of brain. There seems to be no significant compound of brain which does not display a remarkably high rate of turnover (Lajtha, 1961; Palladin, 1964.)

2. Role of RNA. The permanence of memory, therefore, probably cannot be attributed to the establishment of new intra- or extracellular structures laid down by permanent chemical molecules. It seems necessary, then, to explain the stable representation of experience as due to some change in configuration or substance mediated by a chemical system which, although itself not stable, is characterized by the fact that the molecules which break down are resynthesized in a specified way as to maintain the essential features of the change. Such template functions are known to be served by the nucleic acids. Furthermore, instinctive behaviors arise because of the influence of deoxyribonucleic acid (DNA) and ribonucleic acid (RNA) on relationships established between the cells of the nervous system. Since learning can produce long-lasting patterns of behavior as stable as instincts, it seems plausible that analogous mechanisms might be involved.

Deoxyribonucleic acid is almost completely localized to the nucleus. Ribonucleic acid is found not only in the nucleus but is distributed on the microsomes and throughout the cytoplasm of the cell in various forms. Since stimuli impinge on the cell at its outer surface, the chemical systems which permanently alter the cellular response to presentation of such stimuli probably modify substances which are found in proximity to cellular surfaces or in the subjacent cytoplasm. Considerations of this sort have led various workers, both in theoretical formulations and experimental explorations, to turn to the possible role of RNA or proteins in the mediation of long-term information storage. Substances located in the cell nucleus may well mediate such changes. In spite of the nuclear localization and greater resistance of DNA to alteration, there is no reason to assume that modification of DNA action may not also play a role. A central question which must be considered in this general context is whether the hypothetical mechanism is more likely to be *instructional*, so that the structure of the representational molecule is somehow specified by the information to be stored, or *selectional*, so that one of a preexisting set of possible structures is allocated for a given representational function.

Extensive data show that the rate of synthesis of RNA in nerve cells is proportional to the total stimulation received by the cell, and that a chemical concomitant of neural activity is stimulation of ribonucleic acid and of protein synthesis.[1] A number of studies have attempted to demonstrate more directly that there is an increase in RNA synthesis in a learning situation. It has been reported that after establishment of conditioned response, presentation of the conditioned stimulus causes an increased turnover of RNA in regions of the brain related to the modality of the conditioned stimulus but not in adjacent regions. Using the so-called "mirror focus" as a prototype of learning, RNA changes in the apical dendrites have been reported after the mirror focus becomes independent. Using labeled uridine, increased RNA synthesis has been demonstrated in the brains of goldfish learning a conditioned avoidance response, but not in the brains of control fish receiving the same total stimulation

---

[1] This literature is so voluminous that specific citations have not been provided in this section. Abundant documentation is available in John (1967a, 1971), Chapouthier (1973), and Booth (1973).

at random. Similar results have been obtained in mice. Further analysis suggested that the rate of synthesis of a messenger-like RNA in brain increased during learning. Significant changes in base ratios have been found in nuclear RNA in rats in a variety of experimental situations. A reciprocity has been demonstrated between glial and neuronal RNA synthesis. Conditions which cause an increase of certain compounds in neurons seem to cause a decrease of such compounds in glia, and vice versa. This has led some workers to suggest that certain substances might move from glial cells to neurons since the loss of glial RNA during neural activity exactly balances the increase in neuronal RNA with the same base ratio. The glial cells might thus specify part of the neuronal protein synthesis.

A number of studies have explored the effects of interference with RNA synthesis or destruction of RNA on the storage and retrieval of information. Conditioned planaria display retention of the conditioned response following transection and regeneration, whether the tested animals regenerated from head or tail segments. However, conditioned animals who are transected and per-mitted to regenerate in a solution containing a low concentration of ribonuclease behave in a different fashion. Only animals regenerated from head segments can perform the conditioned response following such treatment, while animals regenerated from tail segments perform at the random level. It has been shown that intracerebral injection of ribonuclease, but not deoxyribonuclease, trypsin, or serum albumin, blocks retention of a conditioned defensive reflex in mice. Inhibition of RNA synthesis with 8-azaguanine has been shown to increase the number of errors during maze learning without interfering with retention for previously learned maze habits. Inhibition of RNA synthesis using 8-azaguanine has also been demonstrated to increase the consolidation time for a new pattern of activity in the spinal cord of a rat. Conversely, facilitation of RNA synthesis decreased the consolidation time of this same spinal cord activity pattern appreciably. Using mice, it has been shown that facilitation of RNA synthesis decreases the time following an experience during which erasure of the learning can be accomplished by administration of electroconvulsive shock. These data suggest that chemical stimulation of RNA synthesis accelerates the rate at which consolidation is achieved.

Other experiments exist suggesting that increase in RNA facilitates the storage of information and its retrieval. It has been reported that massive injections of yeast RNA to aged individuals result in marked improvement in memory. Animals injected intraperitoneally with yeast RNA before learning sessions acquired conditioned avoidance responses significantly more rapidly than con-trols who received saline injections and demonstrated greater resistance to extinction. Similar results have been obtained for maze learning in rats. It has been suggested that the injected RNA preempts much of the RNAase available and allows more of the endogenous RNA to survive and accumulate. These effects need not imply direct utilization of yeast RNA for the storage of information in human or animal neural tissue.

II. CHEMICAL TRANSFER OF LEARNING

Planaria that were fed fragments of conditioned worms acquired the conditioned response more rapidly than planaria that ingested fragments of naive worms. The observation of more rapid learning in cannibalistic worms fed conditioned tissue encouraged a number of workers to attempt to extract "trained" RNA from the brains of conditioned animals and to investigate the possible facilitation of learning in animals receiving injections of such extracts. The initial report of positive results in such experiments led a large number of laboratories to undertake similar investigations. This has developed into one of the most controversial areas of research on memory, with large numbers of workers participating, many of whom report positive results and many of whom report complete failure to obtain any indication of transfer of learning by injection of extracts from trained animals (Byrne, 1970).

In the most striking experiment of this genre, Ungar and his associates (1968) reported the extraction, purification, and synthesis of a substance from animals trained to avoid the dark. This substance, called "scotophobin," has been reported to cause dark avoidance in a variety of species when injected intracerebrally. Recently (Ungar, 1972), it has been reported that a similar substance has been extracted from the brains of animals trained to avoid certain colors. While the mechanisms of action of these low molecular weight polypeptides are still obscure, it seems likely that they represent classes of molecules which greatly enhance the probability of certain kinds of behavior. It is still unclear whether these effects reflect the transfer of specific learned responses or nonspecific response propensities. In either case, these results are of great interest.

III. PRESENT UNCERTAINTY OF THE FIELD

Further investigations have indicated a marked difference between cycloheximide and puromycin in their effects on hippocampal electrical activity. Puromycin produces marked electrical abnormalities. Thus, it may be that the effects of puromycin on consolidation should not be attributed to interference with protein synthesis but may be related to its effects on the hippocampus, a structure which has been implicated in the short-term storage of information.

Furthermore, it has been shown (Flexner & Flexner, 1967, 1969) that intracerebral injection of isotonic saline solution or other substances, including water, could restore performance of learned responses long previously suppressed by puromycin. This effect was attributed to the removal of abnormal peptides produced by puromycin. Whether or not this proposed mechanism is correct, it seems impossible to evade the conclusion that the amnestic effects of the initial puromycin injection were not due to blockade of protein synthesis mediating long-term memory. Long-term storage of the information obviously occurred in the presence of puromycin blockade of protein synthesis, but the manifestation of this memory as effective for control of behavior was blocked; that is, *retrieval* could not occur.

Although numerous facts suggest that macromolecular synthesis may well play a crucial role in consolidation, this is an area of research in which many items of data are extremely difficult to reconcile with each other. None of the experimental results which have thus far been obtained warrant the unequivocal conclusion that stable information storage in the brain is mediated by the synthesis of ribonucleic acid or protein. Many of the substances used to interfere with synthetic pathways for these macromolecules also influence synthesis of catecholamines. Of particular importance in generating this uncertainty is the demonstration that the blockade of memory by inhibitors of protein synthesis could be prevented or reversed after long time periods by adrenergic stimulants. Adequate concentrations of catecholamines seem to be necessary for memory to be expressed and can protect against the amnestic effects of inhibitors of protein synthesis (Barondes & Cohen, 1968; Dismukes & Rake, 1972; Flood *et al.*, 1972; Randt *et al.*, 1971; Seiden & Peterson, 1968).

These findings not only cast some doubt on whether protein synthesis was crucial for consolidation in these situations, but highlight the dangers of interpretation of findings in this research area. Although numerous facts suggest that macromolecular synthesis may well play a role in consolidation, this is a research area in which there are abundant quantities of data extremely difficult if not impossible to reconcile with each other, in which laboratories have repeatedly reported the inability to replicate results obtained elsewhere, and in which effects clearly demonstrated on one species for one kind of behavior simply cannot be reproduced using other species or other measures of behavior. In our opinion, none of the experimental results obtained thus far is sufficient by itself to warrant the unequivocal conclusion that stable information storage in the brain is mediated by the synthesis of ribonucleic acid or protein. Yet, it must be conceded that a wide variety of experimental procedures has yielded an impressive quantity of positive findings strongly pointing toward ribonucleic acid and protein synthesis as deeply implicated in the function of information storage in the brain.

IV. THE DEREPRESSOR HYPOTHESIS

In the previous discussion, we argued that the information content in the activity of a neuronal ensemble might consist of the time course of coherence. The memory of such information must be stored as an increase in the probability of the particular temporal pattern of coherence which occurred. We have proposed what we call the derepressor hypothesis to accomplish such storage:

1. In any cell much of the potential for synthesis of specific substances inherent in the DNA structure is repressed.
2. Sustained participation of a neuron in representational activity causes a

critical shift in the concentration of cytoplasmic materials resulting in the derepression of an inhibited synthesis.

3. The resulting alteration in cytoplasmic constituents has two consequences: (a) derepression of that synthesis is thereafter sustained; and (b) the reactivity of the neuron to patterns of stimulation is altered.

The derepressor hypothesis has been discussed in greater detail elsewhere (John, 1967a). There are several essential features of this hypothesis. It is assumed that the postulated changes do not occur as a consequence of mere neural activity, but that the activity must be sustained for a sufficient time and at a sufficient rate to achieve a critical shift in the concentration of a critical substance. Since the sustained activity underlying the critical shift arose from participation in a reverberating loop, the neuronal input presumably involved only a portion of the many synaptic contacts of the cell and was characterized by a particular pattern. Thus, the input possessed particular spatiotemporal characteristics. The resulting neuronal discharge reconciled the various spatial influences at different synapses of the same cell, integrating these into a temporal pattern of neuronal response. There seems to be no compelling reason to require that the change in neural reactivity resulting from the derepressed synthesis necessarily restricts the altered probability of neural response to the arrival of impulses with specified distributions at specific synapses. The integrative nature of neuronal response to multiple synaptic inputs might well result in a "smearing" of the contribution of individual synapses.

A great deal of specification would seem to have been accomplished if the cell becomes effectively "tuned" to some temporal pattern integrated over all of its synaptic inputs, which influence cytoplasmic chemical concentrations during the relevant time period. This tuning might be effective at the level of the axon hillock rather than at the synapse. Such tuning would not require faithful reproduction of the events at each synapse and would also alter response tendencies to partial reproduction of the initial set of input events. Furthermore, it permits alternative inputs to be accepted if temporal constraints are satisfied over the full set of stimuli which occur. Such a formulation would seem to provide advantages of flexibility and overall accuracy while conforming to the probabilistic nature of many of the propositions set forth earlier. The most important consequence of the process envisaged in the derepressor hypothesis is that cells which have been active over a period of time with a particular pattern of activity become more capable of sustaining such a temporal pattern of activity in the future, and thereby the probability is enhanced that a given mode of oscillation will subsequently be displayed by a particular neuronal ensemble.

V. ETHICAL PROBLEMS OF LEARNING ENHANCEMENT

Although this overview of the present status of research on biochemical mechanisms of permanent memory storage makes clear that a final answer is not yet available, it is nonetheless obvious that great progress has been made. While we do not know the exact nature of the crucial chemical reactions, that information might well be forthcoming in the near future. We can expect that the critical reactions will not only be identified, but that practical pharmacological agents for the facilitation of learning and for both the erasure and enhancement of memory will soon become available. This prospect provides enormous therapeutic potentialities which ought to be evaluated systematically. At the same time, such chemical capability poses serious questions which demand consideration. Under what circumstances and to whom should learning-enhancement drugs be administered? When is it permissible to erase or enhance memory? We are on the threshold of being able to control what an individual can learn or remember. It is time to address ourselves to some of the ethical problems raised by these prospects. Although a detailed discussion of these issues is beyond the purview of this textbook, we urge that some classroom time be devoted to their examination.

9
The Localization of Function—
Where Is Memory?

The problem of localization of memory is a special case of the more general problem of localization of function in the brain. Two diametrically opposed viewpoints on this question have long existed. One, the "localizationist" position, views the brain as an aggregate of separate organs with circumscribed regions mediating specific mental processes. The other, the "antilocalizationist" or "gestalt" position, holds the viewpoint that mental activity is the product of the whole brain. A brief historical review of these two positions follows.[1]

I. THE LOCALIZATIONIST POSITION

In the second century B.C., Galen suggested that mental processes were localized in the cerebral ventricles. About 600 years later, Nemesius proposed the posterior ventricle as the seat of the memory, the middle ventricle as the seat of the intellect, and the anterior ventricle as the seat of perception. This idea of three ventricles as the substrate of the major mental abilities was still generally accepted in the Middle Ages, as seen in Fig. 9.1.

In the eighteenth century, a school of psychology emerged which believed that mental processes could be subdivided into separate faculties. As these beliefs became established, a search began for the material substrate of these traits. It was believed that the brain could be subdivided into many organs or centers, each responsible for a separate ability.

[1] I wish to acknowledge my indebtedness to A. R. Luria for his excellent reviews in *Higher Cortical Functions in Man* (1966) and *The Working Brain* (1973), from which I have borrowed heavily in summarizing the development of the localizationist and antilocalizationist positions. The student seeking an unequalled scholarly treatment of these positions is urged to consult those volumes.

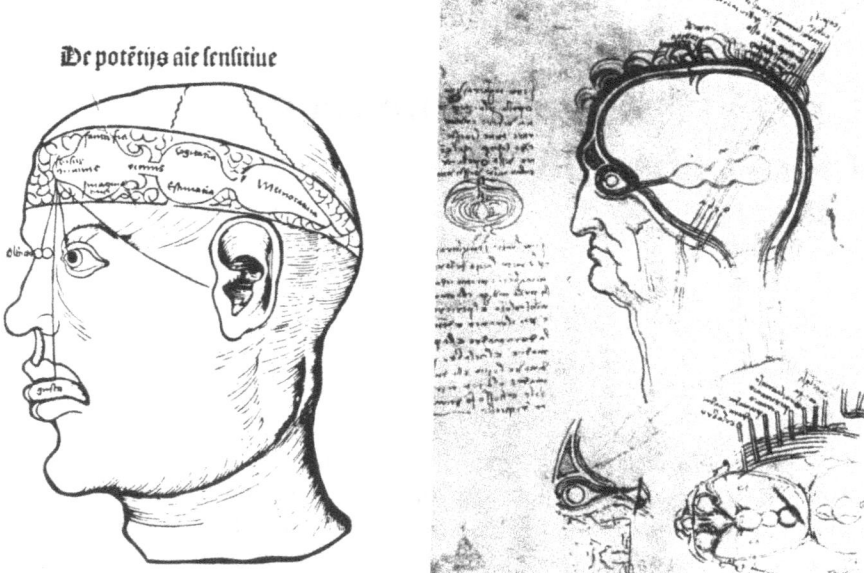

FIG. 9.1 Anatomic drawings from the twelfth (left) and fifteenth (right) centuries still propounded the theory voiced by Nemesius in the fourth century A.D. that perception was located in an anterior ventricle, cognition in the middle ventricle, and memory in the posterior ventricle. This represented a more refined localization of these functions than Galen's hypothesis that mental processes were located in the cerebral ventricles, suggested in the second century B.C.

Anatomic works began to appear which proposed differential mediation of the different mental faculties by various brain regions. Such works suggested, for example, that memory was localized in the cerebral cortex, imagination and reason in the white matter, apperception and will in the basal regions, and integrative processes in the corpus callosum.

Gall (1825) was one of the leading brain anatomists of his time. Early in the nineteenth century, accepting the psychology of faculties, he proposed that each mental faculty was based on a definite group of brain cells and suggested that the whole cerebral cortex was an aggregate of organs, each responsible for a particular faculty. The faculties which he relegated to particular areas of the brain were those identified by the psychological teachings of his time. Gall constructed very detailed phrenological maps in which domestic instincts, destructive instincts, attraction to food, aptitude for education, instinct for continuation of the race, love of parents, self-esteem, and numerous other traits were localized, as illustrated in Fig. 9.2.

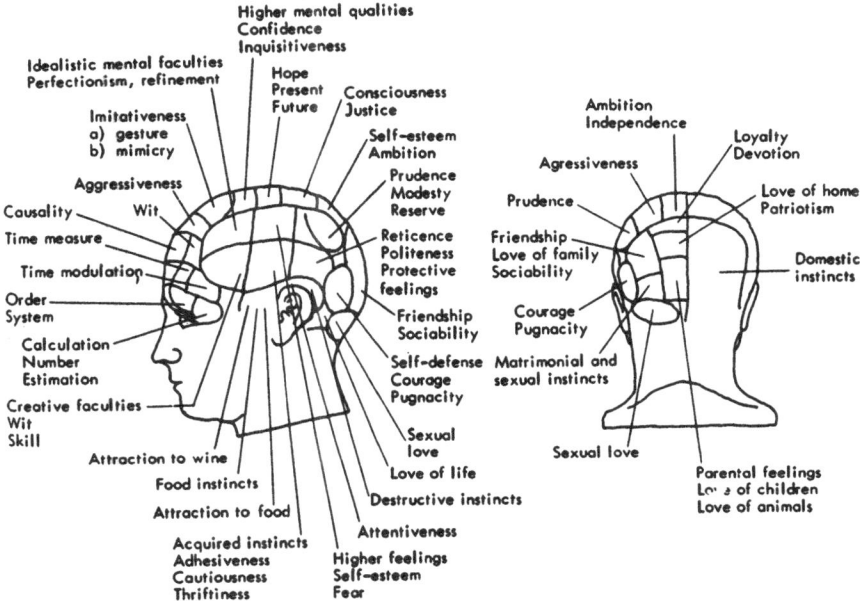

FIG. 9.2   A typical phrenological map of the nineteenth century.

Gradually, scientific evidence supporting the ideas of localization began to come from clinical observations on the effects of local brain lesions, on the one hand, and from neurophysiological and neuroanatomic studies, on the other. Clinicians argued that if the brain did not consist of separate centers, one could not understand how localized defects could appear after damage to particular parts of the brain. In 1861, Broca showed the brains of two patients with disturbance in expressive speech who had lesions in the inferior frontal convolution of the left hemisphere. He concluded from this that expressive speech functions were localized in a center for the motor forms of speech in which cells of a particular area comprised a depot of images of the speech movements. A few years later, Wernicke (1874) found a case with a lesion of the superior temporal gyrus of the left hemisphere and a disturbance in speech comprehension and concluded that the sensory images of speech were localized in this cortical zone.

Encouraged by these discoveries, clinicians in the following years reported the identification of additional centers in the brain. These included areas responsible for visual memory, mind blindness, word blindness, word deafness, writing skills, formulation of logical propositions, conceptual centers, naming centers, and centers of ideas, in addition to the centers responsible for visual, auditory, tactile, and motor function. In 1884, on the basis of many such observations,

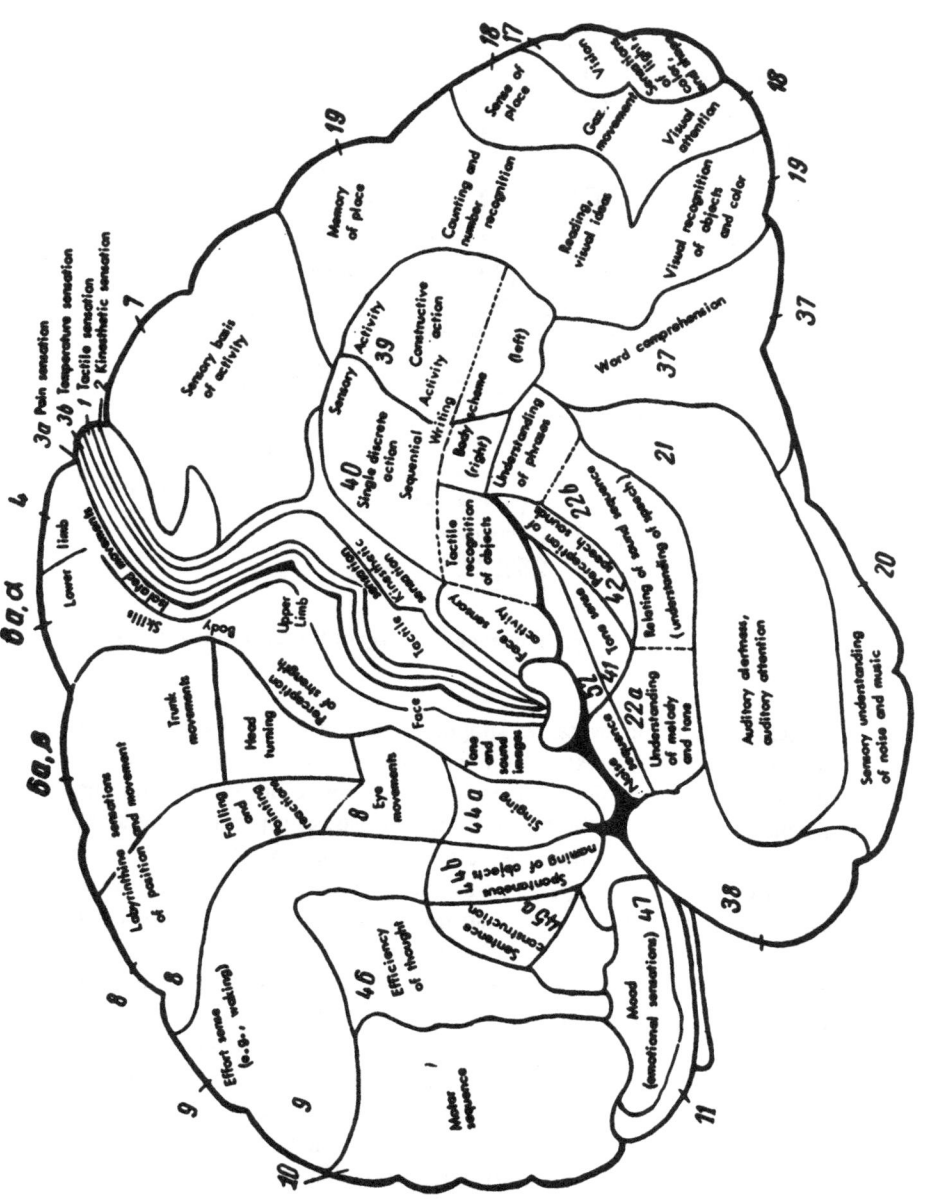

202

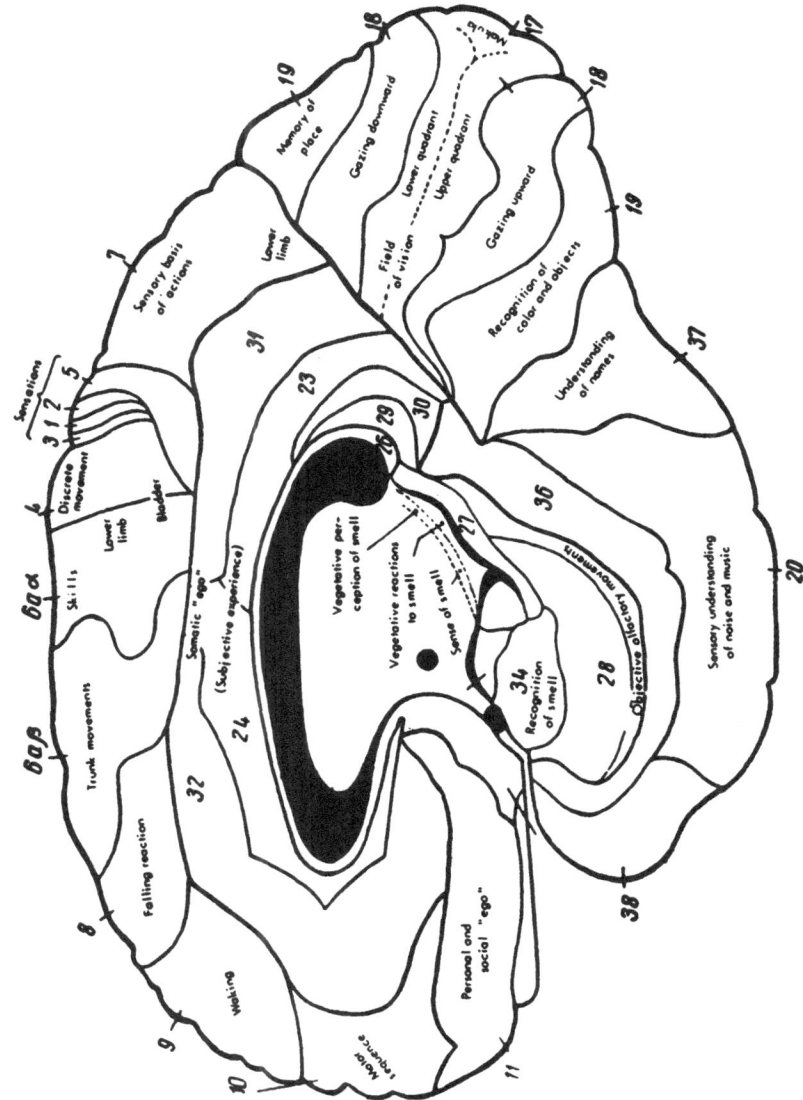

FIG. 9.3 A typical chart of cerebral localization of functions, from a text published in 1934 (Kleist, 1934).

Meynert proposed that cortical cells were carriers of particular mental processes and suggested that each new impression was stored in a new and still vacant cell. This was perhaps the earliest hypothesis that single cells mediated the storage of particular memories.

In parallel with the clinical observations supporting ideas of the localization of function, similar evidence was produced from animal studies. In 1870, Fritsch and Hitzig showed that stimulation of certain cortical areas caused contraction of certain muscles. At about the same time, Betz (1874) found giant pyramidal cells in the anterior central gyrus and associated these cells with motor functions. In one of the earliest experimental studies of lesion effects on memory, Munk (1881) showed that extirpation of the occipital lobes caused the loss of visual recognition. In a series of studies by Hitzig (1874), Ferrier (1876), and Bianchi (1895), it was reported that anterior lesions caused disturbances of attention and intellectual activity. These findings, which were suggestive of centers for various sensory and motor functions, stimulated further search for centers of more complex mental functions. Eventually these ideas led some to the belief that complex mental processes could be localized in circumscribed brain areas. These propositions were formally presented in psychiatric texts which often contained charts of cerebral localization of functions. One such chart is illustrated in Fig. 9.3. Note the localization of "subjective experience."

II. THE ANTILOCALIZATIONIST VIEWPOINT

Throughout the period in which the localizationists were developing their beliefs and the evidence on which those beliefs were based, a diametrically opposed school of thought was marshalling its own arguments. An early representative of this was A. Haller who, writing in 1769, acknowledged that different parts of the brain might well be involved in different functions, but postulated that the brain acted as a single organ composed of parts of equal importance. Damage to a single region can cause disturbance of various functions, and many of these disabilities can be compensated for by the remaining tissue. In what is probably the earliest piece of experimental work supporting the antilocalizationist position, Flourens (1824) showed about the same degree of recovery of function in birds regardless of which part of the brain was damaged. He concluded that the cortex acted as a homogeneous entity. Some years later (1842), he showed that reversing flexor and extensor innervation of the wing in a cock could be compensated, and argued that this was evidence for equipotentiality of function of different regions of the brain. Further experimental studies of a similar sort were done by Goltz between 1876 and 1881. After extirpation of various parts of the cerebral hemispheres, Goltz (1884) reported that dogs showed a variety of marked disturbances of behavior, using general responses as a measure. These disturbances gradually disappeared. The functions were restored, leaving only slight awkwardness. These observations led to the idea that any part of the brain could be associated with ideas and thought, and the defect observed after brain damage was solely related to the size of the lesion.

In 1884, Brown-Sequard provided evidence showing that any region of the cortex could acquire motor properties merely by being associated in its activity over a period of time with the part of the brain which when stimulated caused movement. In 1905, Baer provided a detailed confirmation of Brown-Sequard's observation, voicing again the conclusion that any region of the brain could be made motoric by association over a period of time with the activity of a region causing movement. In 1897, Wedensky published the first of a long series of experiments on the so-called "stability of the motor point" in which the usual interpretation of the observation of Fritsch and Hitzig was challenged. Wedensky showed that the movement which was produced by electrical stimulation of a point on the motor cortex depended on the previous history of stimulation of the animal. If, in a dog under light narcosis shortly after electrical stimulation of the motor point for the extensors of the forepaw (Center A), stimulation is applied to the motor cortex regions of the flexors of the forepaw (Center B), it is possible to observe a paradoxical effect: the result of stimulation of Center B is as if Center A had been stimulated, and extension rather than flexion of the forepaws occurs.

Similar conclusions came from the work of Lashley (1923) who mapped the precentral gyrus of a rhesus monkey using electrical stimulation in a series of tests extending over a period of weeks. In each test motor reactions to stimulation of a particular cortical point were essentially constant, and in different tests the general cortical fields from which movements of face, arm, or leg were elicited tended to remain constant, although the borders of the fields varied somewhat. However, within the arm area stimulation of the same point in different tests produced widely different movements, and at different times the same movement was obtained from widely separated and shifting areas. Such results suggested that within the segmental areas the various parts of the cortex might be equipotential for the production of all the movements of that limb, and that the particular movements elicited in any test depended upon the momentary physiological organization of the area rather than upon any point-for-point correspondence between pyramidal and spinal cells. After lengthy studies of the results of different kinds of ablations and lesions on the performance of conditioned responses, Lashley (1929) concluded that a particular type of disturbance cannot be ascribed to a particular brain lesion. Impairment of function is related to the extent of brain damage (the law of mass action), while complex functions can be mediated by different brain areas (the law of equipotentiality).

In addition to experimental results, clinical observations provided support for antilocalization ideas. The distinguished neurologist Hughlings Jackson formulated a series of principles in sharp opposition to the idea of narrow localization

(1869). Jackson stated that lesions of circumscribed brain areas seldom lead to complete loss of function. While voluntary movements or speech are often blocked, involuntary movements and utterances frequently remain. Jackson argued against the concept of function as mediated by narrowly circumscribed groups of cells in favor of a complex vertical organization with multiple representation at low brain stem levels, middle motor and sensory cortical levels, and high frontal cortex levels. Thus the localization of the cause of a symptom that is the reason for impairment of some function accompanying local lesion could not in any way be interpreted as evidence for localization of that particular function.

At about the same time, other neurologists suggested that speech should be considered as a symbolic function disturbed by any complex damage to the brain. The so-called Noetic School of neurology held that the principal form of mental process was symbolic, and all brain damage caused depression of symbolic function (e.g., Goldstein, 1948). While accepting the cerebral localization of neurological symptoms, sensations, and elementary processes, Monakow and Mourgue (1928) argued against the localization of symbolic activity in any particular brain region. Head (1926) ascribed speech disturbances to lesions of large areas of cortex and attributed these disturbances to the loss of "vigilance." Goldstein (1948) distinguished between the "periphery" of cortex, in which localized lesions disturb the means of mental activity (elementary functions), and the "central part," treated as equipotential, in which lesions cause change in abstract behavior (higher functions) in accordance with the law of mass action. The ideas of this worker combine the principles of narrow localization with those of equipotentiality.

Asratyan (1965) has reviewed the recovery of functions disturbed by ablations in various parts of the central nervous system. He describes two views about such phenomena: (1) the readjustment of behavior which takes place is due to the plasticity of an equipotential system; and (2) the recovery of functions which is observed is due to the stepwise relearning and compensation carried out by various parts of the nervous system, and is not due to the substitution of function.

III. THE SEARCH FOR THE ENGRAM

The controversy between the localizationists and the antilocalizationists dealt with the attempt to identify complex mental processes or psychological concepts with the material structure of the brain. The problem of localization of memory was a particular aspect of the more general question of functional localization. Much of the evidence on which theories of memory localization were based came from studies of conditioned responses in which a sensory stimulus came to elicit a new motor movement after a conditioning procedure. It was assumed that the sensory stimulus affected a particular sensory region of the brain, that the conditioned movement was controlled by the motoric regions of the brain, and that conditioning established a new pathway between the sensory region and the motor region so that the sensory stimulus could influence the activity of the motor region.

Generations of physiological psychologists have attempted to localize the hypothetical pathway mediating the new influence of the sensory stimulus on the motor system by destroying various regions of the brain. The failure of such ablation studies has been thoroughly catalogued by Lashley in his paper "In Search of the Engram" (1950).

IV. RECENT EVIDENCE AGAINST LOCALIZATION OF FUNCTION

Strict correlation of structure with function is difficult, even with respect to such relatively species-constant characteristics as sensory input and motor output regions. The variability of the responses elicited by stimulation of the motor system has already been mentioned. Evidence increasingly accumulates from studies of the unanesthetized unrestrained animal that many conclusions about functional localization must be seriously questioned. Data which contradict expectations based on "classical" cytoarchitectonic or neurophysiological evidence, much of which was summarized in Chapter 1, as well as dramatic evidence of functional compensation, abound in the literature. A few striking examples follow.

1. Evoked potentials (Doty, 1958) or responses of single neurons (Burns, Heron, & Grafstein, 1960) to visual stimuli can be recorded from a widespread extent of cortex, far exceeding the area striata as defined cytoarchitectonically.

2. Visual discriminations can be established after extensive ablations of cortical and collicular regions of the visual system (Urbaitis & Hinsey, 1966; Winans & Meikle, 1966). Ablations that produce severe sensory deficits in adult cats cause no permanent discrimination deficits when performed on kittens (Tucker & Kling, 1966). A similar finding also has been shown for rabbits (Stewart & Riesen, 1972).

3. Cats can perform visual pattern discriminations after destruction of as much as 98% of the optic tract (Norton, Frommer, & Galambos, 1966), even when the lesions are placed at several levels so as to disrupt all retinocortical isomorphism.

4. Cats can relearn an auditory frequency discrimination after bilateral ablation of all cortical auditory areas resulting in complete retrograde degeneration of the medial geniculate body (Goldberg & Neff, 1964).

5. Extensive bilateral lesions of the thalamic and mesencephalic reticular formation do not produce unconsciousness, loss of arousal, inability to acquire

new conditioned responses, or loss of previously acquired conditioned responses if such damage is inflicted in multiple stages (Adametz, 1959; Chow, 1961; Chow & Randall, 1964).

6. The challenge to strict localization of function even extends to the so-called vegetative centers. A series of specialized receptors located near the midline ventricles of the brain stem play an important role in the regulation of respiration, food intake, and other vegetative functions. These receptors correspond to the classical centers for metabolic regulation. Recent data (Coury, 1967) suggest that these centers constitute portions of regulatory systems which are anatomically quite extensive and diffuse and are deployed through the brain at all levels, including the neocortex.

V. IS THE VISUAL CORTEX ESSENTIAL FOR PATTERN VISION?

In man, total destruction of area 17 limited to one hemisphere leads to a unilateral (contralateral) hemianopsia, but compensation by the remaining contralateral visual cortex can be quite successful. Total bilateral destruction of area 17 in man leads to central blindness which is permanent. If unilateral damage includes areas 18 and 19, the hemianopsia which occurs is "fixed"; compensation does not occur, and the patient suffers from "unilateral spatial agnosia," a total unilateral destruction of the visual process (Luria, 1966). These phenomena have long been known, and have also been observed in many other animals. For example, Sprague (1966a, b) has reported that cats with a large, unilateral occipitotemporal cortex ablation develop a permanent hemianopsia.

In view of the functional neuroanatomy of the visual system as well as the single unit studies of "feature extractor" neurons in areas 17, 18, and 19 presented in the initial chapters of this Volume, the reader will probably regard these consequences of destruction of the visual cortex as both obvious and inevitable.

Therefore, it will probably be startling to learn that the hemianopsia produced in cats by extensive ablation of the visual cortex on one side can be virtually abolished by either a lesion of the superior cortex contralateral to the cortical lesion or a transection of the commissure of the superior colliculus (Sprague, 1966b). As a tentative explanation for this remarkable result, Sprague suggested that unilateral ablation of the visual cortex released the ipsilateral superior colliculus from a facilitatory corticocollicular influence, making it more susceptible to inhibitory input from the contralateral superior colliculus. Destruction of the contralateral colliculus or transection of the collicular commissure in turn blocked this inhibition, thereby releasing the ipsilateral colliculus to mediate visually guided behavior. Other data suggest that the ipsilateral medial suprasylvian gyrus (association cortex) assumes an important role in the resulting compensatory "visual" functions.

A convincing confirmation and extension of these findings was recently provided by Sherman (1974b). Not only were Sprague's 1966 findings replicated, but it was shown that animals which appeared *totally blind* on all visually guided tests after bilateral ablation of areas 17, 18, and 19 recovered essentially normal visual placing responses, good following of moving objects in all directions, and a full *binocular* field of view, (although monocular testing showed loss of the contralateral field and behavior was "sluggish") when the collicular commissure was transected in a second-stage operation. Essentially identical outcome was obtained when the cortical ablations and commissural transection were inflicted in a single operation.

If a cat is raised with the lids of one eye sutured together, it develops severe abnormalities in its geniculocortical system. Stimulation of the deprived eye drives very few cortical neurons (Wiesel & Hubel, 1963b, 1965). Also, the lateral geniculate cells innervated by the deprived eye are abnormally small (Guillery & Stelzner, 1970; Wiesel & Hubel, 1963a). These deficits, however, seem to be limited to the binocular regions of the geniculocortical projection system. Wickelgren and Sterling (1969) showed that the deprived eye has apparently normal retinotectal input. Further, they provided electrophysiological evidence that the retinotectal input of the deprived eye is inhibited by the abnormal corticotectal pathway, and that decortication abolishes this suppression. Sherman (1973) showed that on a visual perimetry test, such cats behave as though they can see objects in the monocular (lateral) segment of the visual field but are completely blind in the binocular (nasal) segment. Recently, Sherman (1974b) has shown that visual cortex lesions involving all known projections from the lateral geniculate restore vision to the previously blind field segments of cats who have suffered monocular deprivation almost since birth. These results fit closely with those just described and provide further insight into the mechanisms involved.

These demonstrations that sensory deficit can be alleviated by destruction of certain regions of a damaged sensory system stimulate speculation about their possible implications for rehabilitation after brain damage. Wilson and Wilson (1962) have contributed observations from studies of intersensory facilitation of learning sets in normal and brain-operated monkeys which are very relevant for such speculations. Under the proper conditions, it was possible to demonstrate cross-modal specific transfer of discriminations, based upon a particular stimulus diversion, from tactile to visual and from visual to tactile in normal monkeys. It is known that lesions of inferotemporal cortex selectively disrupt visual discrimination learning in monkeys, while lesions of parietal cortex disrupt tactual discrimination learning. M. Wilson also studied both types of cross-modal transfer in monkeys with either inferotemporal or parietal lesions. She found that monkeys with lesions which disrupt discrimination learning in a particular modality not only fail to benefit from experiences in that modality but are subsequently *hindered* in new learning in another modality. However, *monkeys with such lesions can benefit from discrimination learning involving an intact modality so that their subsequent performance on tasks utilizing cues in the damaged modality approaches that of normal monkeys.*

Taken together, these various findings imply that damaged cortical tissue (and perhaps damage in any part of a sensory system) not only interferes with processing of sensory information in the modality for which the damaged cortex is most relevant, but may interfere with learning in other modalities. This disruptive effect seems to involve inhibitory influences exerted by the cortex on subcortical structures. If such disruption is prevented by ablation of the damaged cortical region or the inhibited subcortical structure, release from inhibition can be achieved with substantial restoration of information processing in that sensory modality. Apparently, the same functional consequence can be approximated by eliminating sensory experience in the damaged modality while carrying out discrimination training in an intact sensory modality utilizing cues which have informational value in specific dimensions or parameters of the stimulus. Subsequent cross-modal transfer to sensory information containing cues along analogous dimensions in the damaged modality then becomes possible. This suggests that during discrimination training in the intact modality, either a neural mechanism is established which subsequently participates in mediation of the analogous discrimination in the damaged modality *or* which somehow releases the inhibitory influences upon subcortical structures in the damaged modality when that analogous discrimination task is presented. Further research will be necessary to clarify which of these alternatives actually operates in such circumstances.

VI. SINGLE- VERSUS MULTIPLE-STAGE LESIONS

A very different line of evidence provides additional support for the suggestions that a substantial part of the functional deficit observed after localized brain damage may be due to the abrupt imposition of inhibitory processes as a result of tissue loss, and that such inhibitory processes can be released by experience or learning, as well as by judicious placement of further lesions, as in the experiments cited in the previous section. A wide variety of different preoperatively learned tasks can no longer be performed if brain regions apparently crucial for the tasks are removed bilaterally in a single operation. Reacquisition of these tasks with intensive training after surgery is often impossible or enormously impeded. Yet, a vast body of evidence shows unequivocally that *complete retention of the same preoperatively learned tasks can often be demonstrated in mammals if the identical brain damage is inflicted in serial or multiple stage operations.* This literature has been critically evaluated in a recent exhaustive review by Stein, Rosen, and Butters (1974). Recovery depends upon the length of the interval between the operations and the size of the lesions.

Of greatest interest to the present discussion is the fact that recovery depends upon the type of sensory stimulation experienced during the interoperative period. For example, animals subjected to two-stage lesions of the visual cortex can retain pattern vision, although such function would be lost after bilateral ablation performed in a single stage. (Note that even the latter deficit is now subject to reinterpretation in view of the findings of Sprague and of Sherman cited in the preceding section.) If animals subjected to two-stage lesions of the visual cortex are kept *in the dark* between the operations, loss of visual function occurs just as though a one-stage bilateral lesion had been inflicted. Also, animals who were exposed for several hours each day to diffuse, unpatterned visual stimulation or who were passively transported through a patterned visual environment for the same period each day between the successive operations failed to retain previously acquired pattern discriminations and could not be retrained on such tasks. However, animals permitted self-produced locomotion in the same patterned environment for several hours each day during the interval between operations showed complete sparing of the pattern discrimination or very rapid reacquisition (Dru, Walker, & Walker, 1975).

VII. A NOTE OF CAUTION

Such findings establish that the localization of function within any area of the brain must be evaluated with great caution, even when apparent physiological correlates of function have been demonstrated in as great detail as has been the case with the elegant studies of "feature extractor" cells of the visual cortex. It should go without saying, but we will nonetheless state for the record, that similar caution must be exercised with respect to our own arguments in this Volume *against* localization. We voice a strong warning against overenthusiastic proponents of either position, localizationist or antilocalization, who ignore or shrug off the contradictions. It is precisely from the reconciliation of the apparent contradictions that we stand to learn most about how the brain processes information.

Yet it seems to us that for at least several decades an overly strong doctrine of localization of function has dominated our thinking. Adherents to this doctrine have strongly influenced the practice of neurology, neurosurgery, and rehabilitation medicine, and have played a critical role in remedial training and education of the handicapped either directly or through the construction of physiological models of sensation, learning, and memory which depend upon specific neuronal pathways connecting neurons or regions mediating specific functions. As a result of the unquestioning acceptance of this doctrine, many individuals showing effects of congenital, traumatic, or surgical brain damage have been regarded as irreversibly handicapped. Localization of function may indeed be more discreet in human beings than in lower mammals, as some workers contend. Yet enough cases exist of striking recovery after loss of brain tissue due to cerebrovascular accident, traumatic injury, disease, or neurological intervention, together with a huge body of experimental data from lower animals which have simply not been exhaustively examined for their possible relevance to human rehabilitation, to make it plausible that a substantial number of functional deficits now considered irreversible in humans could be treated with substantial or complete functional restoration.

VIII. IMPLICATIONS FOR REMEDIATION OF BRAIN DAMAGE

The dramatic results described in the preceding sections raise many questions: What regions guide visual behavior or sustain wakefulness in animals subsequent to these drastic multiple stage lesions? What is the functional uniqueness and significance of the feature extractor cells whose behavior has been so elegantly and meticulously studied? What would be the consequence of inflicting collicular ablations or transection of the intercollicular commissure on a human being made totally permanently blind or hemianopic by damage to the visual cortex? The reader may be intrigued by the perplexing ethical problems confronted by the audacious neurosurgeon who contemplates the latter question.

Obviously a great deal of future research will be required to answer the multitude of questions raised by the examples cited above, but it should at least be clear to the reader that matters are a great deal more complex than might be expected from traditional orientations.

Whatever the details of the neuronal mechanisms involved in the compensatory phenomena revealed by multiple stage lesion experiments, some practical implications ought to be considered: *First,* drastic though it may seem, it may well be that many individuals with hemianopsia or similar sensory deficits due to known cortical damage might benefit from surgery designed to release subcortical structures from inhibition which prevents effective afferent input. *Second,* pending the cautious exploration of such heroic measures, it would seem as though the rehabilitation of patients with damage in a particular sensory modality would be markedly aided by a two-stage remedial procedure. In the initial stage, successive discriminations would be established in an intact sensory modality using cues with well-defined relevant dimensions or parameters, while all sensory input to the damaged modality would be abolished. After substantial overtraining under these conditions, cross-modal transfer to the damaged modality, with gradual "fading" of the information in the intact modality, might substantially facilitate this process of cross-modal transfer. A similar fading procedure has been found to facilitate discrimination learning in human retardates, but within the visual modality alone (Dorry & Zeaman, 1975). These suggestions would seem particularly relevant for the construction of early intervention procedures in children with central damage involving a particular sensory modality, but with functional sensory receptors and adequate primary sensory input to the thalamus. *Third,* whenever possible, human neurosurgical procedures expected to result in functional deficit should be preceded by attempts to establish the preconditions for functional transfer, such as by extensive sensory–sensory training procedures, and should be carried out in multiple stages.

Finally, we suggest that it may not be too ambitious to propose a radical reorientation in our attitude toward those who have suffered catastrophic or congenital bilateral damage in a single stage. Had that damage occurred in multiple stages, with the appropriate events separating the stages, no functional deficit might have ensued. Thus, the brains of such patients contain structures which have the potential capability of mediating the functions that were lost. But these structures have not been exposed to the conditions required to transform their potential capacity into actual function. Systematic study of the parameters influencing recovery after multiple stage lesions, investigation of the neurophysiological processes mediating the function during the period between the lesions and after imposition of the final lesion, and exploration of the possible ways to influence those processes with pharmacological substances and conditioning procedures should provide a far clearer picture than now exists about biologically fixed limitations on remediation. In the meantime, it seems obvious that a far more favorable climate for such systematic studies would arise if neuroscientists candidly admitted that the problem of functional localization has not been resolved.

A. Hippocampus

In spite of the failure of ablation attempts as catalogued by Lashley, there has been a continuing suggestion that certain regions of the brain are importantly involved in memory, particularly the hippocampus and the frontal cortex. A number of experiments, particularly the dramatic observations of Milner and Penfield (1955), have directed attention to the possible role of the hippocampus in the "stamping in" of experience. This proposition seems to receive support from such studies as those of Bureš (Bureš, Burešova, & Weiss, 1960), which show that hippocampal spreading depression can block retention of avoidance learning, and Hunt and Diamond (1957), who reported that bilateral hippocampectomy interferes differentially with performance of avoidance responses to visual and auditory cues and that the effects disappear with overtraining. Yet Grastyán and Karmós (1962) have shown that bilateral removal of the hippocampus in cats does not interfere with the ability to acquire either alimentary or defensive conditioned reflexes. Flynn and Wasman (1960) demonstrated that a defensive reflex could be established during bilateral after discharge of the hippocampus following electrical stimulation. A number of investigators have concluded that various behavioral deficits observed after hippocampal disturbance are not due to recent memory loss but can be attributed to discrimination failure (Cordeau & Mahut, 1964), motivational changes (Grossman & Mountford, 1964), complexity of task (Drachman & Ommaya, 1964), or inability to alter previously established behaviors (Webster & Voneida, 1964). These inconsistencies indicate the necessity of using a variety of response measures and methods of intervention in attempts to assess the anatomic localization of memory processes.

B. Frontal Cortex

Since the work of Jacobsen (1936), there is a long history of belief that the frontal lobe is important for short-term or very recent memory. Recent evidence seems to indicate that the impairment in delayed response observed after frontal resection is not due to a memory defect. Pribram and Tubbs (1967) have shown that the deficiency vanishes if a longer interval is imposed after each pair of trials. Since the performance is restored in spite of the lengthened posttrial period, the defect after frontal lobe injury seems related to the ability to divide a stream of information into its proper segments. After studying the defects observed in man with frontal lesions, Teuber (1959, 1964) concluded that there were no deficits in recent memory.

IX. LATERALIZATION: THE SPLIT BRAIN

A great deal of interest has been elicited recently by the studies of interhemispheric transfer of information in the so-called "split-brain" preparation. This preparation has been extensively studied by Sperry and his pupils (Gazzaniga, 1970). Initial results suggested that learning acquired by one-half of the brain was somehow localized to the structures on that side and not accessible to structures on the other side of the brain. Sperry has summarized the more recent results obtained with the split-brain preparation (1962, 1964). Cats trained with one eye masked were unable to remember with the second eye what they had learned with the first following section of both the corpus callosum and the optic chiasm. The second eye could be trained to do the opposite task with no interference. In numerous studies by various workers, it was found that section of the cerebral commissures prevented the spread of learning and memory from one to the other hemisphere.

Subsequently, extensive studies were carried out on human patients who had undergone surgical separation of the forebrain commissures, including the anterior commissure and the corpus callosum for treatment of neurological disease such as intractable epileptic seizures. The two separated hemispheres in such patients seem to have no functional communication, but each hemisphere is capable of performing certain functions. The fact that these functions are distinctly different establishes that there is marked lateralization of function in man (Bogen & Gazzaniga, 1965; Gazzaniga, 1965). Such patients can describe in words what they see when a picture is presented to the left speech-dominant hemisphere via the right visual field. When the picture is presented via the left visual field, it cannot be identified, presumably because the left hemisphere fails to receive any information. (Interestingly, if the operation is performed early in life, some recovery of normal function occurs, suggesting that a substantial part of lateralization is the result of learning.)

Conversely, spatial perception appears to be better mediated by the right hemisphere, which is nonspeech dominant. Commissurectomy patients recognize spatial information far more efficiently when it is presented to the right hemisphere. For example, objects which have been tactilely explored can be visually selected, or vice versa, very rapidly when the identification is made by the nonspeech-dominant hemisphere, whereas such identification by the speech dominant hemisphere depends upon a slow, tedious verbal attempt to analyze the features of the stimulus (Gazzaniga, 1970; Sperry, 1969).

Evoked potential studies using verbal and spatial stimuli in human commissurectomy patients have provided results consistent with these observations (Gott et al., 1974). Left-hemisphere EPs were larger than right-hemisphere EPs in response to verbal stimuli, while right-hemisphere were larger than left-hemisphere EPs in response to spatial stimuli. Recently, Davis and Wada (1974) have argued that this configuration of observations suggests that the speech-dominant hemisphere possesses superior auditory perception capability, while the non-speech-dominant hemisphere possesses superior visual perception capability, and that this is inherent in the sensory processes carried out in the two hemispheres when normally interconnected. They explored this proposal by studying the coherence of frequency analyses of visual and auditory averaged evoked responses recorded from left and right occipital (visual) and temporal (auditory) scalp electrodes in human subjects. They found significantly greater coherence on the left, speech-dominant hemisphere for responses to clicks, while the right, "spatially-specialized" hemisphere showed significantly greater coherence for responses to flash. Thus, significant interhemispheric differences seem to be present in the adult for simple unstructured nonverbal and nonspatial stimuli in the visual and auditory modalities. These differences in coherence may represent the integrative and associative processes which coordinate sensory and association areas of the cortex into representational systems, and may be important in the processing of visual and auditory (or spatial and verbal) information.

X. LOCALIZATION OF SPEECH FUNCTION

On the basis of the kind of considerations which have been presented, many workers have concluded that memory is not localized to particular brain regions, while at the same time asserting that the functions of speech must be considered an exception and are probably localized. For example, Gerard, writing in 1961, stated that it was highly doubtful that each remembered item was "located at a particular neuron or synapse. . . . Some localization is . . . present . . . as shown by aphasic defects. . . . But even these are hardly cell by cell [p. 27]." Ojemann (1966), after reviewing the deficits in function observed in man after damage to various localized brain regions, concluded that there was little evidence for the localization of memory in the brain except for the fact that speech defects were observed with local lesions. McCleary and Moore (1965) state that in the absence of sensory and motor loss, behavioral deficits in animals following various cortical lesions are partial, sensitive to the amount of cortex removed, generalized in nature, and frequently reversible with time. Except for deficits in speech after brain damage, they believe that the same conclusions hold in man. A basically similar position is voiced by Reitan (1964), who concluded that unknown factors wash out the differences in the effects of lesions in man which vary in type and location.

Speech functions obviously depend on a variety of kinds of memory. In view of the widespread belief that deficits in speech function after brain damage are evidence for the localization of memory, it is worthwhile to take a more careful look at the data bearing on this point. The work of Broca and Wernicke, cited earlier, provided the original basis for the belief that speech functions were localized in particular regions of the temporal lobe of the left hemisphere. The recent studies most frequently cited as providing support for this viewpoint are those by Penfield and Roberts (1959). (Although the split-brain studies cited in the previous section provide evidence of substantial *lateralization* of language function, it must be pointed out that such lateralization is not complete and localization of language function *within* the hemisphere has not been investigated in such studies.)

XI. EFFECTS OF CORTICAL STIMULATION

Penfield has extensively studied the effects of stimulation of various regions of the exposed cortex in man in the course of operations for neurological disorders. Stimulation of primary motor areas causes movement of various peripheral muscles. Excision of these areas does not block voluntary movement subsequently. Stimulation of primary sensory areas produces contralateral sensation. Excision of these areas produces a defect in sensation. Yet the work of Sprague, cited earlier, suggests that these defects may be due analogously to the release of subcortical inhibition. Stimulation of secondary sensory areas causes ipsilateral sensation. Excision of such areas produces little gross defect, although definite albeit subtle symptoms do arise (see Luria, 1966). In 1909 Harvey Cushing showed that stimulation of the postcentral gyrus in man produced sensation in the opposite limbs. Since then, it has been shown by many neurosurgeons that electrical stimulation of other primary sensory and secondary areas causes patients to see, hear, smell, feel, or taste in an elementary way. Patients react to such sensations as imposed by external action and consider the sensation as an artifact.

In 10 cases out of 190 in which Penfield and his colleagues carried out cortical stimulation over a 9-year period, psychical responses to stimulation consisting of experimental illusions, interpreted illusions, or dreamy states were obtained. Each of these ten cases involved stimulation of the temporal lobe. The resulting experiences were recognized as authentic events from the patient's past. Penfield denoted the temporal cortex involved in such experience as the interpretive cortex to distinguish this region from sensory and motor areas and those areas which give no response to stimulation, such as the anterior frontal and posterior parietal regions. The stimulation of interpretive cortex causes an experience described as "a flashback." Similar phenomena appear during epileptic seizures caused by spontaneous discharge in the corresponding temporal regions. Penfield did not conclude that the recording mechanism was actually in the stimulated tissue. Rather, he assumed that the neuronal activity left the area where the electrode was applied and activated the record in some more distant area. He distinguished between experiential responses consisting of specific flashbacks, and interpretive responses consisting of more generalized feelings such as familiarity, novelty, distance, intensity, loneliness, or fear. The flashbacks can be vivid sight and sound in personal interpretations. The person always feels that this is the evocation of a memory rather than a real experience. Experiences evoked in this fashion go forward moment by moment. No still pictures occur, and there is no backward sequence of events reported. Experiences are of a commonplace type, for the most part. There is no crossing between different periods in an individual's past. Experience stops when the electrode is withdrawn, and sometimes it is possible to reactivate the same experience repeatedly beginning at the same point when stimulation begins.

Cortical areas where stimulation caused ideation or speech project to the pulvinar and nucleus lateralis posterior of the thalamus and via centre medianum to Broca's area. Recovery of speech function following cortical lesion and aphasia suggests that the thalamic areas can be used for the ideational mechanisms of speech with the assistance of previously unemployed cortical zones, coordinated with the centrencephalic system. Penfield assumes a central integrating "centrencephalic system" within the diencephalon and mesencephalon, which has bilateral functional connections with the cerebral hemispheres, responsible for integration of the function of the hemispheres and integration of functions between different parts of the same hemisphere. Consciousness accompanies this integrative function and disappears with interruption of function in the centrencephalic system. Penfield considers the mesencephalic reticular formation as part of this system and cites the loss of consciousness after mesencephalic reticular lesion as support for his thesis. However, it must be pointed out that Adametz (1959) showed that the mesencephalic reticular formation can be completely coagulated in multiple stages with no subsequent loss of consciousness. Any area of cortex can be removed without the loss of consciousness.

The lateral temporal cortex is the only brain region from which actual earlier remembered experience can be reinvoked by direct stimulation, *and this only in patients with seizures and probable alteration of function of this part of the brain.* Thus, it appears that a memory can be released by an electrical stimulation to a particular place on the temporal cortex of man. Animal studies show that a conditioned response can be established to local electrical stimulation of the brain and stimulation of the region where the conditioned stimulus has been applied gives access to the memory, the criterion being that the conditioned response is produced (Leiman, 1962). (See Chapter 11 for an extensive discussion of such work.) However, it has been shown that transfer to other regions occurs readily, *and if the primary stimulated region is removed,* conditioned response can nonetheless occur to the stimulation of other regions (Segundo, Roig, & Sommer-Smith, 1959). Such evidence would indicate that the site of stimulation eliciting a response is not necessarily the site of the memory.

XII. EVIDENCE AGAINST STRICT LOCALIZATION OF LANGUAGE FUNCTION

Some insight on contemporary thinking about the localization of the speech function is afforded by studying the transcript of a conference on aphasia (Osgood & Miron, 1963). Most participants in this conference agreed that lesions in specific localities produced definite clinical types of aphasia, yet evidence was presented of the recovery from aphasia after brain surgery. It was proposed that to show the *essential* function of a brain region one would have to demonstrate *first,* that there was no performance after injury, *second,* that no spontaneous recovery took place, and *third,* that no relearning was possible. Evidence was presented that after temporal lobe damage resulting in aphasia, retraining was often possible. In bilinguals, one language frequently returns before the other. The general feeling was expressed that words or concepts are not stored in local regions in the sense of occupying specific locations in the cortex. No electrical stimulation carried out by any of the participants had produced organized speech. Stimulation of the cortex caused vowel-like cries or arrest of speech or other distortions. Such effects were obtained from Broca's area and primary and secondary motor areas on the left hemisphere. Stimulation on the right side was usually ineffective. The vocalization could be produced on either side. The general conclusion reached by the participants in the symposium was that language defects were more probable after damage to the left hemisphere than to the right, consistent with subsequent findings in commissurectomy patients. The more posterior the lesion, the more likely was the development of receptive language difficulties. The more anterior the lesion, the more likely was the development of expressive aphasia. The participants seemed reluctant to go beyond this in allocating localization of speech functions and memories involved in such functions to particular regions of the brain.

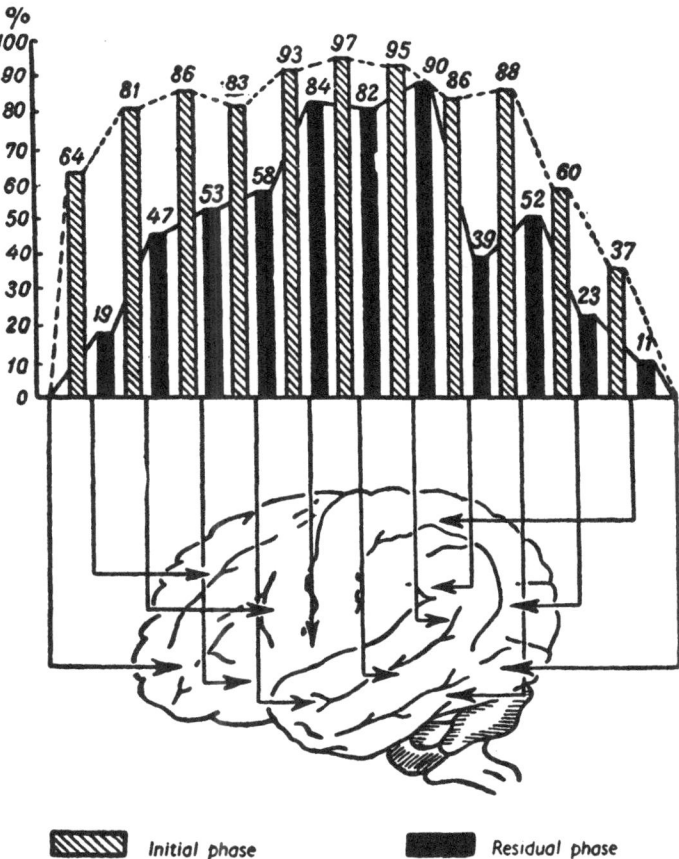

FIG. 9.4 Incidence of speech disorders after damage to different parts of the left hemisphere. Cross-hatched bars show the percentage of cases in which deficits were observed immediately after injury, while solid bars show the percentage of cases in which a long-lasting residual deficit ensued (Luria, 1966).

Evidence has been provided by Penfield that Broca's area can be destroyed with essentially complete recovery from the resulting aphasia, and other workers have provided supporting evidence. There is a case of bilateral destruction of Broca's convolution with retained ability to speak. Roberts (Penfield & Roberts, 1959) has stated that any acute lesion to any gross part of the left hemisphere may produce some disturbance in speech. Similar conclusions are provided by the results of Luria (1966), who has mapped the incidence of speech disorders after wounds to different parts of the brain (FIG. 9.4).

Luria's findings indicate clearly that damage to any convolution of the left hemisphere, from the anterior to the posterior boundaries of the brain, can result in speech defects. Analyzing evidence of this sort, Luria has pointed out that, in order to develop an adequate theory of localization, it is necessary first to revise our concepts of function, and second to reject the idea of centers as cell groups which are responsible for mental processes. Luria suggests that function is the product of a reflex system uniting excited and inhibited areas of the central nervous system into a working mosaic, analyzing and integrating afferent input, and establishing a system of temporary connections which for the moment achieves equilibrium of the organism in its environment.

XIII. FUNCTIONAL PLURIPOTENTIALISM: GRADED LOCALIZATION

Function conceived this way is localized in a network of complex dynamic structures composed of combinations of mosaics of distant points which are united in a common task. Function, particularly higher function, refers to the complex adaptive activity of an organism which is directed toward the accomplishment of some task. Such complex functions are multistage and are mediated by functional systems rather than centers. These functions consist of a set of interconnected steps and are mediated by a complex dynamic constellation of relationships involving different levels of the nervous system. Such a system of functionally united components has a systematic not a concrete structure in which the initial and final links of the system (task and effect) remain stable, and the intermediate system (means of performance of the task) may vary within wide limits. These systems are complex in composition, plastic in their elements, and dynamically self-regulating. Localization of such functions in any circumscribed area is out of the question. Function is accomplished by a plastic system which achieves an adaptive task with a highly differentiated group of interchangeable elements.

Elaborating such ideas, Luria refers to the concept of functional *pluripotentialism* enunciated by Filimonov (1951, 1957). This concept suggests that no formation of the central nervous system is responsible for only one function. A given region may participate in multiple functional systems which mediate the performance of numerous tasks. Filimonov proposed the concept of *graded localization*, by which he meant that functions are mediated by complex systems at multiple levels. Loss or damage at any level leads to reorganization because of feedback and to the restoration of the disturbed act. These notions are related to ideas of dynamic localization expressed by Ukhtomski (1945), who proposed that functions are localized in dynamic systems whose elements are strictly differentiated. Coordination in timing and speed of action creates a momentary functionally unified center from spatially differentiated groups. Luria has used such concepts to explain the variability of the motor point, the achievement of sensory responses from anterior stimulation, and the fact that extirpation does not cause the loss of functions which are elicited by stimulation.

XIV. COMPENSATION BY REORGANIZATION OF FUNCTION

Response elicited by stimulation at a point depends on what has happened before, that is, on the state of the system. Particular areas of the brain are not fixed centers but points in dynamic systems. It is often observed that the total lesion of a cortical center leads to the initial loss of a function that is gradually recovered. If a local cerebral lesion is subsequently inflicted, it is unlikely to cause a secondary loss of the restored function. The recovery of function is not due to the establishment of vicarious performance by new equipotential centers, but to the reorganization of a new dynamic system dispersed in the cortex and lower regions. Thus, the restoration of function is conceived of as reorganization rather than the transfer to equipotential areas. Local brain lesion is hardly ever accompanied by total loss of function, but usually by disorganization and abnormal performance. Particular functions may be disturbed by very different lesions, and a local lesion may lead to the disturbance of a complex of very different functions. The localization of processes such as perception or memory in discrete areas of the brain seems even less likely than the localization of respiration or movement in an isolated brain area. The localization of higher functions seems less plausible than biological functions.

Luria conceives of the wide dynamic representation of experience via synchronously working cells that excite one another mutually. The material basis of higher processes is the activity of the brain as a whole; that is, the brain is a highly differentiated system whose parts are responsible for the different features of the unified whole. In this view, the character of the cortical–intercentral relationship does not remain the same at different stages of development of the function, and the effect of a lesion of a particular part of the brain will differ at different stages of functional development. Some of these ideas are similar to the views of physiological organization upon which speech depends, as expressed by Lord Brain (1965).

XV. THE LOCALIZATION OF ELECTROPHYSIOLOGICAL CHANGES DURING LEARNING

Unique insights into the problem of localization of memory can be achieved by the use of chronically implanted electrodes for study of the electrical activity of the brain in unrestrained animals acquiring and performing learned behavior. In contrast with the static nature of the information obtained by inquiring whether a lesion of a particular anatomic structure of pathway does or does not interfere with acquisition or performance of a specific learned response, such electrophysiological observations provide a picture of the dynamic interactions between numerous brain regions in the mediation of learning and memory.

A. Tracer Technique

A major problem in such electrophysiological studies has been the difficulty in distinguishing between that part of the electrical activity of the brain which reflects the processing of information relevant to the learned behavior (which we will define as "signal") and the other ongoing business of the brain (which we might call "noise"). Since its introduction in the Soviet Union by Livanov and Poliakov in 1945, *tracer technique* has been the most useful method for enhancing the "signal-to-noise" ratio in such studies. Like many crucial methodological innovations, the basic idea of tracer technique is very simple and straightforward. The signal or "tracer conditioned stimulus" (TCS) for the learned behavior to be studied is presented intermittently at a characteristic rate of repetition. The electrical manifestations of the brain mechanisms responsible for processing information about the TCS are expected to appear with reasonable reliability each time this repetitive TCS is presented. In the activity recorded from various brain regions, electrical rhythms which appear at the frequency of the TCS when it is presented are considered to be *"labeled responses"* arising from such processing. These labeled rhythms were discussed briefly in Chapter 7.

Appearance of labeled responses in the activity recorded from a particular brain region is unequivocal evidence that the region is influenced by information about the TCS. Since transformations of this information undoubtedly occur in the brain, regions which fail to show labeled responses may nonetheless be involved in processing the information. Thus, the appearance of labeled responses constitutes sufficient but not necessary evidence that a particular brain region participates in the mediation of a learned response to a TCS. Further, it must be kept in mind that the labeled electrical activity is *not* the actual information which is being processed, but is merely an electrical indication that some neural activity in that region regularly occurs whenever the TCS is presented (John, 1971; John & Killam, 1959). In spite of these reservations, tracer technique provided important information about dynamic processes in the brain during learning and subsequent performance of new behavioral responses to environmental stimuli. Such information is the cornerstone of the picture of the dynamic process of remembering which will be presented in the next chapter. The present discussion is limited to the question of the anatomic localization of memory.

B. Many Brain Regions Participate in Learned Behaviors

An enormous body of evidence shows unequivocally that during learning, widespread changes occur in the distribution of responses evoked in the brain by the conditioned stimulus, as well as the fact that the details of such responses change drastically as stimuli acquire informational significance. Numerous detailed reviews of such evidence have appeared (John, 1961, 1967a, 1971; Morrell, 1961a; Thompson et al., 1972). This evidence can be divided into two categories which complement each other. The first type is concerned with phenomena related to electrical waves, such as labeled responses, while the second type consists of observations of actual neuronal firing patterns. The salient features of these two kinds of evidence are in agreement.

In Livanov and Poliakov's original study, subsequently confirmed by John and Killam (1959), it was noted with surprise that a number of brain regions that showed markedly different electrical responses to presentation of a TCS before conditioning acquired striking similarities in electrical activity during acquisition of a new behavioral response to that stimulus and during subsequent performance. Other workers have since commented upon the same phenomenon (Dumenko, 1967; Galambos & Sheatz, 1962; Glivenko, Korol'kova, & Kuznetsova, 1962; John & Killam, 1959, 1960; John, Ruchkin, & Villegas, 1963, 1964; Knipst, 1967; Korol'kova & Shvets, 1967; Liberson & Ellen, 1960; Livanov, 1962, 1965; Yoshii, Pruvot, & Gastaut, 1957).

These initial observations aroused some controversy because the labeled responses were within the range of electrical frequencies sometimes observed in the spontaneous electrical rhythms of the brain. Although quantitative measurements using electronic frequency analyzers and a variety of control experiments tended to blunt such objections, the issue was not settled decisively until the introduction of the technique of *average response computation*.

C. Average Response Computation

In essence, average response computation is a method for extraction of an estimate of the prototypic electrophysiological response to a specific stimulus from the ongoing electrical activity of the brain. The method assumes that presentation of a specific stimulus is followed by a series of oscillations in the electrical activity of responsive brain regions. These oscillations reflect the arrival of an afferent barrage upon the dendrites of the neurons in that region. The integrated postsynaptic potentials arising in those neurons are due to that input information and the efferent output discharges are caused by that input and the ensuing transactions within that region and between it and other regions. These "phase-locked oscillations" are superimposed upon the ongoing activity reflecting the other transactions in which those neurons are involved.

Since such other unrelated business (noise) is not reproducibly related to the time of presentation of the stimulus, while the response evoked by the stimulus (signal) necessarily is time-locked to the moment at which it occurred, it is possible to obtain an enormous enhancement of the signal-to-noise ratio by merely summating segments of the electrical activity recorded after successive presentations of the same stimulus. The ongoing noise activity in such summated segments tends to average out to zero, since it is in random time relationship to the arbitrary stimulus presentation. The signal-evoked activity, in reproducible time relationship to when the stimulus occurred, yields a nonzero average which is the "typical" evoked response. In this way, not only the amount but the actual waveshape of the electrical response to a particular stimulus can be precisely specified. The improvement in signal-to-noise ratio, that is, visibility of that part of the electrical activity due to the stimulus, is proportional to the *square root* of the number of stimulus repetitions which were averaged. Further details of the averaging technique, as well as its shortcomings, are discussed in Volume 2, Chapter 3 (Principles of Computer Analysis). Diagnostic applications of the average evoked response are discussed in Volume 2, Chapter 2.

D. Increase Similarity of Average Responses in Different Brain Regions

By using the method of average response computation, it was possible to demonstrate unequivocally that during learning, not only do many brain regions which were previously unresponsive to the TCS come to display response, but the detailed features of such evoked responses become markedly similar in different brain regions. This phenomenon is illustrated in Fig. 9.5. The two halves of Fig. 9.5 illustrate the activity caused by presentation of a flicker signal to an unrestrained cat. The left half of the figure shows the rhythmic labeled responses caused by the TCS, while the right half shows averaged evoked responses. For the two kinds of data, the figure illustrates the changes that occur as a meaningless environmental event acquires cue value. In both cases, the initial control data were obtained from a cat performing a previously learned lever pressing response to obtain milk *ad libitum* and the TCS was an extraneous environmental event with no informational relevance. When the TCS was made relevant, either as a cue that food was not available (left) or that electric shock was imminent (right), the response of many brain regions to the TCS became markedly enhanced and many different regions displayed basically similar electrical responses (John, 1972).

Inspection of the waveforms in Fig. 9.5 makes it obvious that the presence or absence and the detailed morphology of an evoked response in a particular brain region is not uniquely determined by the anatomic connectivity of the region or the physical attributes of the stimulus, but also reflects the informational relevance of the signal. These disparate factors both contribute to the electrical response of the brain to a stimulus, and are discussed in greater detail in the next chapter.

E. Widespread Neuronal Involvement in Learned Behaviors

The appearance of labeled rhythms and of evoked responses of similar wave-shape in many different anatomic regions suggests that neurons in widespread regions of the brain are involved in the mediation of learned behaviors. This inference can be drawn from the correlation between spontaneous or evoked activity on the one hand and neural discharge on the other, which has been amply documented (Calvet, Calvet, & Scherrer, 1964; Creutzfeldt *et al.*, 1966a, b; Fox & O'Brien, 1965; Fromm & Bond, 1967; Frost & Gol, 1966; Fujita & Sato, 1964; Gerstein, 1961; Green *et al.*, 1960; Kandel & Spencer, 1961; Klee & Offenloch, 1964; Klee, Offenloch, & Tigges, 1965; Pollen & Sie, 1964; Purpura & Shofer, 1964; Robertson, 1965; Stefanis, 1963; Vasilevs, 1965).

Thus, one would expect that changes in single unit activity during conditioning should be very marked and should occur in many regions. That such is indeed the case can be seen from studies of changes in activity of single neurons during conditioning, which consistently report that a large proportion of the cells observed (10–60%) change their response during conditioning, either altering their quantitative response to the CS or becoming functionally multimodal (Adam, Adey, & Porter, 166; Buchwald, Halas, & Schramm, 1965; Bureš, 1965; Bureš & Burešova, 1965, 1967, 1970; Hori & Yoshii, 1965; Jasper, Ricci, & Doane, 1960; Kamikawa, McIlwain, & Adey, 1964; Morrell, 1961b, 1967; O'Brien & Fox, 1969a, b; Olds & Hirano, 1969; Olds & Olds, 1961; Travis, Houten, & Sparks, 1968; Travis & Sparks, 1967, 1968; Travis, Sparks, & Hooten, 1968; Woody, Vassilevsky, & Engel, 1970; Yoshii & Ogura, 1960). Many of these studies have been reviewed in detail by Leiman and Christian (1973). The fact that such a large percentage of neurons in so many different anatomic regions alters behavior during the most simple conditioning procedure consti-tutes an irrefutable argument against connectionistic theories of learning, which localize the memory in synaptic alteration manifesting itself as facilitation of neural firing. If firing per se represented information about prior experience such representation would be destroyed or made hopelessly ambiguous a few experi-ences later, since so many cells alter their responsivity in a single effective learning session. How can memories be protected and isolated in such a system?

In a series of recent studies (Olds, Segal, Hirsch, Disterhoft, & Kornblith, 1972), an extensive mapping endeavor has shown that such changes in neuronal

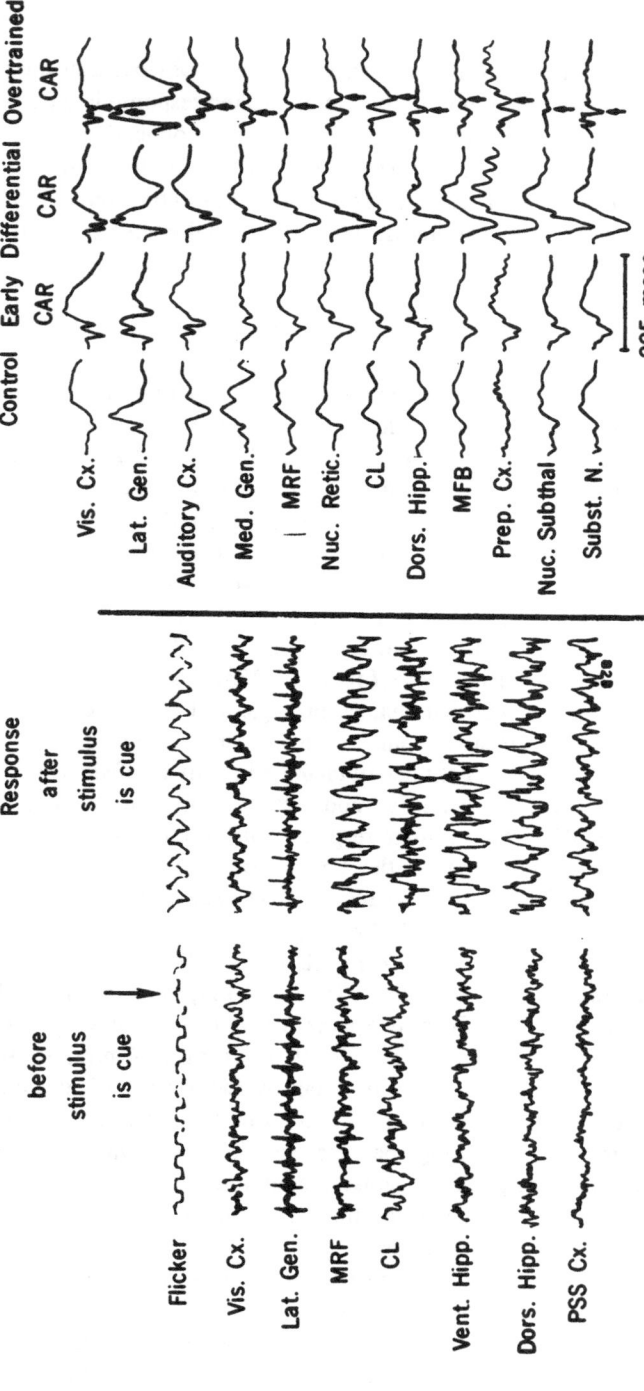

FIG. 9.5

226

response to a stimulus occur in most regions of the brain as conditioning proceeds. Using multiple chronically implanted movable microelectrodes, we have studied the characteristics of neuronal responses to discriminated stimuli (John & Morgades, 1969a–c). Two important observations resulted from these studies. First, we found that across an extensive anatomic domain traversed by the movable microelectrodes, both the waveshape of the average evoked response and the firing pattern of the neurons within the region (average post-stimulus histograms) were basically homogeneous. Gradients were constructed for each component of the evoked response and each peak in the firing pattern across the mapped region, and no evidence for localization was obtained. Typical gradients obtained from such mappings are illustrated in Fig. 9.6.

The second observation is illustrated in Fig. 9.7. At each point in the traverse, evoked responses and average neuronal firing patterns (poststimulus histograms) were computed to two discriminated signals differing in frequency and serving as cues for two different behaviors. An analysis of variance of the electrophysiological activity was then carried out comparing the variance in responses to the two different signals *within* the recordings obtained at a given electrode location to the variance in responses to the same signal *between* the recordings obtained across many electrode positions. The results were quite clear. At any location,

FIG. 9.5 (Left) Response before stimulus is cue shows the effect of photic (6-Hz flicker) stimulation on a cat after it had learned that milk could be obtained whenever a lever was pressed. The flicker had no signal value at this stage. Little "labeled" activity was elicited by the flicker except in the lateral geniculate. The response disappeared in the lateral geniculate because of the internal inhibition that occurred as the cat pressed the lever and waited for milk. Response after stimulus is cue shows records during presentation of flicker (6 Hz) after the cat had learned a frequency discrimination. The cat was responding correctly by not pressing the lever down. There is marked enhancement of labeled responses at the stimulus frequency (Flicker, 6-Hz photic stimulation; Vis. Cx., visual cortex; MRF, mesencephalic reticular formation; CL, nucleus centralis lateralis; Vent. Hipp., ventral hippocampus; Dors. Hipp., dorsal hippocampus; PSS Cx., posterior suprasylvian cortex). (From John, 1967a.)

(Right) Evolution of visual evoked responses. Control: average responses evoked in different brain regions of a naive cat by presentation of a novel flicker stimulus. Several regions show little or no response, and different regions display differing types of response. Early CAR: responses to the same stimulus shortly after elaboration of a simple conditioned avoidance response (CAR). A definite response with similar features can now be discerned in most regions. Differential CAR: changes in the response evoked by the flicker CS shortly after establishment of differential approach–avoidance responses to flicker at two different frequencies. As usual, discrimination training has greatly enhanced the response amplitude, and the similarity between responses in different structures has become more marked. Overtrained CAR: after many months of overtraining on the differentiation task, the waveshapes undergo further changes. The arrows point to a component usually absent or markedly smaller in behavioral trials on which this animal failed to perform (Nuc. Retic., nucleus reticularis; MFB, median forebrain bundle; Prep. Cx., prepyriform cortex; Nuc. Subthal., nucleus subthalamus; Subst. N., substantia nigra).

AMPLITUDE AND LATENCY GRADIENTS

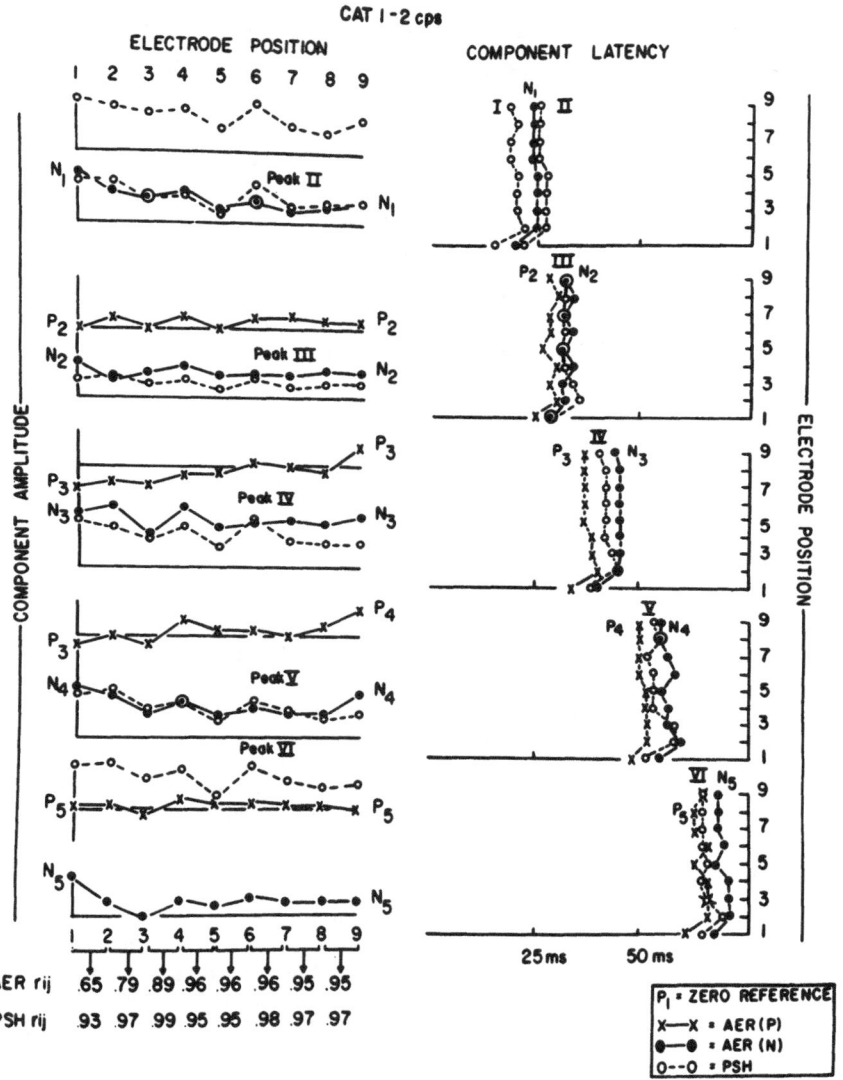

FIG. 9.6 Graphs on the left side illustrate the amplitude gradients of positive (P) and negative (N) AER components and PSH peaks (see key), computed for Cat-1, 2-Hz. Successively later components are depicted from top to bottom graph. In each graph, amplitude of the component is plotted as vertical displacement, while depth of electrode penetration is plotted as horizontal displacement. At bottom are presented the correlation coefficients rij between response waveshapes at adjacent electrode positions.

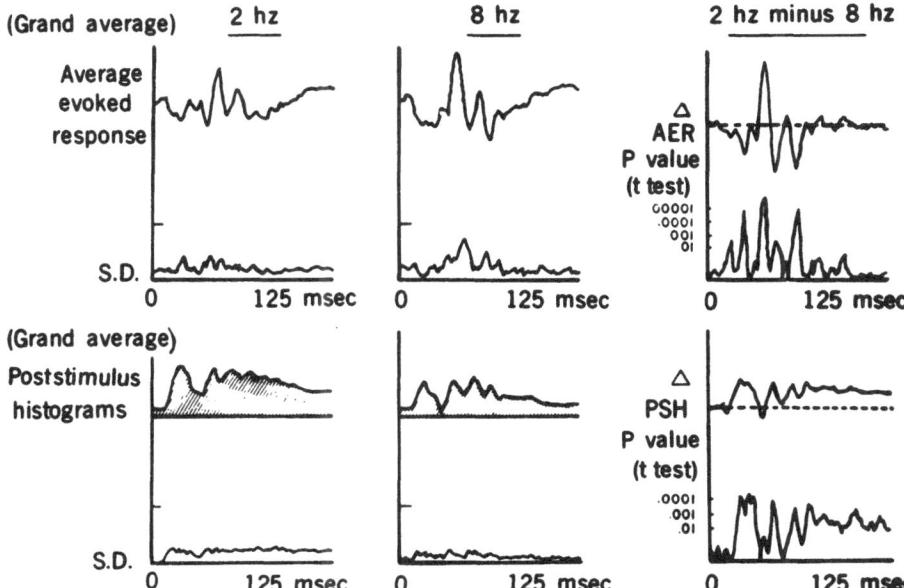

FIG. 9.7 (Top left) The top curve shows the grand average of the AERs elicited by the 2-Hz CS across all electrode positions in the mapped region, while the lower curve shows the standard deviation (S.D.) of the groups of AERs. (Bottom left) The top curve shows the grand average of the PSHs elicited by the 2-Hz CS across the same electrode positions and the lower curve shows the S.D. (Top center) The grand average of the AERs elicited by the 8-Hz CS and the corresponding S.D. (Bottom center) The grand average PSH elicited by the 8-Hz CS, and its S.D. (Top right) The top curve shows the difference waveshape resulting from the subtraction of the grand average AER elicited by the 8-Hz CS from the grand average AER elicited by the 2-Hz CS. The lower curve shows the *p* value as computed by the *t* test for each point of difference wave. (Bottom right) The top curve shows the difference waveshape resulting from the subtraction of the grand average PSH elicited by the 8-Hz CS from the grand average PSH elicited by the 2-Hz CS. The lower curve shows the *p* value for each point of the difference. (From John & Morgades, 1969b.)

the difference in response to the two different signals was greater than the variability in response to a particular signal observed at many different locations.

In other words, not only the waveshape of evoked responses but the firing pattern of neuronal groups or "ensembles" in different brain regions displayed a characteristic pattern whenever a particular learned signal was presented. This observation indicated that different anatomic regions were *qualitatively*

Each graph on the right side shows the latency distribution of the positive and negative AER component and the PSH peak whose amplitude gradients are found in the graph at the same level on the left side of the figure. Component latency is plotted along the abscissa, while depth of penetration is represented along the ordinate. (From John & Morgades, 1969a.)

homogeneous in their representation of a particular learned stimulus. At the same time, different patterns of response were elicited by signals with different significance. The data also suggested that the quantitative impact of a specific signal varied from region to region.

Thus, electrophysiological data strongly suggest that the response of many different anatomic regions of the brain to a signal is drastically altered by learning. If we ask where is *the* change brought about by learning, we must conclude that change occurs in most if not all of the brain regions responsive to the signal. The data suggest strongly that the function of memory, if defined as change in response to an event because of prior experience with that event, is broadly distributed in the brain. Nonetheless, it is possible that this widespread change in response to a familiar event is initiated by changes localized in some functional system capable of influencing the activity of many brain regions.

In the next chapter, we will examine further details of the process of remembering. Certain of those details provide a quantitative evaluation of the relative involvement of different brain regions in the mediation of memory. Consideration of that quantitative evaluation suggests a way to reconcile the localizationist and antilocalizationist theories of memory.

10
How Do We Remember?

I. EXOGENOUS AND ENDOGENOUS COMPONENTS OF EEG ACTIVITY

A. Assimilation of the Rhythm

In their initial studies using tracer technique, Livanov and Poliakov (1945) noted a phenomenon which they termed "assimilation." By this term they referred to the observation that when animals were being trained to perform some behavioral response to an intermittent sensory TCS, during intertrial intervals the spontaneous EEG became dominated by electrical rhythms at the frequency of the absent conditioned stimulus. This observation was confirmed by John and Killam (1959) and assimilation has since been observed in rat, rabbit, cat, dog, monkey, and man, in a wide variety of experimental situations (John, 1961, 1967a; Morrell, 1961a). Remarkable specificity of the frequency of assimilated rhythms in the firing patterns of single cells and small groups of cells in the cortex of unrestrained trained cats during intertrial intervals has recently been observed by Ramos and his associates (1974, 1976).

Analogous reproduction of the frequency of an intermittent stimulus in its absence has been observed in a variety of other circumstances. If intermittent stimuli are presented at regular intervals, frequency-specific repetitive responses will often be observed after the appropriate time interval even though the actual stimulus is withheld, suggesting that this electrical pattern represents a conditioned response established by the cyclic procedure (John & Killam, 1960; Yoshii & Hockaday, 1958). In so-called "cortical conditioning," a steady CS which has been repeatedly paired with an intermittent US comes to elicit a frequency-specific electrical rhythm before the US is presented (Morrell & Jasper, 1956; Yoshii et al., 1957). Accurate reproduction of the firing patterns characteristically elicited by syncopated as well as rhythmic stimuli has been demonstrated in the discharge of single neural units (Morrell, 1961b, 1967; Morrell, Engel, & Bouris, 1967; Yoshii & Ogura, 1960) in the absence of the conditioning stimuli.

B. Functional Significance of Assimilated Rhythms

As this evidence accumulated, there have been numerous attempts to assess the functional significance of this possible representational process. Soon after assimilated rhythms were first observed, it was reported that such rhythms were absent while the animal was in his home cage, but appeared as soon as he was placed in the familiar training environment (Yoshii & Ogura, 1960). It was also observed that spontaneous performance of the conditioned response was often preceded by the appearance of bursts of assimilated rhythms at the frequency of the absent TCS (Yoshii, 1962). In an ingenious extension of such observation, steady tone was associated with a flicker which had served as a TCS for an avoidance response. After such cortical conditioning, presentation of the tone often elicited a burst of EEG waves at the frequency of the absent flicker, with concomitant performance of the behavioral response learned to the flicker cue (Schuckman & Battersby, 1965). Thus, it seemed that presentation of the tone activated a neural system representing the flicker with which the tone had been associated, releasing the electrical rhythms appropriate to the actual presence of the flicker, which in turn induced performance of the behavior for which flicker was the learned cue. In a conceptually related experiment, Livanov and Korol'kova reported that performance of a conditioned response could be elicited reliably and with minimal latency if the motor cortex was stimulated electrically at the frequency of assimilated rhythms that had appeared there during training with a peripheral TCS (Livanov, 1965).

Another line of evidence suggested the functional significance of temporal patterns of electrical activity originating within the brain. As stated previously, many different brain regions come to display markedly similar waveshapes of evoked responses as well as frequencies of electrical rhythms during learning. However, we observed that when animals who had been highly overtrained in the performance of differential conditioned responses to discriminated stimuli at two different frequencies committed errors, certain brain regions often displayed electrical rhythms inappropriate to the actual stimulus, but corresponding to the frequency of the absent stimulus which would have been the appropriate signal for the observed behavior (John, 1963, 1967a, 1972; John & Killam, 1960). Similar findings have been reported by Lindsley, Carpenter, Killam, & Killam (1968). This phenomenon is illustrated in Fig. 10.1.

The similar electrical patterns observed in different brain regions of trained animals, then, come from two origins. One, which we have termed *exogenous*, reflects afferent input due to external events. The other, which we have termed

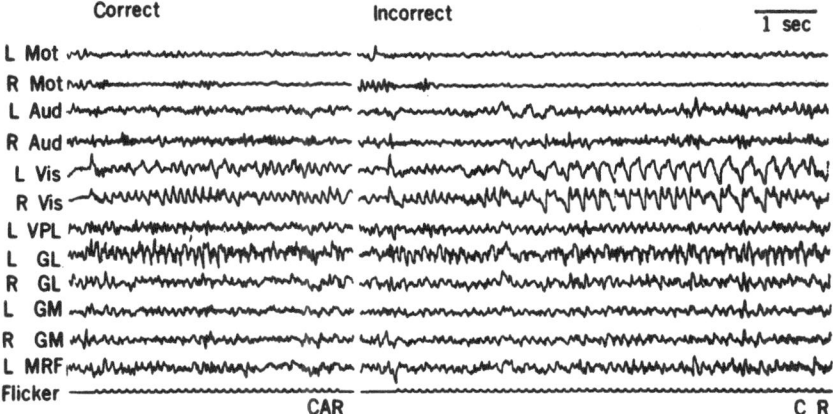

FIG. 10.1 Comparison of electrophysiological activity elicited by the same visual signal, interpreted in two different ways. Records on the left were obtained as the differentially trained cat responded correctly to a negative CS (7.7-Hz flicker) by performing a conditioned avoidance response. Records on the right were obtained on the next trial when the cat erroneously responded to the same flicker frequency by performing the conditioned approach response appropriate to the positive stimulus (3.1-Hz flicker). (Mot, motor cortex; Aud, auditory cortex; Vis, visual cortex; VPL, ventroposterolateral nucleus; GL, lateral geniculate nucleus; GM, medial geniculate nucleus; MRF, mesencephalic reticular formation.) All records bipolar. (From John, 1972. Copyright © 1972 by The American Association for the Advancement of Science.)

endogenous, reflects the release of electrical patterns from some internal representational system. The appearance of two different electrical patterns in the brain of an animal as it commits a behavioral error indicative of a misinterpretation of a signal suggests that the exogenous activity reflecting the actual stimulus somehow activated inappropriate endogenous activity reflecting the significance normally attributed to a different stimulus. A mismatch has occurred between stimulus input and retrieval from memory, or "readout."

C. Generalization

The release of endogenous activity is well illustrated during generalization when an animal performs a behavioral response previously learned to some repetitive TCS in response to a new test stimulus. Under such circumstances, behavioral generalization has been found to be regularly accompanied by the release of electrical rhythms appropriate to the TCS rather than to the test stimulus (John, 1963, 1967b; John & Killam, 1959; John, Leiman, & Sachs, 1961; Lindsley *et al.,* 1968; Majkowski, 1958, 1967; Thatcher, 1975; Thatcher & Cadell, 1969). Similar findings have been observed in the firing patterns of neurons (John & Morgades, 1969c).

Numerous controls have been provided which reassure that these released electrical patterns are not due to unspecific factors such as arousal, attention, orientation, movement, motivation, changes in receptor position, and so on, factors that might activate brain mechanisms associated with the production of particular electrical rhythms or waveshapes (John, Bartlett, Shimokochi, & Kleinman, 1973). A particularly elegant demonstration that these released rhythms do not reflect such unspecific factors has been provided by Majkowski (1967). This worker studied the electrical activity which occurred during training to an intermittent flicker CS and subsequent generalization to flicker at a different repetition rate, using a population of chronically implanted split-brain cats. During generalization to the test stimulus, inappropriate rhythms corresponding to the previously used CS rather than to the new stimulus were found only on the trained side of the split-brain cat. After differential training, the rhythms elicited by the test stimulus (now the differential stimulus) corresponded to the actual stimulus. Similarly, assimilation of the rhythm in intertrial intervals during the initial training was only observed in structures on the side of the split brain being trained. These results strongly support the suggestion that the rhythm of the released temporal patterns reflects the significance attributed to the release of specific memories rather than the unspecific activation of anatomic systems producing hypersynchronous rhythms.

II. EXOGENOUS AND ENDOGENOUS COMPONENTS OF EVOKED RESPONSE

A. New Components Appear during Learning

The release of temporal patterns of electrical activity reflecting the activation of specific memories can be more precisely explored using evoked potential (EP) methods. In numerous papers, it has been noted that as conditioning to a particular CS proceeds, a new late process with an onset latency of about 60 to 70 msec appears in the EP recorded from many brain structures (Asratyan, 1965; Begleiter & Platz, 1969b; Galambos & Sheatz, 1962; John, 1963, 1967a, b; John & Killam, 1959; John & Morgades, 1969b; Killam & Hance, 1965; Leiman, 1962; Lindsley et al., 1968; Sakhuilina & Merzhanova, 1966). Further, when the signal failed to elicit performance of the conditioned response, the late process which had appeared during learning failed to occur.

B. Readout to Absent but Expected Events

A body of evidence has accumulated which shows that certain aspects of the EP may reflect previous experience rather than responses to afferent input and are thus of endogenous rather than exogenous origin. One important line of such evidence comes from studies primarily carried out on human subjects and is particularly important in assessing the likelihood that these released electrical patterns actually correspond to the activation of specific memories, because it has been possible to establish unequivocally that there is a subjective correlate to the appearance of these released potentials. These studies show that when an *expected* event does not occur, a cerebral potential appears at a latency similar to that of potentials usually evoked by the expected stimulus. Evoked potentials elicited in man by absent but expected events have been reported (Barlow, Morrell, & Morrell, 1967; Klinke, Fruhstorfer, & Finkenzeller, 1968; Picton, Hillyard, & Galambos, 1973; Riggs & Whittle, 1967; Rusinov, 1959; Sutton, Tueting, Zubin, & John, 1967; Weinberg, Grey-Walter, Cooper, & Aldridge, 1974; Weinberg, Grey-Walter, & Crow, 1970). Similar findings in the cat were reported by John (1963). These cerebral events, termed *readout* or *emitted potentials,* have been interpreted by Weinberg *et al.* (1974) to reflect the generation of processes corresponding to the memory of past or imaginary stimuli.

C. Readout during Generalization

In our laboratory, we have concentrated upon studies of EP waveshapes during generalization in differentially trained cats. When generalization occurs upon presentation of a novel stimulus to such animals, the EP waveshape closely resembles the waveshape usually evoked by the CS used to train the animal. However, if generalization fails to occur, the new stimulus waveshape differs radically from the typical response to the CS. The same phenomenon has been observed in the firing pattern of neuronal ensembles during generalization (John, 1963; John & Morgades, 1969b; Ruchkin & John, 1966). This finding is illustrated in Fig. 10.2.

D. Difference Waveshapes

As can be observed in Fig. 10.2, certain late components were present in the EP when the novel stimulus elicited generalization but were absent if generalization failed to occur. By subtraction of the averaged EPs obtained during trials in which no behavioral response was elicited by presentations of a novel stimulus from averaged EPs computed during trials resulting in generalization, it was possible to construct the *difference waveshapes* which showed the form of the readout process released (John, Ruchkin, Leiman, Sachs, & Ahn, 1965). The result of this computation is shown in Fig. 10.3.

Examination of Fig. 10.3 shows that readout processes were found in many brain regions and displayed a general similarity of waveshape. Marked latency differences can be seen from structure to structure. The readout process seems to be generated in a corticoreticular system, from which it extends to involve other brain regions in a temporal sequence.

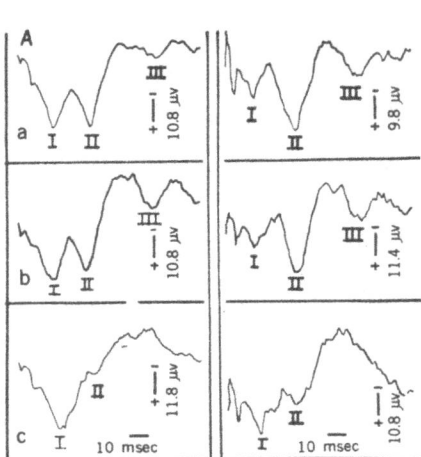
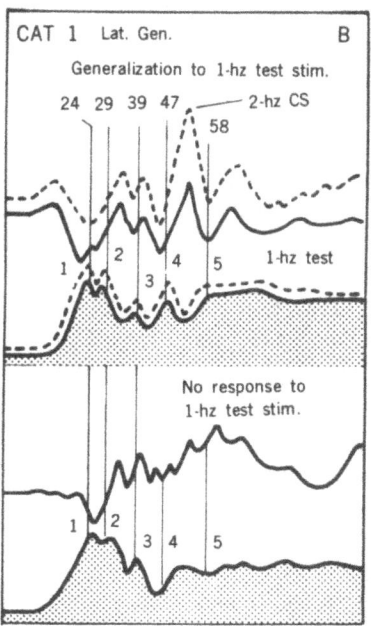

FIG. 10.2 (A) Computations of average responses obtained from the lateral geniculate nucleus and nucleus reticularis of the cat under various conditions during the same experimental session. First row of averages (a) is based upon 100 repetitions, second (b) and third (c) rows are based upon 42 repetitions of the same stimulus applied during a number of behavioral trials. Analysis epoch was 90 msec. (a) Average responses evoked in structures by the 10-Hz CS (flicker) actually used in training, during repeated correct behavioral performances. (b) Average responses evoked by a novel 7.7-Hz CS, during repeated generalization behavior. Test trials with the 7.7-Hz stimulus were interspersed among trials with the actual 10-Hz CS, and were never reinforced. (c) Average responses evoked by the 7.7-Hz flicker on presentations when no generalization behavior was elicited. The waveshape elicited by the actual CS is similar to the response evoked by the novel stimulus during generalization behavior. Notice the absence of the second positive component in the evoked potential when generalization behavior failed to occur. (From Ruchkin & John, 1966.) (B) (Top) Records of AERs and PHSs obtained during 18 trials that resulted in CR to the 2-Hz CS (dotted curves) and during 32 trials that resulted in behavioral generalization in response to a 1-Hz flicker used as a test stimulus (solid curves). The test stimuli were randomly interspersed between presentations of 2-Hz (dotted curves) and 8-Hz flickers in a long experimental session. (Bottom) Records of AERs and PSHs obtained during 17 trials that resulted in failure to elicit generalization behavior in response to the test stimulus. Note change in late components. Analysis epoch, 100 msec. (From John & Morgades, 1969b.)

236

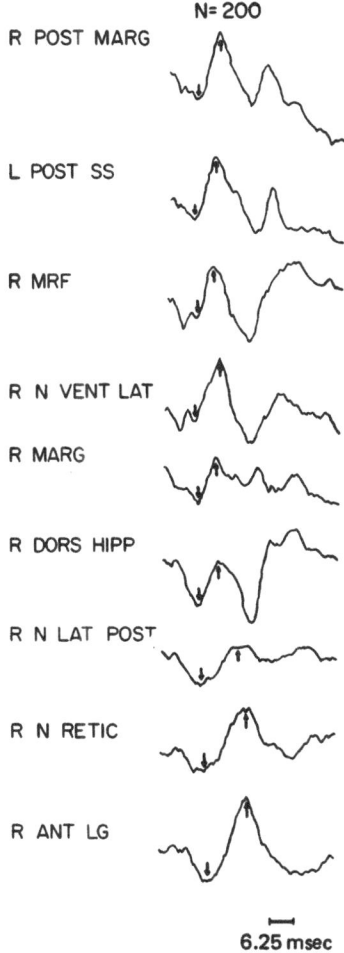

N= 200

R POST MARG

L POST SS

R MRF

R N VENT LAT

R MARG

R DORS HIPP

R N LAT POST

R N RETIC

R ANT LG

6.25 msec

↓ TIME OF FIRST DIFFERENCE COMPONENT
↑ TIME OF SECOND DIFFERENCE COMPONENT

FIG. 10.3 "Difference" waveshapes constructed by subtraction of averaged responses evoked by 7.7-Hz test stimulus during trials resulting in no behavioral performance from averaged responses evoked by the same stimulus when generalization occurred. Each of the original averages was based on 200 evoked potentials providing a sample from five behavioral trials. Analysis epoch was 62.5 msec. These difference waveshapes begin 10 msec after the stimulus. The onset and maximum of the difference wave has been marked by two arrows on each waveshape. The structures have been arranged from top to bottom in rank order with respect to latency of the difference wave. Note that the latency and shape of the initial component of the difference wave is extremely similar in the first four structures, and then appears progressively later in the remaining regions. (Post Marg, posterior marginal gyrus; Post SS, posterior suprasylvian gyrus; MRF, mesencephalic reticular formation; N Vent Lat, nucleus ventralis lateralis; Marg, marginal gyrus; Dors Hipp, dorsal hippocampus; N Lat Post, nucleus lateralis posterior; N Retic, nucleus reticularis; Ant Lg, anterior lateral geniculate; R, right side; L, left side. (From John, 1967.)

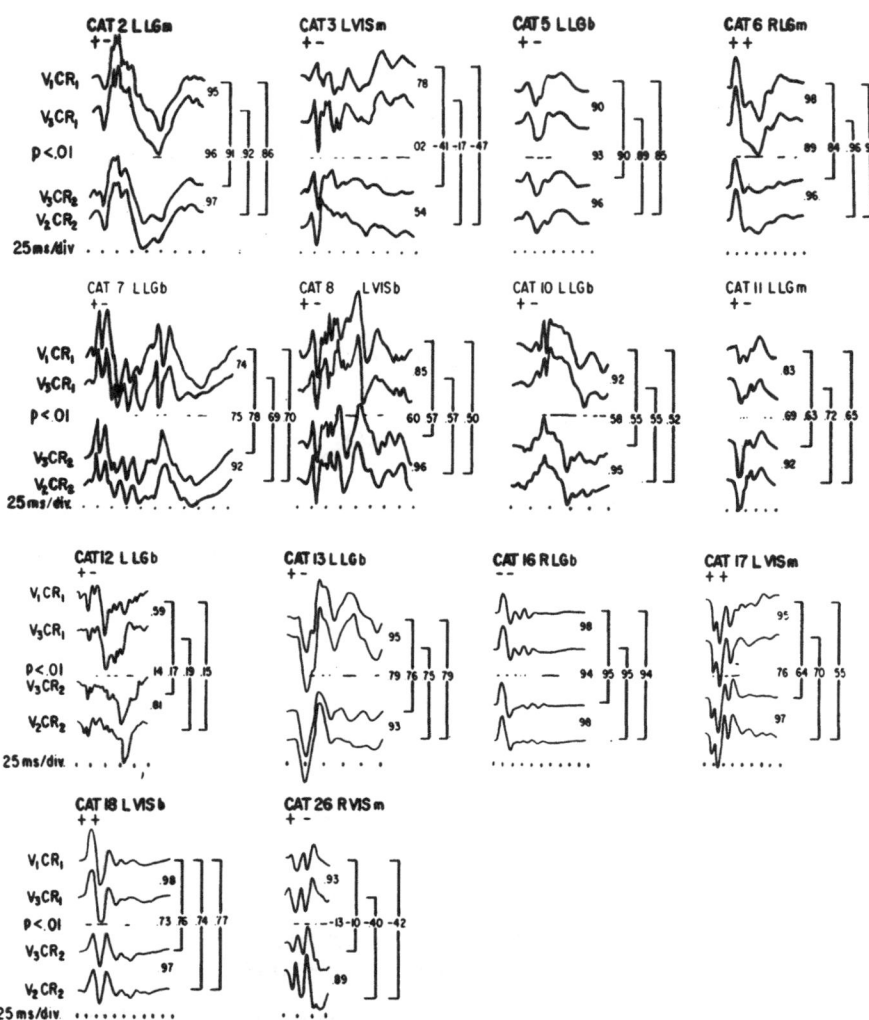

FIG. 10.4 Examples of similar waveshapes evoked by different visual stimuli when they elicit the same behavioral response, from 14 different trained cats. In each set of four waveshapes the top curve shows the waveshape elicited by flicker stimulus V_1 during trials resulting in correct performance of CR_1, and the bottom curve shows the response elicited by flicker stimulus V_2 during trials resulting in correct performance of CR_2. The second and third curves were elicited during differential generalization to flicker at a frequency, V_3, midway between V_1 and V_2. The second curve shows the evoked response waveshape during trials when V_3 elicited performance of CR_1, appropriate to V_1, while the third curve shows the waveshape evoked by V_3 during trials resulting in CR_2, appropriate to V_2. The intermittent line between the second and third curves in each set shows latencies at which the difference between data averaged in V_3CR_1 and in V_3CR_2 reached significance at better than the .01 level, assessed by the t test. Numbers to the right of each set of curves represent

E. Differential Generalization

In order to provide more adequate controls for possible contributions of unspecific factors to the "go–no go" alternatives of the procedure just described, we developed a technique which we called differential generalization. In this method, animals were conditioned to perform one behavioral response, CR_1, to a flicker stimulus at one frequency, V_1, and a different behavioral response, CR_2, to a second stimulus at another frequency, V_2. After thorough overtraining, a third stimulus, V_3, was occasionally introduced into random sequences of V_1 and V_2. The frequency of V_3 was midway between V_1 and V_2. Sometimes the cat treated V_3 as equivalent to V_1 and CR_1 was performed, while sometimes V_3 was judged to be like V_2 and CR_2 was elicited. It was found that the waveshapes elicited during $V_3 CR_1$ were significantly different from those evoked in $V_3 CR_2$ trials. Further, when V_3 presentation resulted in CR_1 performance, the average response evoked by V_3 closely resembled the usual response to V_1. Conversely, when V_3 presentation resulted in CR_2 performance, the potentials evoked by V_3 closely resembled those usually elicited by V_2 (John, Shimokochi, & Bartlett, 1969). These findings are illustrated in Fig. 10.4.

The results obtained in this go–go paradigm have established that previous findings were not merely attributable to gross differences in excitability between trials resulting in no behavioral performance as opposed to those in which a behavioral response occurred. It was also established that the same physical stimulus could elicit different evoked potential waveshapes in differential generalization whether CR_1 and CR_2 were both approach responses, both avoidance responses, or differential approach–avoidance responses. Differences observed in waveshape were, therefore, not restricted to behavioral responses with a particular motivation. These differences were not unique to particular types of instrumental behaviors, training situation, stimulus frequencies, or sensory modality,

---

the Pearson product moment correlation coefficients between the bracketed curves. Inspection of data shows that in every case, $V_3 CR_1$ closely resembles $V_1 CR_1$, while $V_3 CR_2$ closely resembles $V_2 CR_2$. In most cases this visual evaluation is corroborated by the values of the correlation coefficients. In some cases (e.g., cats 2, 5, 7) the correlation between $V_3 CR_1$ and $V_3 CR_2$ is higher than the correlation between $V_3 CR_1$ and $V_1 CR_1$, contradicting the impression given by visual evaluation of similarity. These examples illustrate the inadequacy of the correlation coefficients as a pattern–recognition procedure. EPs combined into these averaged waveshapes were selected using subjective evaluation of the experimenter in half the cases (cats 2, 3, 7, 8, 11, 12, 26) and using computer-sorting methods in the other half (cats 5, 6, 10, 13, 16, 17, 18). Sample sizes ranged from 6 to 30 EPs for the subjectively selected samples and from 25 to 200 for the computer-sorted data, and were composed from several behavioral trials in most cases. This variability was dictated by choices made by the animal in differential generalization, a factor beyond control of the experimenter. The cat number, recording derivation, and type of discrimination performed are indicated above each set of data. The signs + − signify approach–avoidance; + +, approach–approach; − −, avoidance–avoidance; subscript m denotes monopolar, b, bipolar derivation. Time scale, 25 msec/division. (From John et al., 1973.)

since a wide variety of such variables was explored. Thus, a number of the more obvious unspecific factors were ruled out which might contribute to differences in the waveshapes of potentials evoked by the same physical stimulus in trials resulting in different behavioral outcome.

In about two-thirds of the 96 electrode placements in the cats initially subjected to the differential generalization procedure, evidence of readout components was found. A film which illustrated the readout potential in a series of differential generalization trials was made and is available for demonstration purposes (John, Shimokochi, & Bartlett, 1969). Audiences, when asked to predict the cat's behavior on the basis of the evoked potential waveshapes, generally can do so. (This 40-minute teaching film has been shown on numerous occasions to both naive and sophisticated audiences.)

F. Differing Facets of the Same Experience

Similar results were obtained in differential generalization using auditory stimuli. Furthermore, in certain brain regions, the readout processes discernible in differential generalization tests conducted using either auditory or visual novel stimuli displayed similar features. Such findings showed that when similar behaviors are elicited by signals in two different sensory modalities, the readout processes in certain structures display common features, as illustrated in Fig. 10.5.

This observation would seem intuitively to be a neurophysiological correlate of the fact that the same memory can be elicited by stimuli in different sensory modalities. The recollection of an incident can be activated by a picture or by a

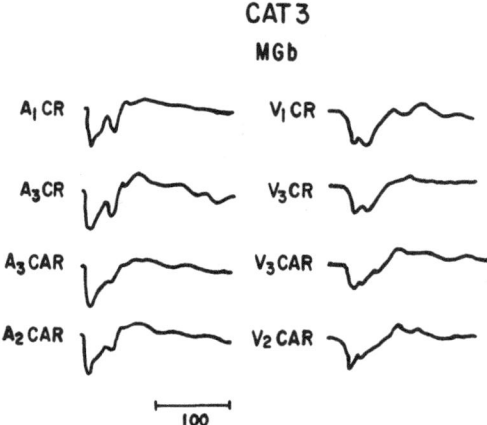

FIG. 10.5 Similar features of waveshapes evoked by visual and auditory stimuli that elicited the same behavioral response. Data from cat 3, medial geniculate, bipolar. (From John *et al.*, 1973.)

verbal statement. Necessarily, the neural representation of that incident must possess certain basic invariance no matter how it becomes activated. However, the same stimulus can elicit a variety of different waveshapes, or *response modes,* upon successive presentations, as seen in Fig. 10.6.

III. "MODES" OF RESPONSE

Examination of Fig. 10.6, showing that many different waveshapes can be elicited by the same physical stimulus, clearly illustrates that indiscriminate averaging of these many different processes can produce a seriously misleading result. The "average" evoked response not only may not correspond to *any* of the different types of response elicited by a particular stimulus, but failure to recognize this may obscure relationships between evoked response waveshapes and behavior.

Careful examination of the sequence of single evoked response waveshapes in many behavioral trials from individual cats convinced us not only that a wide variety of waveshapes appeared, but also that certain waveshapes reliably predicted the occurrence of particular subsequent behaviors. Yet, because these waveshapes with high predictive value might be few in any single behavioral trial, their contribution was often so slight as to be indiscernible in the "whole trial" average. Further examination of the fine structure of individual EPs, that is, the small voltage oscillations superimposed upon the large major components, convinced us that this fine structure was often exquisitely reproduced in a subset of EPs elicited within the behavioral trial. Similar fine structure could sometimes be observed across trials, as if EPs with this common fine structure corresponded to a certain functional phase through which the representational system necessarily passed on successive performance of similar decision processes. We have published numerous examples of this phenomenon (John, 1972, 1973a; John, Bartlett, & Shimokochi, 1973).

Once we became aware of the dangers of whole-trial averaging, it became necessary to develop appropriate pattern recognition procedures to permit the accurate classification of individual EP waveshapes. *If EP fine structure was potentially meaningful, then it was possible that the presumed "noise" eliminated from the set of EPs by the averaging process was actually an important, perhaps even essential, part of the signal.* The first computer procedure devised to deal with this problem was reported by Ruchkin (1971). We have further described this procedure and the results obtained with it in behavioral studies (John, 1973a; John *et al.,* 1973).

Recently, a powerful statistical pattern recognition procedure has been implemented which offers the potential of even more precise and objective classification of single evoked potential waveshapes (Kruskal, 1964; Schwartz, Ramos, & John, 1974, 1976a).

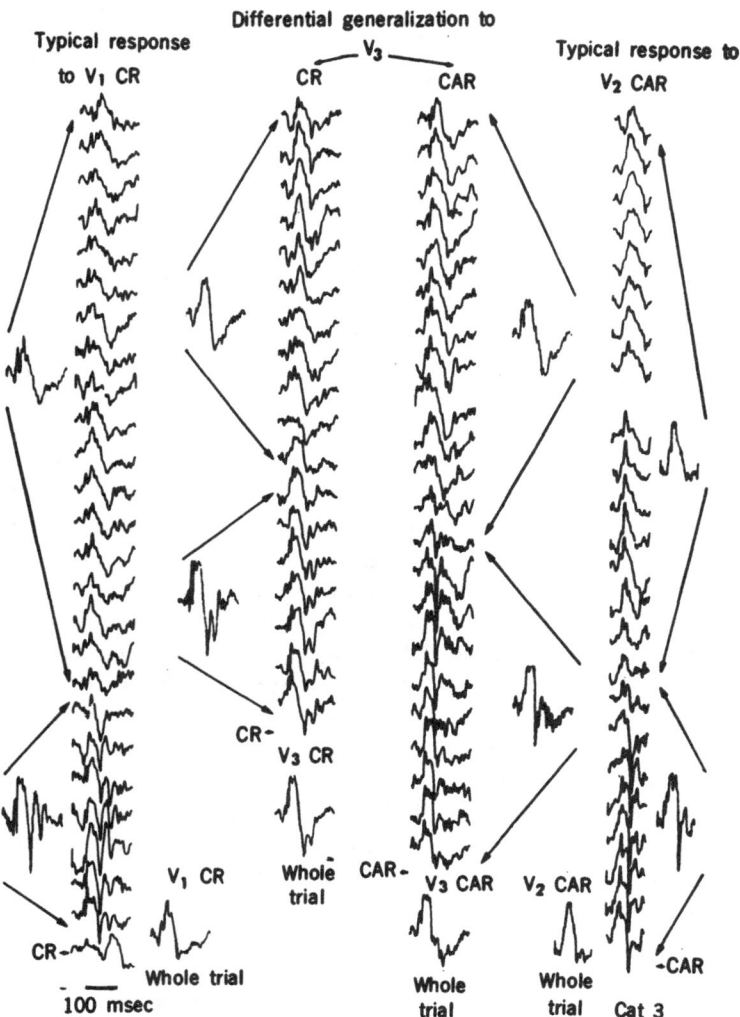

FIG. 10.6 Each column contains the series of evoked potentials that occurred in a behavioral trial, arranged from CS onset (top of column) to behavioral performance (bottom of column). Sequential (bipolar) potentials evoked from the lateral geniculate during a V_1 CR trial are displayed in the first column, during a V_3 CR trial in the second column, a V_3 CAR trial in the third column, and a V_2 CAR trial in the fourth column. Each line represents the sweep of the oscilloscope triggered by occurrence of a single flicker CS, at the respective repetition rates. The waveshapes shown beside or below each column represent average evoked responses computed over the whole behavioral trial or across the subset of evoked potentials indicated by the brackets. In the V_1 CR trial, a radical change in the waveshape of the evoked potential (EP) can be seen on the 19th EP, indicated by the upper arrow of the lower bracket. The sharp new, single downward deflection assumes a double shape on the 20th EP; it is clearly evident on the 21st EP and remains visible until the end of the trial. In

For further details, the reader should consult the chapter on computer data analysis methods in Volume 2. The essential point to be comprehended at this juncture is that as the brain changes state, the same physical stimulus elicits a variety of different evoked potential waveshapes. Sometimes the EP classes are so few and the numbers of events in each class so large that whole trial averages nonetheless reveal clear differences in average response waveshape correlated with different behavioral outcomes. Sometimes the EP classes are greater in number and the critical events relatively few, or the feature correlated with behavior may be the *order* in which particular waveshapes appear (that is, the sequence of *states* entered by the brain). In such cases, perfectly definite relationships between electrophysiological and behavioral events can be hopelessly obscured by whole-trial averaging and success can only be obtained by methods which permit qualitative evaluation of individual evoked potentials.

IV. BEHAVIORAL PREDICTION
BY PATTERN RECOGNITION METHODS

A large body of evidence has been presented which establishes that the waveshape of the electrical oscillation caused by presentation of a novel stimulus resulting in differential generalization reflects the activation of specific memories. It has been possible to program computers for pattern recognition so that the behavioral performance of differentially trained animals responding to ambiguous test stimuli can be reliably predicted (Bartlett & John, 1970, 1973; John, 1972; John *et al.*, 1973; Ruchkin, 1971). This successful prediction depends upon the fact that the behavior which is eventually performed depends upon the shape of the readout process released by the meaningless test stimulus. Computer prediction of behavior in differential generalization is illustrated in Figs. 10.7 and 10.8.

---

the trial shown in the fourth column, the first eight EPs have approximately similar shape. A sharp, short upward deflection caused by a late component appears in the 9th to 16th EP. On the 17th EP, at the upper arrow of the lower bracket, a downward deflection appears, which becomes quite sharp on the 18th EP and remains as a single, sharp downward deflection until the end of the trial. Clear differences are evident in the EPs of the second and third columns. On the 12th EP, marked by the upper arrow of the lower bracket, the double downward deflection characteristic of the last potentials in the V_1 CR trial appears. Its precursors can be seen in the 4th and 8th EPs. The remaining EPs of the V_2 CR trial display this double downward deflection. The first eleven EPs of the V_3 CAR contain a few single or double downward deflections at about the latency of the downward deflection seen in the 19th to 25th EPs of the V_1 CR trial. However, in the 12th EP, a sharp, short downward deflection appears, and the remainder of the EPs in this trial closely resemble the final EPs of the V_2 CAR trial. (From John, 1972. Copyright © 1972 by The American Association for the Advancement of Science.)

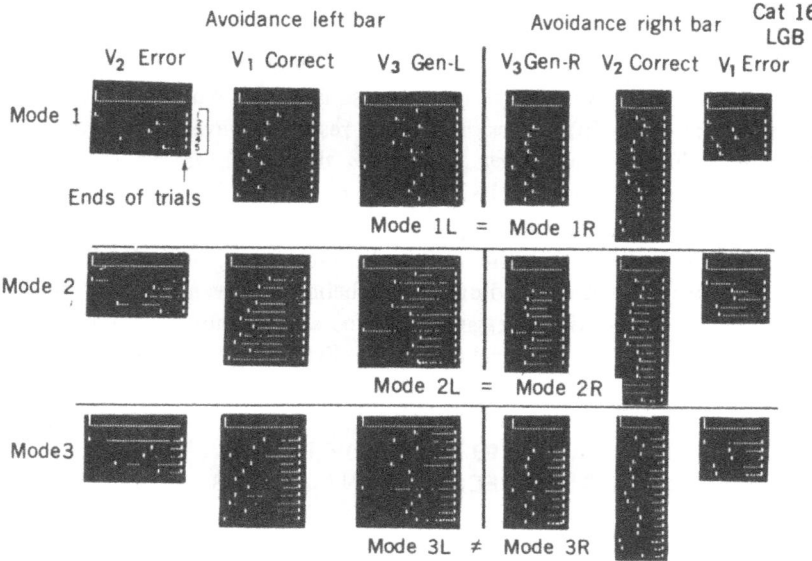

FIG. 10.7 Distribution of modes of evoked potentials. Each column represents a set of trials in which the same behavioral response was performed to a particular stimulus, as indicated in the heading. The V_1 (4.0-Hz flicker) was the CS for pressing the left bar; V_2 (2.0-Hz flicker) was the CS for pressing the right bar, and V_3 (3.0-Hz flicker) was a test stimulus used to elicit differential generalization. The 59 trials included five errors in response to V_2, 11 correct responses to V_1, and 12 generalization-L (Gen-L) responses to V_3, all resulting in left bar responses; there were also 10 generalization-R (Gen-R) responses to V_3, 16 correct responses to V_2, and five errors in response to V_1, all resulting in right bar responses. Both left and right bar responses were to avoid electric shock from a floor grid. The top row of white dots in each black rectangle represents the occurrence of successive CS (flicker) in the behavioral trials. Each row of dots corresponds to a separate trial, beginning at the left vertical bar and ending at the right bar. The occurrence of a dot indicates that the potential evoked for that CS was classified as an example of the indicated mode of evoked response. Three modes were identified within all six types of trials, and are represented by the three sets of waves illustrated in Fig. 10.8.

Results of this sort establish the fact that the brain has the capability to reproduce a remarkably accurate electrical facsimile of an absent event. This facsimile, which appears in many brain regions, must be the product of complex transactions between neuronal ensembles in those diverse anatomic structures. It must also associate into a functional representational unit as a result of the temporal contiguity of activity which occurred in those various regions in a systematic and sustained fashion during the consolidation of memory after the learning experiences established the conditioned behavior.

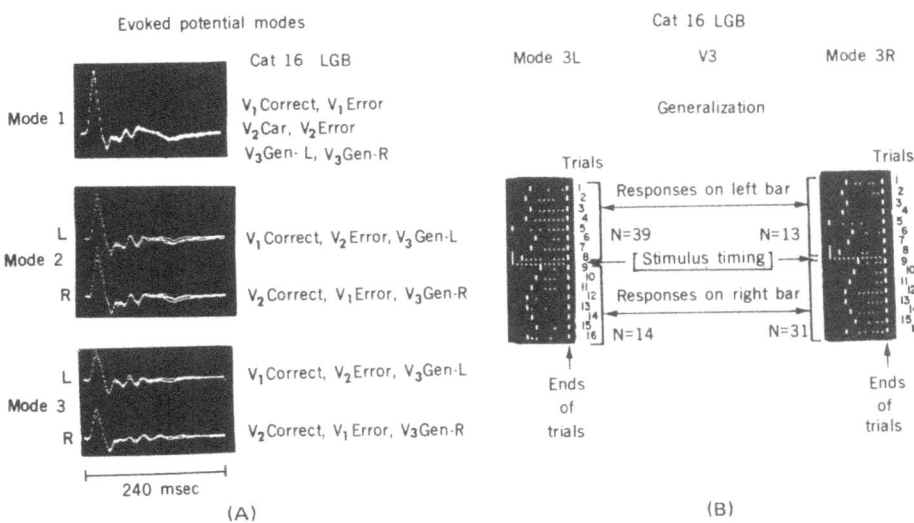

FIG. 10.8 (A) Mode 1 consists of the initial evoked potential in 46 of the 59 trials, plus a small number of additional evoked potentials, all of which occurred early in the trials. The mode 1 potentials from the six types of trials are superimposed in the top row of data. They are essentially identical. Mode 2 consists of the evoked potentials that dominated the early portions of all types of trials. The second row of data, mode 2 left, shows the superimposed waveshapes of this mode for the three sets of trials that resulted in the cat pressing the left bar; while the third row, mode 2 right, illustrates the corresponding data for responses in which the right bar was pressed. Mode 3 consists of the evoked potentials that dominated the final portions of all types of trials. The fourth row of data, mode 3 left, shows the superimposed waveshapes of this mode for the different sets of responses in which the left bar was pressed; while the corresponding data for responses in which the right bar was pressed are illustrated in the bottom row. The three different mode 3 left waveshapes are closely similar, independent of whether they were elicited by V_1, V_2, or V_3. Similarly, the three mode 3 right waveshapes are essentially identical. However, mode 3 left waveshapes are markedly different from mode 3 right. All data were recorded from the lateral geniculate, bipolar derivation.

(B) Correlation between modes of evoked response and behavior. Each rectangle contains data for eight V_3 trials that resulted in left bar behavioral generalization (trials 1–8), and eight V_3 trials that resulted in right bar behavioral generalization (trials 9–16). The rectangle on the left shows the incidence of mode 3 left waveshapes in 16 generalization trials, while the right rectangle shows the incidence of mode 3 right waveshapes in the same trials. The trials were truncated so that only the late portion dominated by mode 3 waveshapes was subjected to this further analysis. These results showed a significant deviation from randomness in the probability that a particular waveshape mode would occur in a trial with a particular behavioral outcome ($p = .00001$). (From John, 1972. Copyright © 1972 by The American Association for the Advancement of Science.)

V. ANATOMIC DISTRIBUTION OF THE ENGRAM

We have hitherto considered only two possibilities: that the received opinion may be false, and some other opinion, consequently, true; or that, the received opinion being true, a conflict with the opposite error is essential to a clear apprehension and deep feeling of the truth. But there is a commoner case than either of these; when the conflicting doctrines, instead of being one and true and the other false, share the truth between them . . . [John Stuart Mill, 1859].

A. Algebraic Treatment of Evoked Potential Processes

The method developed in such studies offers the basis to obtain a definitive answer to the problem of whether memory is localized or distributed. The evidence thus far presented shows unequivocally that the evoked potential recorded from a given brain region during differential generalization is composed of two independent processes, an exogenous process which reflects the physical reality of sensory input, and an endogenous process which represents release of a specific memory. If the exogenous process is represented by the symbol X and the endogenous process by the symbol Y, the evoked potential EP can be represented as follows: $EP = X + Y$.

In the differential generalization procedure, three different stimuli are presented: CS_1 which is the signal for behavior 1, CS_2 which is the signal for behavior 2, and the ambiguous test stimulus. Since errors are occasionally performed either to CS_1 or CS_2, and the test stimulus may elicit either behavior, it is possible to obtain many instances in which the two different behaviors are performed to any one of the three different stimuli.

Let us consider the case in which two different behaviors are elicited by CS_1. If behavior 1 occurs, the EP presumably contains the exogenous process X_1 representing CS_1 and the endogenous process Y_1 representing the learned relation that CS_1 is the cue for behavior 1. Thus, if behavior 1 occurs,

$$EP\ (CS_1, \text{behavior } 1) = X_1 + Y_1.$$

However, if behavior 2 is erroneously performed when CS_1 is presented, the wrong memory must be presumed to have been activated. That is, the EP when such an error occurs presumably contains the exogenous process X_1, representing CS_1, and the inappropriately activated endogenous process Y_2, representing the learned relation that CS_2 is the cue for behavior 2. Thus, if behavior 2 occurs,

$$EP\ (CS_1, \text{behavior } 2) = X_1 + Y_2.$$

If the EP obtained when CS_1 correctly elicits behavior 1 is subtracted from the EP obtained when CS_1 erroneously elicits behavior 2, the following expression is the result:

$$EP\,(CS_1,\text{behavior 1}) - EP\,(CS_1,\text{behavior 2}) = (X_1 + Y_1) - (X_1 + Y_2)$$
$$= Y_1 - Y_2$$

Note that *the residual* $Y_1 - Y_2$ *is the difference between the endogenous processes representing two different memories, independent of the afferent stimulus which released the memory.* Since behavior 1 or behavior 2 could be elicited either by CS_1, CS_2, or the test stimulus, three independent estimates of the difference, $Y_1 - Y_2$, between the two memory readouts are available in the data yielded in differential generalization.

In similar fashion, the difference between the exogenous processes produced by two different sensory stimuli can be obtained. When CS_1 correctly elicits behavior 1, we assume

$$EP\,(CS_1,\text{behavior 1}) = X_1 + Y_1.$$

When CS_2 erroneously elicits behavior 1, we assume

$$EP\,(CS_2,\text{behavior 1}) = X_2 + Y_1.$$

By subtraction, we obtain

$$(X_1 + Y_1) - (X_2 + Y_1) = X_1 - X_2.$$

This residual represents the difference between the exogenous processes representing CS_1 and CS_2.

B. Computation of Residuals

It should be obvious from the above that a number of independent estimates of both exogenous and endogenous processes can be obtained by appropriate manipulation of the EPs available in the body of data yielded from differential generalization studies. If stable exogenous processes are contained in the EPs elicited by particular sensory stimuli no matter what behaviors they elicit, the residuals obtained by such manipulations should have similar waveshapes. Whether or not such similarities of residual waveshapes exist can be estimated by computing the correlation coefficients between residuals obtained via independent measurements. Similarly, if stable endogenous processes are contained in the EPs elicited when similar behaviors are elicited no matter what stimuli were presented, the residuals should have similar waveshapes which can be estimated by computing correlation coefficients.

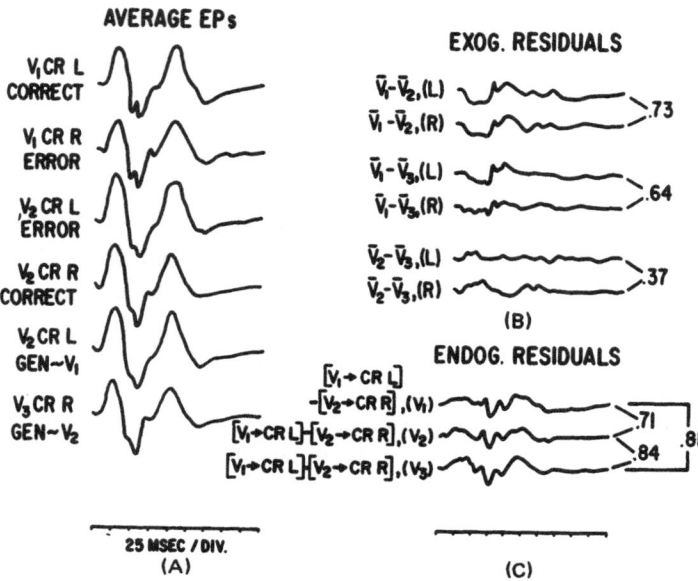

FIG. 10.9 (A) Average evoked potentials obtained from the left lateral geniculate body (bipolar derivation) of a cat subjected to differential behavioral generalization during trials in which six different stimulus–response to contingencies were represented: V_1, CR L, the cat pressed the left lever in correct response to stimulus V_1; V_1 CR R, the cat pressed the right lever in erroneous response to V_1; V_2 CR L, the cat pressed the left lever in erroneous response to V_2; V_2 CR R, the cat pressed the right lever in correct response to V_2; V_3 CR L, the cat pressed the left lever in response to the novel stimulus V_3 during test of differential generalization; V_3 CR R, the cat pressed the right lever in response to V_3. (B) Waveshapes representing pairs of exogenous residuals obtained by manipulating the appropriate averages as described in text. (C) Waveshapes representing independent estimates of endogenous residuals, obtained by subtraction of the appropriate average response waveshapes. Numbers at the right refer to Pearson product moment correlation coefficients between the bracketed waves. The expressions to the left of each residual identify its hypothetical meaning and the responses or stimuli which yielded the residual.

The algebraic manipulations of EPs described above, designed to establish whether stable exogenous and endogenous processes imbedded in EP waveshapes could be separated and identified, have been carried out on an enormous body of data consisting of thousands of EPs obtained in hundreds of trials in 18 cats, each bearing 34 electrodes implanted in numerous anatomic regions of the brain (Bartlett & John, 1973).

Figure 10.9 illustrates that the waveshapes of residuals reflecting similar exogenous and endogenous processes are in fact markedly similar.

The distribution of correlation coefficients between residuals of exogenous or endogenous processes theoretically expected to be similar was found to be significantly skewed toward positive correlations, indicating that the residuals did in fact have similar waveshapes.

C. Anatomic Distribution of Exogenous and Endogenous Residuals

For each brain region studied, the contribution of exogenous processes to the EP was plotted versus the contribution of endogenous processes. The results are illustrated in Fig. 10.10.

Examination of Fig. 10.10 reveals that a systematic relationship exists between the contribution of exogenous and endogenous processes to the EP. *The amount of exogenous process in a region is roughly proportional to the logarithm of the amount of endogenous process in that region.* The absolute contribution of both the exogenous and endogenous processes varied greatly from region to region. Clearly, all regions were not equivalent with respect to the signal-to-noise ratio of the signal or their participation in the representation of experience. This finding supports the contention of adherents of the localizationistic theory. Sensory stimuli affect some brain regions far more than others, and some brain regions participate in the representation of experience far more than others. At the same time, it is clear that the effects of a sensory stimulus are qualitatively distributed throughout the brain, and all brain regions participate in the representation of experience to an extent which is proportional to the afferent input to those regions. This finding supports the antilocalizationist position. The discussions in Chapter 8 are relevant to this finding in that the critical shift postulated for the critical substance during consolidation would necessarily be proportional to the amount of exogenous activity.

These data suggest how to resolve the century-long dispute between localizationist and antilocalizationist theory. Both positions are correct, but neither position is *accurate*. In fact, it appears that *different anatomic regions differ quantitatively in the extent to which they participate either in the direct mediation of afferent sensory experience or the representation of past experiences, yet they all participate qualitatively in both kinds of process.* This reconciliation of the two viewpoints would account for why lesions in certain brain regions produce more severe performance deficits than lesions in other areas, while accounting for the fact that memory can survive damage to any area provided that sufficient sensory input remains to the system as a whole.

The evidence summarized suggests that information, past or present, is represented in the brain by a statistical process: the average spatiotemporal pattern of activity in anatomically extensive neuronal populations. The activity of the single neuron is not informationally significant except insofar as it contributes to the activity of the ensemble. The same information can be represented in diverse regions, with a varying signal-to-noise ratio (S/N). In any region, some cells

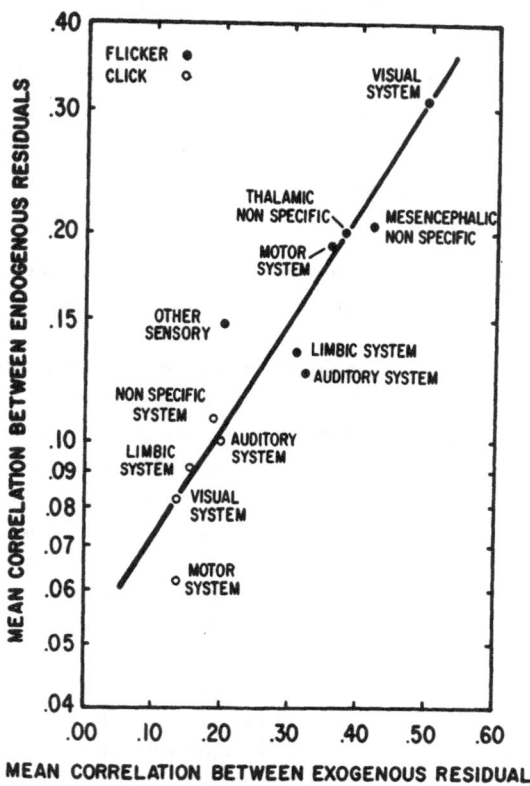

FIG. 10.10 Plot of mean correlation coefficients between exogenous residuals for different neural systems and for different cue modalities. Closed circles: flicker frequencies as stimuli. Auditory system: $N = 305$; Aud Cx (16 cats), Med Genic. (16), Brach Inf Coll (1). Limbic System: $N = 303$; Hippocampus (16), Dentate (3), Cingulate (5), Septum (5), Prepyriform (6), Med Forebrain Bundle (6), Mamm Bodies (5), Hypothalamus (7). Mesencephalic nonspecific: $N = 158$; Retic Form (18), Cent Gray (1), Cent Teg Tract (1). Motor system: $N = 146$; Motor Cx (4), Subs Nigra (10), Nuc Ruber (4), Nuc. Vent. Ant. (9), Subthal (5). Other sensory: $N = 54$; Sensorimotor Cx. (4), Nuc. Post. Lat. (1), Nuc. Vent Post Lat (5), Nuc. Vent. Post. Med. (1). Thalamic nonspecific: $N = 139$; Cent. Lat. (13), Nuc Retic (6), Nuc Reuniens (1), Med Dors (5), Pulvinar (1). Visual system: $N = 394$, Visual Cx (18), Lat Genic (18), Sup Coll (2). Open circles: click frequencies as stimuli. Auditory system: $N = 48$; Aud Cx (5 cats), Med Genic (5). Limbic system: $N = 69$; Hippocampus (5), Dentate (3), Cingulate (3), Septum (3), Prepyriform (2), Med. Forebrain Bundle (3), Mamm Bodies (3), Hypothalamus (2). Motor system: $N = 37$; Motor Cx. (1), Subs Nigra (4), Nuc Ruber (1), Nuc Vent Ant (5), Subthal (2). Nonspecific system: $N = 50$; Mesen Retic Form (6), Cent Gray (1), Cent Teg Tract (1), Cent Lat (3), Nuc Retic (3). Visual system: $N = 55$; Visual Cx (6), Lat Genic (6), Sup Coll (1).

Data from monopolar and bipolar derivations were combined. Replications varied across cats and structures. (From Bartlett & John, 1973.)

appear to be stimulus-bound, displaying the same average firing pattern to stimuli independent of how they are perceived, although they may display different response patterns to different stimuli. Such cells would appear to be relatively reliable reporters of sensation in terms of their ability to construct reproducible *average* firing patterns characteristic for each different stimulus, in spite of the short-term variability of their responses. Other cells in the same regions display average patterns of response to the same stimulus which are reactive, depending upon the meaning attributed to the signal. These latter cells seem to be involved in perceptual processes and in the storage of memories about the stimulus–response contingencies. The mixture of these two types of cells varies from region to region, producing variable S/N for both exogenous and endogenous processes, that is, making different relative contributions to sensation and to perception.

VI. CELLULAR PARTICIPATION IN MEMORY

Thus, memory is not localized in any particular brain region, but is more or less diffusely distributed across numerous brain regions. It remains to establish whether, within any of these brain regions, a memory is localized in particular cells or synaptic pathways such that the firing of specific cells represents a unique memory, or whether a memory is distributed across a population of numerous cells such that the average pattern of activity in the ensemble represents a unique memory independently of the firing of any particular neuron.

Evidence presented in earlier chapters (Chapters 4 and 6) indicates that single cells in primary sensory cortex may display remarkable specificity of response to stimuli possessing certain features. The subsequent processing of topographically specific activity patterns propagating from primary cortex to other regions is not yet clear. Some workers believe that similar progressively more complex mechanisms of feature extraction operate successively so that the output of neurons extracting lower order features converge upon neurons at a higher level which serve to extract features of ever greater complexity, culminating in what Sherrington (1947) and Fessard (1954) epitomized as "pontifical neurons" or what Barlow (1969) called "cardinal cells," which encode percepts, memories, or abstract ideas. As Freeman (1975) has pointed out, this hypothesis at first seems logically incontrovertible because the brain contains enough neurons with enough possible connections so that if each connection were regarded as an on–off switch, the brain could perform almost all conceivable logical operations by binary switching. Among the difficulties with this hypothesis is the fact that the neuronal signal is not binary. Rather, it is a pulse train with graded attributes.

As pointed out elsewhere (John, 1972), the single neuron is an unreliable reporter. It fires with no identifiable input on some occasions, is refractory on others, responds to a wide variety of inputs, and has a wide variety of responses to any specified input. Freeman considers the evidence for feature extraction properties to be relevant to the function of primary cortical regions, which extract information about the local organization of a sensory event, but inappropriate as a model for the general forebrain. Since space does not permit a more detailed rebuttal of this concept here, the interested reader is urged to consider the fuller expositions noted above as well as the sections which follow below.

A. Statistical Features of Neuronal Activity

Numerous other publications have pointed out that information processing and storage of memories about particular items of information may depend upon characteristics of the activity of populations of neurons rather than the firing of single cells (Bullock & Quarton, 1966; Burns, 1968; John, 1967a; MacKay, 1969b; Whalen, Thompson, Verzeano, & Weinberger, 1970). Numerous recent neurophysiological studies have revealed that neuronal activity involves a probabilistic element (Moore, Perkel, & Segundo, 1966). Most cells in the brain are in incessant activity and can only be described as responsive to peripheral stimuli in a statistical sense. Sequential responses of a neuron to the same stimulus may well be markedly different, the same cell responds to many different stimuli, and cells occasionally fire spontaneously. The firing pattern of almost any arbitrarily selected cortical cell can be shown to change upon local stimulation of almost any other cortical area (Burns & Pritchard, 1964; Burns & Smith, 1962).

Burns and Smith have stated:

> ... During one second a single neuron does not provide the rest of the brain with sufficient information to identify the presence and nature of the stimulus. Our results suggest that sensory inputs to the brain set up a spatial and temporal pattern of activity which probably involves most of the cells in the cerebral cortex. It would appear that differential of the effects of a stimulus from the "noise" of continual or "spontaneous" activity is only made possible by the simultaneous weak response of many neurons [1962].

As the elegant early experiments which suggested that single cells served as "feature extractors" (Hubel, 1959; Hubel & Wiesel, 1962, 1963; Lettvin, Maturana, McCulloch, & Pitts, 1959) were pursued further, it was found that the activity induced in supposed "feature extractor" cells depended not only upon the afferent input but upon the configuration of activity elsewhere in the brain (Chow, Lindsley, & Gollender, 1968; Lindsley, Chow, & Gollender, 1967; Spinelli, Pribram, & Weingarten, 1966; Spinelli & Weingarten, 1966; Weingarten & Spinelli, 1966). A more detailed review of evidence against the representation of particular stimulus characteristics by the deterministic firing of single cells has recently been presented elsewhere (John, 1972).

B. Unit Activity in Conditioned Responses

A large number of studies have been carried out on changes in unit activity during conditioning (see Chapter 9). These studies all agree that a high population of the cells observed (from 10% to as high as 60%) change their responses during conditioning. Examination of data from such studies further indicates that units respond to the CS before as well as after establishment of the new conditioned response, but the features of response to the stimulus are altered. Such findings provide further evidence that the firing of a particular cell cannot

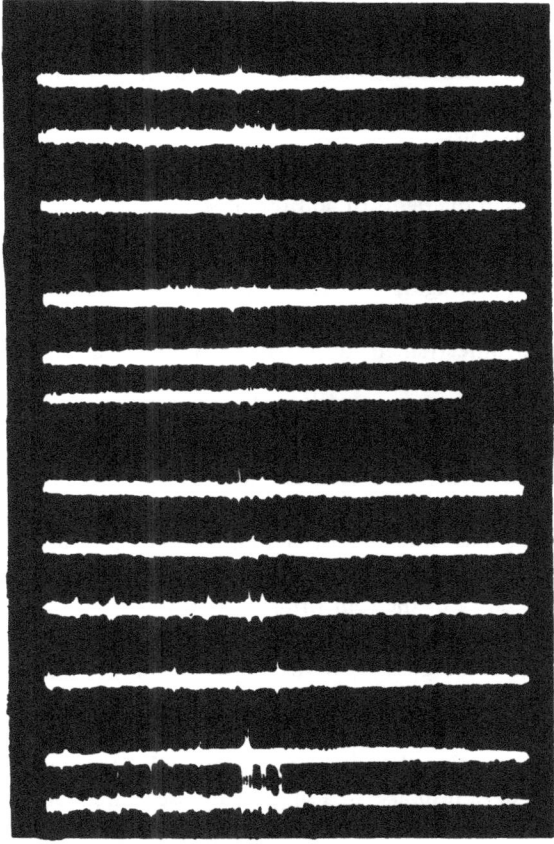

FIG. 10.11 Responses of cortical neuron to successive presentations of the same conditioned stimulus. Each row of data shows activity during 250 msec immediately following stimulus onset. The vertical lines are deflections of the oscilloscope beam caused by spike discharges. Note the great variability of response from stimulus to stimulus.

per se represent the arrival of unique afferent information or the activation of a unique memory.

Some of our recent experimental observations illustrate the comments above and support the general conclusions reached in that discussion. Figure 10.11 shows a series of unit discharges, recorded from a chronically implanted microelectrode in an unrestrained cat performing differential conditioned responses, to many successive presentations of the same conditioned stimulus. Note the variability of unit response to the repeated stimuli.

These responses come from a few neurons in the vicinity of the electrode tip. Each of these different neurons produces a voltage spike of characteristic amplitude. By use of pulse height discriminators, such an assortment of spikes of different voltages can be electronically separated into several channels, each containing only spikes of a given size. In this fashion, the activity of a neuronal ensemble, recorded from a microelectrode surrounded by a number of cells, can be fractionated so that the firing of separate neurons in the ensemble can be identified and analyzed.

C. Similar Patterns in Elements of an Ensemble

Using so-called pulse height discrimination methods which classify discharges according to their voltages, we have studied the average responses of the neuronal components of ensembles thus fractionated as discriminated signals were presented to unrestrained, differentially trained cats. We found that the activity of the ensemble as a whole showed one characteristic temporal pattern to one CS and a different temporal pattern to the other CS. The neurons in the ensemble, separated by the electronic discriminator, showed the same average firing pattern as the ensemble and as each other. This homogeneity of the averaged neuronal response is illustrated in Fig. 10.12, showing the averaged poststimulus histograms obtained from four neurons producing spikes of four characteristically different amplitudes as the same CS was repeatedly presented. Note the basic similarity of the four firing patterns.

However, this similarity of *averaged* patterns did not arise because the cells were synchronized or because their responses to single stimulus presentations displayed the pattern seen in the averages. Figure 10.13 shows that the four cells showed radically different *momentary* firing behavior, with great variability. These findings indicate that the neurons in an ensemble show marked differences from one to another in response to a single CS presentation, and great variability in their responses to successive stimuli. Statistically, however, they all converge to a *common* or shared average firing pattern which is characteristic for a particular environmental signal.

These findings illustrate the difficulty of extracting unique informational significance about an environmental stimulus from the firing of a single cell. However, *since each cell in the ensemble converges to the common average*

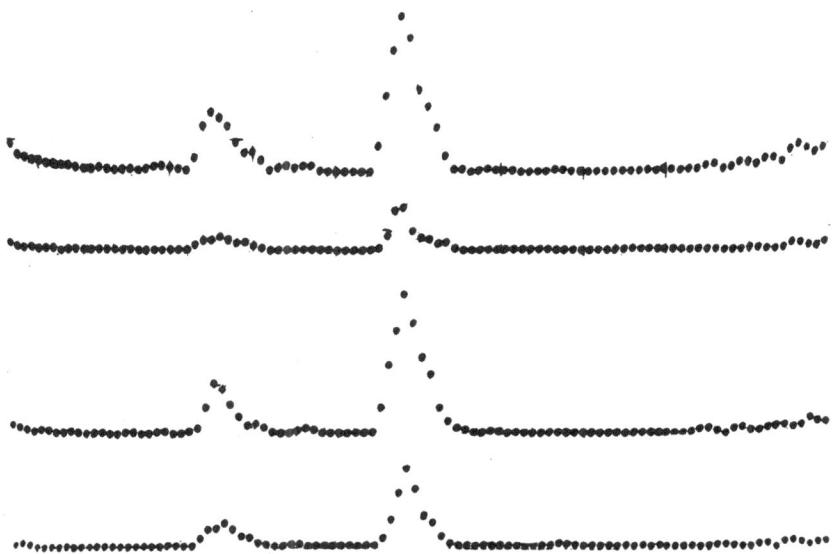

FIG. 10.12 Each line of data shows vertical deflections representing the probability of firing of a different single neuron or small subgroup of neurons, extracted from the activity of a neuronal ensemble by discrimination according to spike voltage (poststimulus histograms). Time after stimulus presentation is indicated along the horizontal axis, with the full line representing 250 msec. Each line is an average summating neuronal responses to 256 stimuli. Note the basic similarity of firing patterns in the different components of the ensemble (graticule of oscilloscope can be dimly seen).

FIG. 10.13 Same neurons as in Fig. 10.12. Each line of data shows the output of a discriminator detecting the firings of a different neuron to a single stimulus presentation. Note that although the average firing patterns of these different cells was extremely similar, as illustrated in Fig. 10.12, their response to a single stimulus presentation was extremely different and not at all synchronized.

pattern if its responses are sampled across many presentations of the stimulus, the same information would be available if the responses to a single stimulus presentation were averaged across many neurons in the ensemble. By this reasoning, we are led to the conclusion that the information being processed in a neuronal ensemble is represented by the firing pattern averaged across the population, *the temporal fluctuation of coherent activity,* rather than by the responses of any individual cell(s) in the ensemble.

VII. MASS ACTION REVISITED

In a series of thoughtful and provocative papers, Freeman has proposed by far the most detailed quantitative formulation of a theory of information processing by mass action of populations of neurons (Freeman, 1972a, b, 1974, 1975). He proposes that each neuron in the central nervous system is a nonlinear device, typically receiving from many neurons in its surrounding and transmitting to many other neurons, including some or all of those from which it receives input. The output is a pulse train. The functional interconnections of populations of these nonlinear elements define interactive *sets* of neurons, with properties belonging to the sets and not to the component neurons. He conceives of information processing in terms of a hierarchy of so-called "K sets." The lowest level is the K_0 set, which is noninteractive and which can be described as an "average" neuron. Elementary neural information, for example the response to an afferent signal, is defined as a space–time pattern in a K_0 set. At this level the representation is discrete and can be treated by the concepts of pulse logic. The next level is the K_1 set, which consists of interactive neurons in excitatory or inhibitory relationships. If the space–time pattern consists in the cooperative activity of a K_1 set, the concepts of mass action apply. Interaction of excitatory and inhibitory K_1 sets forms a K_2 set. Freeman's hypothesis of information coding in K_2 sets is that the ensemble average over an anatomic domain is the manifestation of a carrier frequency. *The signal is conceived to be carried in the spatial patterns of phase differences from the average.* This formulation is supported by detailed analyses of neural and EEG activity in the olfactory system during adaptive behavior of cats and rabbits. In contrast, the formulation which we have proposed earlier in this chapter considers such phase differences as random deviations from the average.

These theories both concur in treating information processing in the central nervous system as mediated by cooperative processes in neural masses, to which the contribution of any individual neuron is of but statistical significance. Further analysis will be required to decide whether the temporal pattern of ensemble activity is the primary vehicle of information representation or merely the "carrier wave," with the information defined by local phase differences in this statistical process. No matter which will prove to be the more correct hypothesis, it seems highly probable that Freeman's quantitative formulations will be of fundamental importance in unraveling this problem. Since further explication of this work is beyond the scope of this volume, the interested reader is urged to examine the original papers.

VIII. NEURONAL ACTIVITY DURING READOUT OF SPECIFIC MEMORIES

The readout process in differential generalization was studied on the level of neuronal activity, using the methods described above (John, 1974; Ramos, Schwartz, & John, 1974, 1976a). Evoked potentials and firing of neuronal ensembles were simultaneously recorded from the same microelectrode, using appropriate electronic filters to separate the slow evoked potential waveshapes from the fast neuronal discharges. The evoked potential from a set of differential generalization trials, in which the same test stimulus elicited performance of two different kinds of conditioned responses, were then classified using sorting techniques. Two readout modes of the EP were identified, one predictive of one behavior and the other predictive of a different behavior.

POSTSTIMULUS HISTOGRAMS

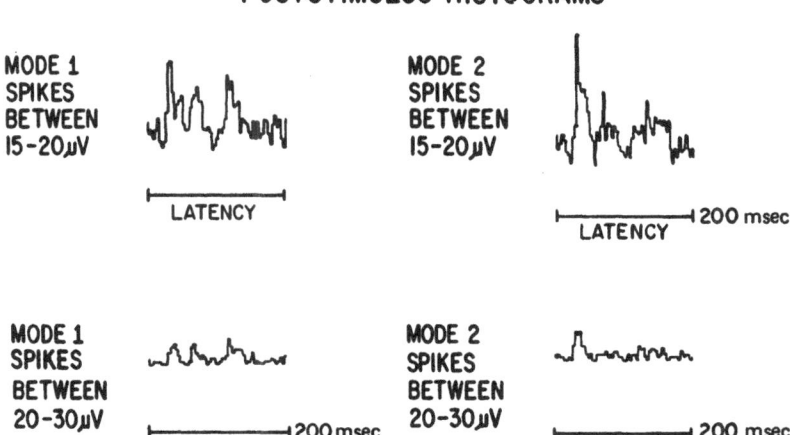

FIG. 10.14 Different firing patterns displayed by the same cells during activation of two different memories. Pattern recognition methods like those illustrated in Fig. 10.7 were utilized to classify evoked potentials elicited by a novel stimulus during differential generalization. Two different readout modes were found, predicting two different behavioral outcomes. Neurons in the lateral geniculate body were discriminated into two groups, on the basis of spike voltage, and poststimulus histograms were constructed. (From Ramos, Schwartz, & John, 1974, 1976a.)

The activities of the neuronal ensembles in response to each stimulus presentation were then classified according to the readout mode of the evoked potential which had been simultaneously recorded. Thus, two groups of neuronal responses were constructed: (1) those which had occurred during readout mode 1, and (2) those which had occurred during readout mode 2. According to the evidence presented earlier in this chapter, *those two groups of neuronal responses corresponded to firing patterns released when two different memories were activated.*

The results of fractionation of the discharges of the neuronal ensemble during activation of the readout of the two different memories are shown in Fig. 10.14 (Ramos *et al.*, 1974, 1976a). Examination of the data shows clearly that *the same cells in the ensemble fired during both kinds of readout activity. One temporal pattern of activity was released when one readout was activated, and a different pattern appeared when the other readout was activated.*

Recently, evidence has been obtained that two different kinds of cells exist, which have been observed in several brain regions: (1) "stable" cells, which display the same average firing pattern to a particular stimulus, independent of the subsequent behavioral response, and (2) "plastic" cells, which display a characteristic average firing pattern during trials culminating in a particular behavior, independent of the physical stimulus. Stable cell activity would seem to be related to exogenous EP processes, while plastic cells are related to endogenous processes (Ramos, Schwartz & John, 1976b, c). The evidence thus far obtained does not provide any support for the expectation that the memory of particular events is represented by the occurrence of discharge in specific cells or pathways reserved for the unique mediation of a single memory. Rather, cells seem to be involved in the mediation of multiple memories, each represented by a particular temporal pattern of firing discernible only in the statistical features of the activity of a population of neurons. Memory appears to be a distributed, cooperative process of a statistical nature.

IX. GENERALITY OF CONCLUSIONS
BASED ON TRACER TECHNIQUE

The electrophysiological studies which have provided the major source for the picture of brain mechanisms of learning and memory we have constructed were largely based upon *tracer technique;* the study of electrical rhythms, evoked potential waveshapes, and unit activity observed in the brain as animals learned to discriminate between signals at different repetition rates. Such studies have revealed a number of phenomena which enable a fairly detailed description of how information is coded, stored, and retrieved in the brain. Before we accept this description and consider its implications for related problems, it is important to evaluate the extent to which the processes that have been revealed have generality.

The tasks utilized in the vast majority of the studies cited above on electro-physiological changes during learning and performance required the animals to discriminate between *intermittent* sensory stimuli presented at *different repetition rates.*

Two important questions must be discussed:

1. Repetition rates of stimulation are not an important feature of everyday life. Are the processes in the brain which mediate discriminations between repetition rates relevant for the processing of other kinds of information?

2. Most environmental stimuli are continuous rather than intermittent. What is the relevance of mechanisms that process intermittent stimuli to the process-ing of information as it occurs naturally?

A. Relevance to Different Modalities of Information

Information about the repetition rate of flashes, clicks, or electrical stimulation of different brain regions seems to be represented by the rapid fluctuation of brain potentials with a particular temporal pattern and by the occurrence of corresponding fluctuations in the probability of firing in neuronal ensembles. Activation of memories about such information causes the release of electrical potentials and neuronal discharges with the corresponding pattern.

Fundamentally similar processes have been revealed for many different types of information. For example, it has been demonstrated that the color and intensity of light, the pitch and intensity of sound, the frequency and duration of stimulation, the size and shape of geometric forms, uncertainty about the nature of a forthcoming stimulus, and affective connotations can all change the waveshape of the evoked potential (Begleiter, Gross, & Kissin, 1967; Buchsbaum & Silverman, 1968; Burkhardt & Riggs, 1967; Cavonius, 1965; Clynes, Kohn, & Gradijan, 1967; Davis, 1966; Davis, Hirsch, Shelnutt, & Bowers, 1967; Fields, 1969; John, Herrington, & Sutton, 1967; Keidel & Spreng, 1965, 1970; Begleiter & Platz, 1969a; Pribram, Spinelli, & Kamback, 1967; Rapin & Graziani, 1967; Rapin, Ruben, & Lyttle, 1970; Regan, 1966; Shipley, Jones, & Fry, 1965, 1966; Sutton, Braren, Zubin, & John, 1965; Vaughan & Hull, 1965; Vaughan & Silverstein, 1968; Wicke, Donchin, & Lindsley, 1964). From what is known about the relationship between EPs and neuronal activity, such changes in EP waveshape reflect changes in neuronal firing pattern. Thus, it appears that a wide variety of stimulus attributes cause characteristic EP waveshapes and neuronal firing patterns, presumably reflecting the encoding of these different kinds of information.

An observation that may afford basic insight into this process was made by John and Ruchkin (John, 1967a). A brief electrical stimulus was delivered in turn to each of a large number of brain regions. Factor analysis of the EP waveshapes recorded from other brain regions revealed that the *EPs which propagated throughout the brain from a local disturbance in any particular region possessed a typical waveshape, which constituted the "key signature" of the region from which the disturbance emanated.*

Activation of relatively localized regions by afferent input or changes in internal state might be expected to cause propagation, from those regions to many other regions in the brain, of an evoked potential whose waveshape reflected both its *region of origin* (the *modality* of the stimulus) and the detailed *nature of the disturbance* (the *quality* of the stimulus). Therefore, it seems reasonable to conclude that the details of EP waveshape and the temporal pattern of neuronal firing represent general processes relevant to the coding, storage, and retrieval of many, if not all, kinds of information in the brain.

B. "Readout" of Other Types of Information

Readout processes have been demonstrated for other kinds of information than stimulus repetition rate. For example, presentations of geometric forms of the same shape but different size elicit EP waveshapes containing invariant features (Clynes, Kohn, & Gradijan, 1967; Hudspeth, personal communication; John *et al.*, 1967). This phenomenon is illustrated in Fig. 10.15.

The observation that geometric forms of the same shape but different size could elicit the same evoked potential waveshape shows that the brain can produce an invariant temporal pattern of voltage independent of the physical size of a visual form. Upon reflection, it seems apparent that this invariant EP

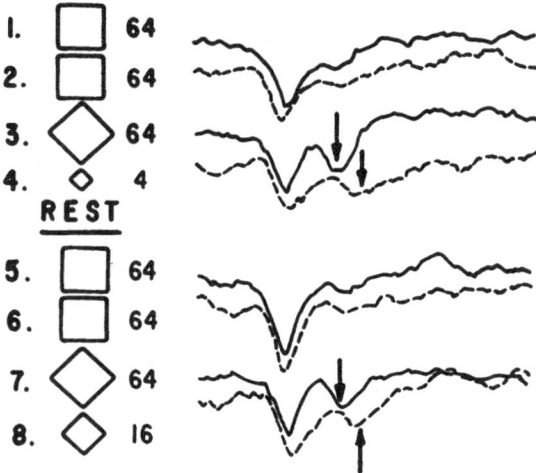

FIG. 10.15 Averaged responses from two sessions separated by 30 minutes with the same subject. All averages based upon 100 repetitions of the stimulus, and a 500-msec analysis epoch. Negative deflections are upward. Responses 1, 2, 5, and 6 were to squares with an area of 64 in² (412.8 cm²), response 3 and 7 to diamonds 64 in² (412.8 cm²) in area, response 4 to a diamond of 4 in² (25.8 cm²) and response 8 to a diamond of 16 in² (103.2 cm²).

"I" AS A NUMBER VS "I" AS A LETTER

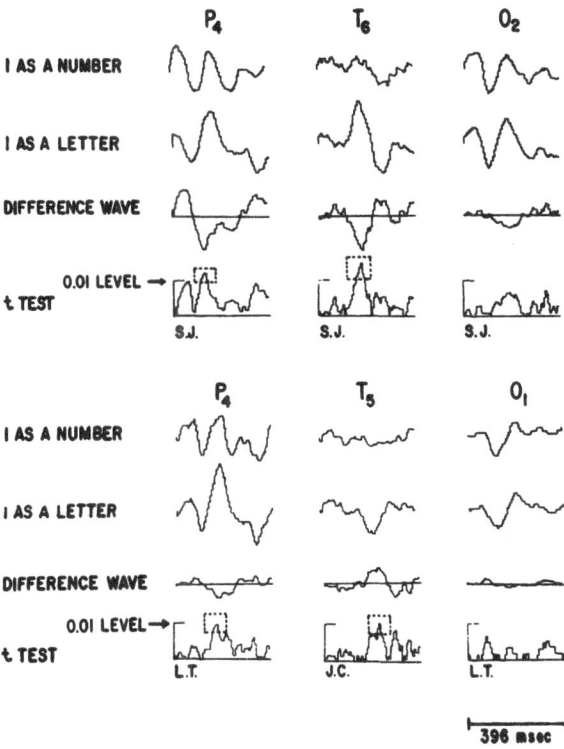

FIG. 10.16 Depiction of results obtained in two subjects when "I" as a number and "I" as a letter were presented. The first and second rows show samples of average evoked potentials recorded from parietal, temporal, and occipital derivations. The third row depicts the difference wave; and the last row, the t test results. Notice that the parietal and temporal leads reach significant differences at a latency of approximately 160 msec.

waveshape must be of endogenous rather than exogenous origin, and that it corresponds to the release of a readout pattern representing the *concept* of a shape independent of its size. Similar conclusions can be drawn from our observations (Fig. 10.16) that the EP waveshape elicited by presentation of the visual form, "I," alters dramatically depending upon whether the subject is instructed to look at the numbers "1" and "2" or the upper case *letters* "I" and "K" (Grinberg & John, unpublished observations). These studies indicate that the same physical stimulus can release two different readout processes depending upon the language meaning which the subject attributes to the visual signal.

Quite the same phenomenon has been described for auditory stimuli. Brown *et al.* (1973) recently demonstrated that the waveshape of the auditory evoked potentials caused by hearing the word "rock" varied very markedly depending upon whether the work was imbedded in a context causing it to be interpreted as a *noun* or a verb.

Another example of the correlation between information about stimuli which were not frequency coded and differences in evoked potential waveshape has been provided by Begleiter, Porjesz, Yerre, and Kissin (1973), and is of particular importance because it constitutes a direct extension of our differential generalization paradigm into the domain of nonfrequency-coded stimuli. These workers trained human subjects to judge between the *intensities* of bright and dim visual stimuli, and found that different flash intensities elicited characteristically different EP waveshapes, a finding in which Tepas, Guiteras, and Klingaman concur (1974). Begleiter and his colleagues went on to intersperse flashes of intermediate intensity among the familiar discrimimanda. It was found that the potentials elicited by these test flashes of intermediate intensity were of differing waveshapes, depending upon the subjective flash intensity (bright or dim) attributed to these test stimuli. The responses to the test flashes corresponded in waveshape to those usually elicited by flashes of the intensity erroneously attributed to the test flash.

C. Readout to "Imaginary" Stimuli

As noted earlier, the release of endogenous potentials by human subjects during the absence of a visual or auditory stimulus in a variety of situations in which there is a created expectation of a stimulus has been reported by a number of workers (Barlow *et al.*, 1967; Dicheva, Atsev, & Popivanov, 1971; Haider, Spong, & Lindsley, 1964; Klinke, Fruhstorfer, & Finkenzeller, 1968; Picton *et al.*, 1973; Riggs & Whittle, 1967; Rusinov, 1959; Spong, Haider, & Lindsley, 1965; Sutton, Tueting, Zubin, & John, 1967; Weinberg *et al.*, 1970, 1974). The last authors in particular commented upon the good correlation between the shape of the emitted or readout potential and the potentials elicited by the expected stimulus when it was present. In fact, such studies show the readout process accompanying an imaginary stimulus. Herrington and Schneidau (1968) attacked this problem more directly. They first ascertained the shape of the EPs usually elicited by presentation of a flash illuminating either a square or a circle. Subsequently, the subject was asked to *imagine* a square or a circle on the screen when a blank flash was presented. In many cases, it was possible to *tell* the subject which form he had imagined because the blank flash produced an EP which was a good facsimile of that usually caused by actual presentation of the imagined stimulus.

These studies indicate that the readout phenomenon observed in studies of frequency discrimination, upon which we based so much of our earlier analysis of memory mechanisms, is not unique to frequency-coded information but has been found in a variety of other situations involving coding of geometric forms, letters of the alphabet, and verbal stimuli. Further, in numerous studies, readout processes have been observed to expected but absent visual or auditory stimuli and have resembled the responses usually elicited by those stimuli when actually experienced. Inquiry from these human subjects established that these readout processes were correlates of an anticipated imagined event. The evidence cited in the final section of this chapter suggests that the mechanisms previously discussed are probably involved in the storage and retrieval of many different kinds of information.

D. Relevance to Continuous Environmental Stimuli

The final question to be considered is the relevance of findings derived from the use of intermittent stimuli for the encoding, storage, and retrieval of information about the vast majority of environmental events which, though their duration may be relatively brief, are present continuously during that period. There exists a body of evidence that perception is not continuous, but occurs in overlapping intervals called "perceptual frames," which has been discussed in Chapters 7 and 11. The sequence, onset and termination of different events occurring within the same perceptual frame cannot be distinguished from each other. Apparently, such subjectively congruent events are brought into the brain as unitary packets or samples of environmental information.

The brain mechanism responsible for the chopping of environmental input into sequentially sampled or moving perceptual frames is not known definitely, but some workers have suggested that this function may be accomplished by fluctuations in excitability related to the alpha rhythm (see Section IV.B, Chapter 3). The rhythmic fluctuation from positive to negative and back to positive potentials is interpreted as an alternation between a phase of relatively high excitability and a phase of inhibition, constituting an "excitability cycle." Each phase of this excitability cycle begins with the gating open of afferent pathways, proceeds with processing of the incoming information, and ends with closing the input paths and terminating the sample of the environment corresponding to that perceptual frame.

So it seems that the apparent continuity of subjective experience of the natural environment is deceptive. We sample our environment in brief, sequential intervals. There seems to be no reason to assume that the processing of such intermittent sequential samples of information differs in any essential fashion from the processing of the intermittent stimuli used in the evoked potential studies which have provided much of the basis for our present ideas about how information is coded, stored, and remembered.

11

Activation of Memories by Electrical Stimulation of the Brain: A Direct Test of Statistical Theory

In Chapters 9 and 10, we have presented a body of evidence about the changes in evoked potential and unit activity observed during learning and remembering. We proposed that information was coded and retrieved in the brain on the basis of the statistical behavior of large neuronal ensembles rather than by conduction along specific synaptic pathways established or facilitated by a learning experience. It may be worthwhile to summarize the major features of the evidence on which that argument was based in order to clarify our rationale for the experiments to be described in this chapter, which we believe constitute a critical test of statistical theory.[1] Since those experiments involve direct electrical stimulation of the brain, we will also provide a review of the present status of stimulus generalization studies in which the informational utility of electrical stimuli has been examined.

I. RATIONALE FOR THE USE OF ELECTRICAL STIMULATION TO TEST STATISTICAL THEORY

In numerous studies, it has been reported that during conditioning, evoked responses to the conditioned stimulus become anatomically more widespread and similar (Dumenko, 1967; Galambos & Sheatz, 1962; Glivenko, Korol'kova, & Kuznetsova, 1962; John, 1967a; John & Killam, 1959, 1960; John, Ruchkin, & Villegas, 1963, 1964; Knipst, 1967; Korol'kova & Shvets, 1967; Liberson & Ellen, 1960; Livanov, 1962, 1965; Livanov & Poliakov, 1945; Yoshii *et al.*, 1957). New late components appear in the evoked response (Asratyan, 1965; Begleiter & Platz, 1969b; Galambos & Sheatz, 1962; John, 1963, 1967a, b; John

---

[1] The original studies reported in this chapter were carried out in collaboration with Dr. David Kleinman, now at the University of Durham, England.

& Morgades, 1969c; Killam & Hance, 1965; Leiman, 1962; Sakhuilina & Merzhanova, 1966). After acquisition of a conditioned response, generalization often occurs when a novel stimulus is presented. When an animal performs such a generalized behavioral response, the evoked potential waveshape elicited by the novel stimulus is a good facsimile of that usually evoked by the conditioned stimulus (John, 1963; Ruchkin & John, 1966). If the animal has been differentially conditioned to discriminate between two stimuli, the waveshape evoked by a novel stimulus intermediate to those two cues depends upon which behavior the animal displays during differential generalization, and is then a facsimile of that waveshape usually elicited by the differential conditioned stimulus which is the appropriate cue for the performed behavior (John, Shimokochi, & Bartlett, 1969).

Since the waveshape of the evoked response under these conditions was not determined by the physical characteristics of the stimulus, but appeared to correlate with the behavior subsequently performed, we speculated that these phenomena might reflect the activation of specific memories. Further studies, involving the investigation of many possible nonspecific factors which might have been involved in such observations and utilizing computer pattern recognition techniques to classify single evoked response waveshapes, supported the conclusion that the brain could produce a facsimile of an absent event (John *et al.*, 1973). The evoked potential was found to contain exogenous processes, determined by the nature of the stimulus, and endogenous or readout processes, released by the stimulus from memory. By appropriate computer manipulations of large quantities of evoked potential data obtained under a variety of stimulus–response contingencies, it was possible to achieve separation of these exogenous and endogenous processes. Although the absolute contribution of these processes varied greatly from region to region, both exogenous and endogenous activity was demonstrated in most brain regions (Bartlett & John, 1973).

Microelectrode studies indicated that characteristic firing patterns in neural ensembles were correlated with the different waveshapes evoked by the differential conditioned stimuli in such experiments. Movement of chronically implanted microelectrodes revealed that those same characteristic firing patterns to a particular stimulus were found throughout anatomically extensive regions. However, at any point in these regions, two markedly different patterns of discharge were elicited by the two discriminated signals. The response of single cells to individual stimulus presentations was highly variable, but poststimulus histograms to repetitions of that stimulus always converged to the shape characteristic of the response to that signal (John & Morgades, 1969b, c).

In view of these findings and a body of related considerations, we proposed that the information about a conditioned stimulus was represented by the time course of nonrandom firing in anatomically extensive neuronal populations rather than by the occurrence of activity in any particular set of synaptic pathways. Activation of the memory of that sensory event and associated behavioral contingencies was assumed to reproduce the same statistical firing pattern (John, 1972).

These findings suggest that a sensory stimulus influences widespread regions of the brain, which become organized into a representational system, integrated by the reticular formation (RF) and the associated diffuse projection system, and store information about that experience. Whether entering or being retrieved from the different portions of that system, information was proposed to be represented by the time sequence of deviations from random or baseline firing, averaged across large neuronal ensembles. Afferent input into these regions caused the exogenous processes, which in turn activated the release of particular readout processes.

This theory, based upon the presumed informational significance of ensemble firing patterns and associated evoked potential waveshapes observed to correlate with discriminative behaviors, is amenable to direct test. Specifically, if the meaning of sensory cues were indeed encoded as the average firing pattern of anatomically widespread neuronal ensembles rather than as firing in selected synaptic pathways, it should be possible to elicit performance of previously learned discriminative behaviors by using electrical stimulation of the brain to fire large numbers of neurons in the appropriate patterns.

Significant occurrence of stimulus generalization between differentiated sensory stimuli and analogous electrical stimuli delivered directly to various brain regions, or rapid transfer of differentiated response from peripheral to central stimuli or from central site to central site would constitute support for statistical theory because it would be implausible that such gross electrical excitation could fortuitously duplicate some hypothetical precise pattern of synaptic activation elaborated to represent the effects of the earlier learning experiences.

If it proved possible to produce differential behaviors by stimulating the brain with different temporal patterns of electrical input, it would be necessary to ascertain whether such electrical signals merely mimicked the sensations caused by peripheral discriminative stimuli or whether they actually simulated the activation of a memory arbitrarily selected by the experimenter. Selective retrieval of a specific memory might be inferred if direct stimulation of a particular brain region could successfully contradict concurrent conditioned stimuli independent of sensory modality. In this chapter we report the results of experiments designed to test these propositions.

II. PRIOR STUDIES OF STIMULUS GENERALIZATION

A body of earlier work on stimulus generalization involving direct electrical stimulation of the brain was relevant to this undertaking. Such studies are conveniently divided into the following five categories.

A. Stimulus Generalization to Brain Stimuli after Peripheral Training

The results of such studies have been contradictory. Where stimulus generalization (SG) has been reported, initial training has either been to a minimal criterion or no differential training has been involved (Doty & Rutledge, 1959; Kitai, 1965, 1966; Neider & Neff, 1961). When high criterion levels and discriminative training have been involved, negative SG has been obtained (Leiman, 1962; Schuckman, 1966; Schuckman & Battersby, 1966). One of the few recent studies to address this problem was reported by Kelly and his co-workers (1973) who trained cats to discriminate between a visual stimulus plus tone versus a tone alone, using a conditioned avoidance paradigm. It should be pointed out that the discrimination involved was between the presence and the absence of a visual stimulus rather than between the qualities of two visual stimuli. Stimulus generalization to central stimuli after peripheral training was obtained from the optic chiasm and lateral geniculate at high levels. Intermediate stimulus generalization was obtained from stimulation of the optic radiation and the superior colliculus, and very low levels of SG were obtained in response to stimulation of the visual cortex. Using high criterion levels but no discrimination training, Clark and his associates (1972) studied the effects of central stimulation after conditioned avoidance response (CAR) training to tones of various frequencies. Good SG was observed from tone to stimulation of the cochlea, and some SG occurred when lateral lemniscal or auditory cortex sites were stimulated.

Probably the most direct antecedent to our own studies was the work of Livanov and Korol'kova (1951) who stimulated the motor cortex at 3 per sec after a conditioned limb flexion response had been established to 3 per sec flicker, and demonstrated that such cortical stimulation elicited limb movement with the shortest response latency.

B. Stimulus Generalization after Training to Subcortical Stimulation

Most studies of generalization between subcortical loci have produced negative results (Knight, 1964; Nielson, Knight, & Porter, 1962; Schuckman & Battersby, 1966). Nielson and his co-workers found some SG between the mesencephalic reticular formation (MRF) and center median or the superior colliculus, and between two levels of the medial lemniscus, but used a very low criterion (60%). Buchwald and his associates (1967) found SG between the contralateral caudate nucleus and the ventrolateral nuclei of the thalamus after caudatal training. Stutz concluded in 1968 that SG does not occur between brain structures which are not functionally interrelated. This worker reported SG between parts of the limbic system. In 1970, Schuckman and his associates found no SG to stimulation of the contralateral lateral geniculate after training of the other geniculate, nor was there transfer to striate cortex. Conversely, no SG was observed in the lateral geniculate (LG) after training of striate cortex. However, SG was obtained when any striatal region was stimulated after striate cortex training. We have confirmed Schuckman's observation of failure to obtain SG to stimulation of the contralateral LG after differential training of one lateral geniculate. Interestingly enough, prompt SG was obtained when we shifted the site of stimulation from the right MRF to homotopic contralateral MRF electrodes. This difference may pertain to our discussion of the different abilities of MRF and LG stimulation to contradict peripheral stimuli. Pusakulich and Nielson (1972), Nielson (1968), and Gerken (1971) studied changes in threshold to subcortical stimuli as a function of drugs, ECS, or various stimulus parameters, but did not explore the questions of transfer which are of special concern to this study.

C. Stimulus Generalization to Cortical Sites after Training to Cortical Stimulation

There is general agreement that SG does not occur when cortical regions other than the site of training are stimulated (Doty, 1969; Grosser & Harrison, 1960; Schuckman & Battersby, 1966). There is some controversy as to whether SG occurs between different sites within the same cortical area. Doty (1965, 1969), Schuckman (1966), and Schuckman and Battersby (1966) agree that SG occurs when other striate cortical sites are stimulated. Schuckman, Kluger, and Frumkes (1970) concur in this finding. Freeman (1962) failed to obtain SG to stimulation of the contralateral prepyriform cortex after training to prepyriform stimulation. Similarly, Woody and Yarowsky (1972) failed to obtain SG to nearby points on the coronal–precruciate cortex after training of stimuli delivered to interspersed electrodes. It is noteworthy that the CR was not dependent, in this study, upon changes in the "fine structure" of the stimulus when stimulus parameters were changed.

D. Stimulus Generalization to Subcortical Sites after Cortical Training

Schuckman and his colleagues (1970) have confirmed our earlier finding (John, 1963) that there is no SG to stimulation of the lateral geniculate after training of the visual cortex.

E. Stimulus Generalization to Cortical Sites after Subcortical Training

Neider and Neff (1961) reported good transfer from the inferior colliculus to the auditory cortex. Similarly, Doty (1965) reported SG from lateral geniculate to optic radiations (in one monkey subjected to five test trials!). Both of these studies used low criteria and no discrimination training. Leiman (1962) reported SG to stimulation of the marginal gyrus after lateral geniculate training. Schuckman and co-workers (1970) reported no SG from lateral geniculate to the contralateral lateral geniculate or the striate cortex, and conversely, no SG to lateral geniculate stimulation after striate training. SG was found after striate training only in other striatal areas.

These various findings suggested relatively difficult access to the mechanisms established during learning which mediated conditioned responses to stimuli of one modality when stimuli were delivered via another modality. However, in most of these earlier studies the conditioned stimulus was a sensory event of some particular quality. Little or no effort was made to construct generalization stimuli which were analogs of the original conditioned stimulus along some stimulus dimension. Under such circumstances, two possible outcomes of the search for stimulus generalization might be predicted. On one hand, if no discrimination training was involved and low criteria of learning were accepted, a broad generalization gradient might be established such that high levels of stimulus generalization were obtained. Such apparently high levels, however, might well be spurious, arising from nonspecific factors such as changes in the overall level of excitation or arousal. On the other hand, were differential conditioning established to high criteria, no stimulus generalization might be obtained because of the establishment of sharp generalization gradients. Previous results in studies of this genre seem to correspond to this pattern.

III. SENSORY–SENSORY TRANSFER OF FREQUENCY DISCRIMINATION

Since the purpose of the brain stimulation studies reported herein was primarily to provide a critical test of our theoretical formulations, it seemed preferable to bias our procedures against any possibility of spurious generalization so that positive results could be construed as strong support for the theory. In the studies to be reported, accordingly, animals were trained to very high levels of discrimination between auditory or visual stimuli presented at two different repetition rates. The central stimuli utilized in tests of stimulus generalization were brief trains of electrical pulses delivered at the same repetition rates as the peripheral discriminanda, thus constituting a set of analogous signals designed to mimic the neuronal excitation postulated to represent the sensory stimuli used during initial training.

In six cats, approach–approach or avoidance–avoidance discriminations were established to either flicker or click at two different repetition rates, indicated by subscripts 1 and 2. All procedures were carried out in a 2 × 2 × 2 foot apparatus with a work panel bearing pedals and dippers on the left and right sides and with a shock grid floor placed inside a soundproof room. Sensory stimuli were delivered from sources in the roof of the apparatus. Training to auditory and visual cues was first carried out using conventional shaping procedures, with counterbalanced sequence of sensory modalities. Results of initial training and transfer to the second peripheral sensory modality are shown in Table 11.1.

IV. PERIPHERAL–CENTRAL TRANSFER
OF FREQUENCY DISCRIMINATION

After substantial overtraining of the discriminations A_1 versus A_2 and V_1 versus V_2, stimulus generalization and transfer to direct electrical stimulation of various brain structures was studied. Brain stimuli were delivered to bipolar electrodes, chronically implanted into a wide variety of brain structures by means of a flexible cable connected to a subminiature plug mounted on the skull (John, 1973a; Kleinman & John, 1975). Complete details about training methods, stimulation procedures, thresholds, stimulus generalization, and transfer between visual, auditory, and central stimuli are available elsewhere (John & Kleinman, 1975).

Using bursts of electrical pulses delivered at the rates corresponding to the peripheral signals, the effects of the direct reticular formation stimuli, RF_1 and RF_2, were first explored. The brain stimuli were occasionally introduced into random sequences of A_1, A_2, V_1, and V_2, using a sequence counterbalanced for modality and frequency of the previous stimulus. High initial levels of stimulus generalization were displayed by all six cats, with a mean discrimination accuracy of 61.7% on the first day, as seen in Table 11.2.

Of the 300 initial trials of stimulus generalization to reticular input performed by this group (first 50 trials for each cat), 235 resulted in performance of one or another conditioned response (78%). Seventy-two percent of the conditioned responses were correct discriminations. For three of these animals the probability of obtaining the observed discrimination levels by chance was less than the .001 level, a level achieved by the group as a whole. This confirms previous reports of rapid transfer of training from peripheral to RF stimuli (Leiman, 1962) and extends those results to differentiated behavior, providing a control for nonspecific effects.

V. CONFLICT BETWEEN SIMULTANEOUS SENSORY
AND CENTRAL STIMULI

After criterion performance was achieved in response to differential RF stimulation, requiring two to ten days of further training, peripheral stimuli at either repetition rate were combined with RF stimuli at the other rate (conflict). The various compound conflict stimuli, A_1RF_2, A_2RF_1, V_1RF_2, or V_2RF_1, were

TABLE 11.1

Initial Training, Overtraining, and Transfer to Other Sensory Cues

	Cat 1	Cat 2	Cat 3	Cat 4	Cat 5	Cat 6
Response[a]	--	++	--	++	--	++
Initial discriminanda	Tone (600 vs. 1200 Hz)	Flicker plus click (4 vs. 2)	Flicker plus click (5 vs. 1.8)	Flicker plus click (5 vs. 1.8)	Click only (5 vs. 1.8)	Click only (4 vs. 2)
Trials to discrimination criterion (2 days > 85%)	820	840	1380	500	460	540
Trials overtraining	650	380	640 (4 vs. 2)	360	640	660
Second discriminanda	Flicker only (4 vs. 2)	Flicker only (4 vs. 2)	Flicker only (4 vs. 2)	Flicker only (5 vs. 1.8)	Flicker only (5 vs. 1.8)	Flicker only (4 vs. 2)
Trials to discrimination criterion	620	120	680	340	360	40
Percent of original training trials required to achieve transfer	76	14	49	68	78	7.5
Trials overtraining before RF transfer	460	560	680	1480	700	720

[a] --, avoidance–avoidance training; ++, approach–approach training.

TABLE 11.2
Transfer to Various Brain Regions after Peripheral Training

	Cat 1	2	3	4	5	6	Average
Initial transfer to RF							
1st day (%)	40	96	28	76	50	80	61.7
1st 50 trials							
% CR	80	92	86	92	54	66	78
% correct	85^a	87^a	47	63	70^b	85^a	72^a
discrimina-							
tion							
Subsequent transfer to other regions							
1st day (% correct)							
Visual cortex	42	80	44	64	0	–	46.0
Lateral geniculate	90	0	50	50	0	13	33.8
Medial geniculate	0	13	60	59	–	–	33.0
Intralaminar–	100	95	0	75	0	60	55.0
midline thalamus							
1st 50 trials (% CRs)							
Visual cortex	26	88	60	80	33	–	58
Lateral geniculate	54	90	54	88	56	64	68
Medial geniculate	30	86	60	80	–	–	64
Intralaminar–	98	94	6	92	–	58	70
midline thalamus							
1st 50 trials (% correct)							
Visual cortex	54	80^a	63	53	46	–	63^a
Lateral geniculate	48	88^a	44	48	68^b	41	58^b
Medial geniculate	60	58	53	80^a	–	–	64^a
Intralaminar–	76^a	83^a	100	63	–	62	74^a
midline thalamus							

$^a p < .001.$
$^b p < .05.$

inserted in counterbalanced fashion into a random sequence of the individual auditory, visual, and electrical signals. Compound concordant stimuli $A_1 RF_1$, $A_2 RF_2$, $V_1 RF_1$, and $V_2 RF_2$, provided controls for unspecific interaction effects. Data were discarded if performance to individual or concordant cues fell below criterion levels.

In each conflict session, RF current was varied parametrically above and below the usual training intensity. At each current level, several conflict trials of each type were presented and sessions usually included an ascending and descending series. Figure 11.1 shows the results of visual-RF conflict (top) and auditory-RF conflict (bottom). As stimulus current increased, RF input completely controlled the outcome in three of four studies of both kinds of conflict. In the two exceptions, a significant control of behavior by RF stimuli was apparent but limited by disruption of discriminative responses at higher current levels.

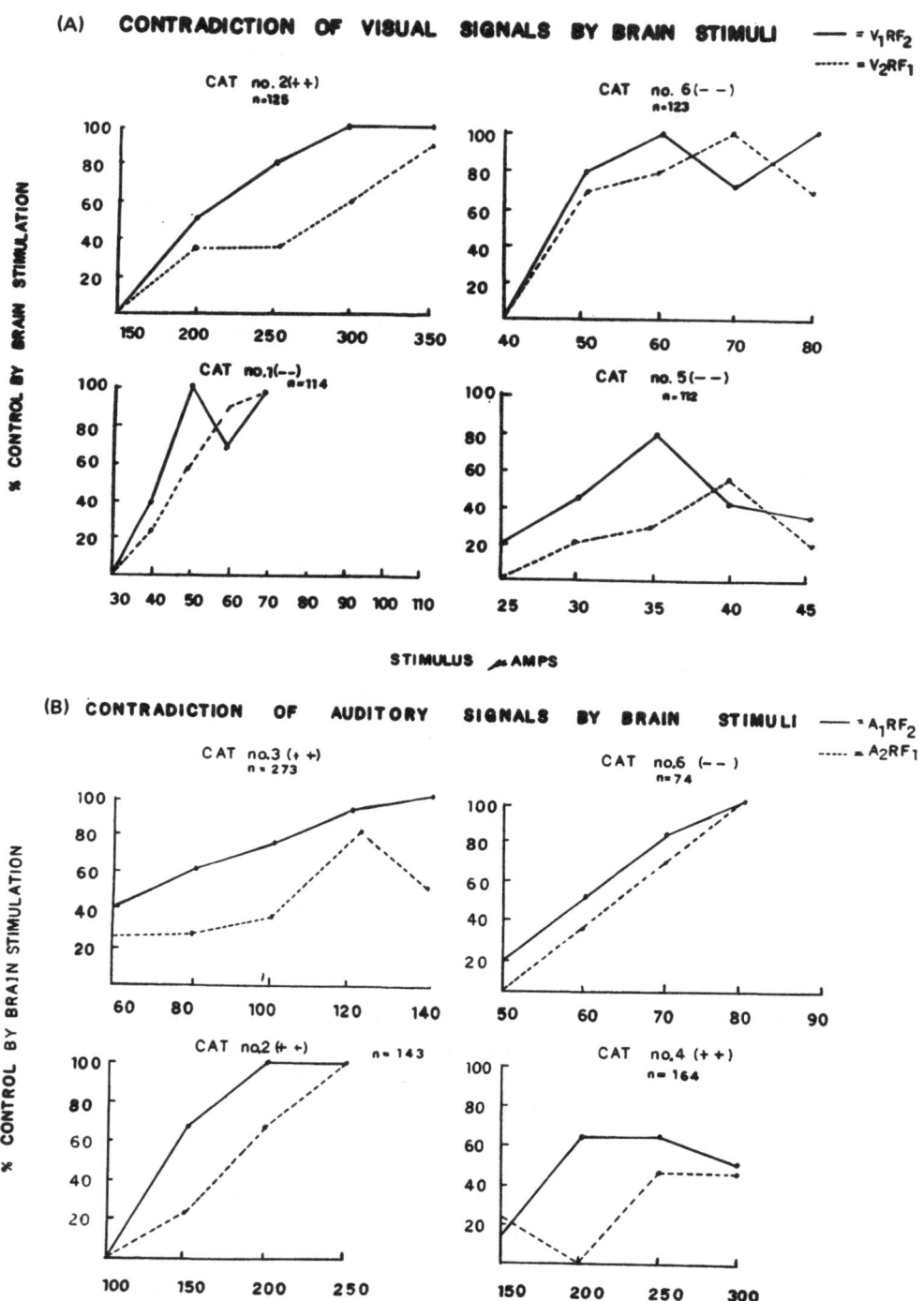

(A) CONTRADICTION OF VISUAL SIGNALS BY BRAIN STIMULI — = V_1RF_2 ⋯⋯ = V_2RF_1

(B) CONTRADICTION OF AUDITORY SIGNALS BY BRAIN STIMULI — = A_1RF_2 ⋯⋯ = A_2RF_1

FIG. 11.1

VI. CENTRAL–CENTRAL TRANSFER
OF FREQUENCY DISCRIMINATION

After completion of these peripheral versus RF conflict trials, transfer of training was initiated to the visual cortex, lateral geniculate, medial geniculate, and the intralaminar nuclei of the thalamus (INT). The order in which these structures were trained was permuted daily. In spite of the substantial experience with electrical stimuli and further overtraining received during RF training and conflict studies, the transfer to these brain regions was slower than the initial transfer to RF. Nonetheless, good stimulus generalization was observed in four cats to INT stimulation, confirming previous reports (Leiman, 1962), while one or two cats showed high levels of stimulus generalization and discrimination ($p < .001$) to each of the other stimulated structures (Table 11.2).

After achievement of criterion performance to lateral geniculate (LG) stimulation, peripheral versus LG conflict was carried out in three cats. Occasional instances were obtained in which an LG at 4 per sec successfully contradicted flicker at 2 per sec, while LG stimuli at 2 per sec seldom prevailed over 4 per sec flicker. Lateral geniculate stimulation was uniformly ineffective to control behavior in auditory-LG conflict.

Our findings of high levels of differentiated stimulus generalization and rapid transfer to RF stimulation provide strong support for the contention that discriminations such as these are mediated by the average temporal patterns of firing in extensive neuronal ensembles rather than by discharges in particular synaptic pathways representing a specific experience. These RF stimuli cannot conceivably reproduce a unique and intricate topology of synaptic discharges corresponding to those normally excited by particular peripheral signals. Undoubtedly, gross electrical stimuli merely impose a corresponding temporal pattern upon masses of cells. The stability of performance when the fine structure of RF stimuli was altered, as well as the stimulus generalization obtained so readily when other brain regions were stimulated, provided further

---

FIG. 11.1 Each graph shows the effectiveness with which stimulation of the mesencephalic reticular formation at either of two frequencies (RF$_1$ and RF$_2$) contradicted simultaneously presented visual (A) stimuli (V$_2$ and V$_1$) or auditory (B) stimuli (A$_2$ and A$_1$), plotted as a function of increasing current intensity. For cats 1, 3, and 6, frequency 1 was 4 per sec and frequency 2 was 2 per sec. For cats 2, 4, and 5, frequency 1 was 5 per sec and frequency 2 was 1.8 per sec. The solid line shows the outcome when peripheral stimulation at the higher frequency (subscript 1) was pitted against RF stimulation at the lower frequency (subscript 2), while the dotted line shows the outcome when the higher frequency stimulus was delivered to the RF. Cats 1, 5, and 6 were trained to perform an avoidance–avoidance discrimination (– –), while cats 2, 3, and 4 were trained to perform approach–approach (+ +). N refers to the total number of conflict trials carried out in each cat, accumulated in 3 sessions for cats 2, 5, and 6 and 4 sessions for cat 1 (visual–RF conflict) and in 3 sessions for cat 2, 4 for cat 6, 5 for cat 4, and 7 for cat 3 (auditory–RF conflict). (From Kleinman & John, 1975.) Note that cat 1.0 numbers are not the same as used in Tables 11.1 and 11.2.

proof that these discriminations do not depend upon activation of specific synapses or pathways (see Fig. 11.2). These results cannot be attributed to nonspecific factors because they require correct discrimination between two different patterns of stimulation applied to the same site.

Lateral geniculate stimulation successfully contradicted visual cues only when the rate of central stimulation was more rapid than the flicker, but completely failed to contradict auditory cues at either rate. Visual cues were hardly ever found successful in contradicting auditory cues. These results suggest that LG stimulation simulates visual sensation. The ability of RF stimuli to preempt control of behavior whether in conflict with visual or auditory cues shows that RF input does not merely simulate the sensations caused by ordinary sensory events, but seems to provide unique access to the brain mechanism which interprets sensory events of whatever modality. These findings suggest that the organized firing of anatomically extensive neuronal ensembles accomplished by patterned RF stimulation simulates the activation of specific memories (John & Kleinman, 1975; Kleinman & John, 1975).

VII. PERCEPTUAL INTEGRATION OF STIMULI SIMULTANEOUSLY DELIVERED TO DIFFERENT SITES

In view of the widespread appearance of both exogenous and endogenous processes throughout the representational system and the ease of transfer of training to direct stimulation of many different anatomic regions in the system, further experiments were undertaken to evaluate the functional implications of those observations. Since animals in this study had been trained to discriminate between stimuli at two different repetition rates (4 and 2 per sec), whether delivered as light flashes, clicks or electrical pulses to RF, visual cortex (VIS), LG, medial geniculate (MG), or midline thalamus (IMT), it became feasible to ask whether the activity in any part of the representational system was the exclusive determinant of perception or whether the brain was capable of integrating activity in disparate brain regions into a perceptual whole.

For this purpose, we carried out an experiment which explored whether the behavioral response to delivery of two trains of 2 per sec stimuli to each of two regions, with one train delayed 250 msec with respect to the other, was appropriate to a 4 per sec or 2 per sec CS (see Fig. 11.2). Although six cats were trained in these discriminations, the majority of the central–central summation data was forthcoming from two animals. The training necessary to achieve high levels of discrimination to so many different signals required much time and the other members of the population succumbed to systemic diseases before providing systematic data. Of the two survivors, one was trained to press the left lever on a work panel to obtain food whenever a 4 per sec stimulus was presented via any modality, and to press the right lever to get food when a 2 per sec stimulus

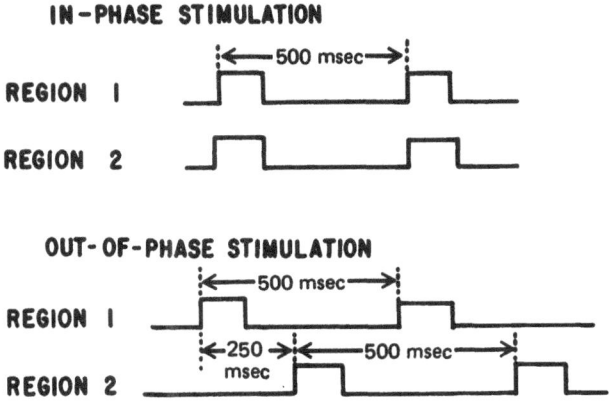

FIG. 11.2 Diagrammatic illustration of in-phase and out-of-phase 2 per sec stimulation of two different brain regions. Note that the stimulus current was not actually a square wave, as schematized, but a high-frequency pulse train (200-nsec biphasic square waves every 5 msec). After training to this "carrier" pulse train, other stimuli such as 100-nsec biphasic square waves every millisecond were utilized. As long as the *intertrain interval* remained constant, the animals performed correctly, indicating that the behavioral response did not depend upon the particular cells stimulated. It is known that pulse trains differing in characteristics of their constituent individual pulses activate different subpopulations.

occurred. The other cat learned the same discrimination but was motivated by shock avoidance. Basically similar results were obtained from both animals.

The postulated structure of the visual decision-making system of the brain is illustrated in Fig. 11.3.

Each of these input modalities (Fig. 11.3, 1–5) was trained to a high level of discrimination between 4 per sec and 2 per sec stimuli. Summation experiments were then carried out in which 2 per sec stimuli via one modality were systematically combined in- and out-of-phase with 2 per sec stimuli of every other modality. During in-phase combinations, the brain as a whole received 2 stimuli per sec, while during out-of-phase combinations, the brain received 4 stimuli per sec. In either case, any single region received only 2 stimuli per sec. These summation trials were randomly interspersed in long sessions in which random sequences of 4 per sec and 2 per sec stimuli were delivered individually to those sites being tested. Data were discarded if responses to individual stimuli or to concordant, in-phase stimuli fell below criterion level.

The results, summarized in Table 11.3, show that responses appropriate to a 4 per sec stimulus occurred when 2 per sec stimuli were alternately delivered as follows: flicker + LG, flicker + RF, flicker + VIS, RF + LG, RF + IMT, LG + IMT, LG + VIS, IMT + VIS. Inputs to any two levels of the "visual decision system" could be integrated into a unified message. Although the relative ease of summation varied within the different portions of the system, all of these interactions must be accepted as indicative of significant summation if we take

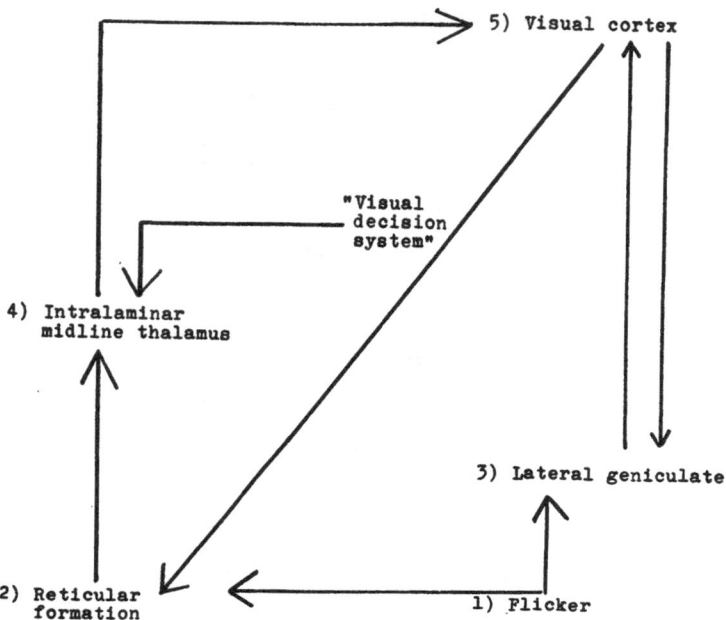

FIG. 11.3 A postulated structure of the visual decision-making system of the brain. Peripheral flicker stimulation. This figure illustrates the heterarchical, as opposed to hierarchical, organization of brain parts capable of integrating portions of messages arising in disparate anatomic regions. The numbers and arrows reflect relative sequencing and communication arrangements rather than a hierarchy of control systems.

into account that these cats normally discriminated almost perfectly between 4 per sec and 2 per sec stimuli delivered to any site. Although interactions within the analogous "auditory decision system" have been less thoroughly studied, they seem more difficult to achieve. The reason for this is not understood.

These data indicate that the brain is capable of integrating portions of messages arising as nonrandom patterns of neuronal activity in many disparate anatomic regions into a conceptual whole. Although the ease of such integration seems to vary somewhat from region to region, there seems little justification for the belief that activity in any single region is decisive for perception. Rather, the evidence indicates that behavior is guided by a synthesis of information from a variety of simultaneously activated regions, forming an overall "gestalt."

VIII. CONFLICT STUDIES BETWEEN SIMULTANEOUS CENTRAL STIMULI

In the hope that these methods might yield insight into whether there existed a critical brain locus for "decision-making," central–central conflict studies were carried out. In such studies, two different brain loci were simultaneously

TABLE 11.3
Behavioral Outcome of Compound Stimulations at 2 per sec[a]

Stimuli	Outcome at 4 per sec (%)	N
Flicker + VIS cortex	92	13
	59	17
Flicker + LG	80	20
	39	33
Flicker + IMT	67	15
	65	20
RF + IMT	66	58
	33	39
RF + LG	64	22
	9	53
Flicker + RF	62	65
	36	31
LG + IMT	54	24
	43	40
VIS cortex + IMT	45	20
Flicker + click	43	40
	12	16
	8	27
VIS cortex + LG	25	20
Click + LG	25	20
Click + RF	25	8
	7	24
	0	10
	0	5

[a]For every brain region, each line represents data from a different animal.

stimulated at two contradictory frequencies in order to ascertain whether any single region always succeeded in preempting the behavioral outcome.

Although the summation technique produced consistent and interpretable results within an animal, central–central conflict provided inconsistent data. Formidable psychophysical problems are involved in equating the effective intensity of disparate brain stimuli, and must be resolved before the method of central–central conflict becomes practical.

The decision to use a particular current level for stimulation of a brain region in conflict is arbitrary. One might select the intensity which is threshold for accurate discrimination, the value usually used in training, or the value which preempts behavior when contradicting some standard reference signal. Each of these choices would be defensible, but each will produce different results. In fact, the outcome in conflict is a parametric function of stimulus intensity in many cases. Often, if stimuli are really balanced in intensity, a decided bias in favor of the higher frequency emerges. After substantial exploration of this method, we decided to set it aside in favor of the more readily interpretable summation technique.

IX. SUMMARY OF CONFLICT STUDIES

A substantial amount of conflict data was gathered. Table 11.4 presents the results of our conflict studies. In *all* the data here tallied, *stimulus current was that usually used during training.*

Bearing in mind the reservations stated above, the sequence of prepotency seems to be visual cortex > RF > flicker. However, flicker > LG, while LG > visual cortex and RF > LG. This suggests a *heterarchical* rather than a *hierarchical* organization. Analogously, click > flicker, click > RF, flicker > LG, but RF > LG. Here, the result is consistent with a hierarchical structure. Part of the problem may be attributed to differences in outcomes of conflict from animal to animal, but some of these internal contradictions were obtained within data from single animals. The method therefore was considered to be inappropriate for the analytical purposes for which it had been intended.

These methods provide a way to evaluate the functional significance of electrophysiological findings which indicate that a particular anatomic region participates in the representational system mediating a particular memory. A particular behavior may occur due to some unknown cause which also produces the observed electrophysiological correlates, or the behavior may actually result from the same neural activity which is directly reflected as electrophysiological readout patterns. Once it had been established that identifiable, anatomically extensive electrophysiological events correspond to the readout from memory, it was necessary to shift to other levels of analysis in addition to the correlative level.

X. ELECTROPHYSIOLOGICAL FINDINGS

Once the summation experiment began to yield interesting behavioral results, it became important to study the evoked potentials propagating to other regions of the brain from the two stimulated central structures whose interactions were being evaluated behaviorally.

Figures 11.4–11.7 illustrate some of these findings in one animal. Each of these figures shows the activity recorded from the intralaminar–midline thalamus under two conditions: *top*—when presentation of flicker stimulus at 2 per sec plus brain stimulation at 2 per sec *delayed 250 msec* resulted in performance of the behavior appropriate to a 4 per sec CS; *bottom*—when the same compound stimulus resulted in behavior appropriate to a 2 per sec CS. In Fig. 11.4,

TABLE 11.4
Outcome of Various Conflict Studies

Conflict (S₁ ys. S₂)			Wins (%)		X²	N Trials	N Cats	Outcome
			S₁	S₂				
Click vs. flicker	Wins	Freq. 1	55	35	7.19	149	3	Click > flicker
	(%)	Freq. 2	65	45	p < .01			
Click vs. RF	Wins	Freq. 1	58	55	14.97	766	6	Click > RF
	(%)	Freq. 2	45	42	p < .001			Freq. 1 wins
Flicker vs. RF	Wins	Freq. 1	46	61	16.68	649	4	RF > flicker
	(%)	Freq. 2	39	54	p < .001			
Click vs. LG	Wins	Freq. 1	83	40	27.40	116	3	Click > LG
	(%)	Freq. 2	60	17	p < .001			Freq. 1 wins
Flicker vs. LG	Wins	Freq. 1	82	67	21.82	80	2	Flicker > LG
	(%)	Freq. 2	33	18	p < .001			Freq. 1 wins
RF vs. LG	Wins	Freq. 1	56	61	5.24	164	2	RF > LG
	(%)	Freq. 2	39	44	p < .03			Freq. 1 wins
LG vs. IMT	Wins	Freq. 1	100	75	5.00	8	1	Freq. 1 wins
	(%)	Freq. 2	25	0	p < .03			
LG vs. MG	Wins	Freq. 1	100	100	12.00	12	1	Freq. 1 wins
	(%)	Freq. 2	0	0	p < .001	12	1	
Click vs. cent. lat.	Wins	Freq. 1	97	43	26.67	60	1	Click > CL
	(%)	Freq. 2	57	3	p < .001			Freq. 1 wins
Vis. Cx. vs. LG	Wins	Freq. 1	33	67	3.33	30	1	LG > vis. cx.
	(%)	Freq. 2	33	67	p = .10			
Vis. Cx. vs. RF	Wins	Freq. 1	60	35	2.80	45	1	Vis. cx. > RF
	(%)	Freq. 2	65	40	p = .10			
Vis. Cx. vs. flicker	Wins	Freq. 1	50	38	2.00	54	1	Vis. cx. > flicker
	(%)	Freq. 2	62	50	p = .20			

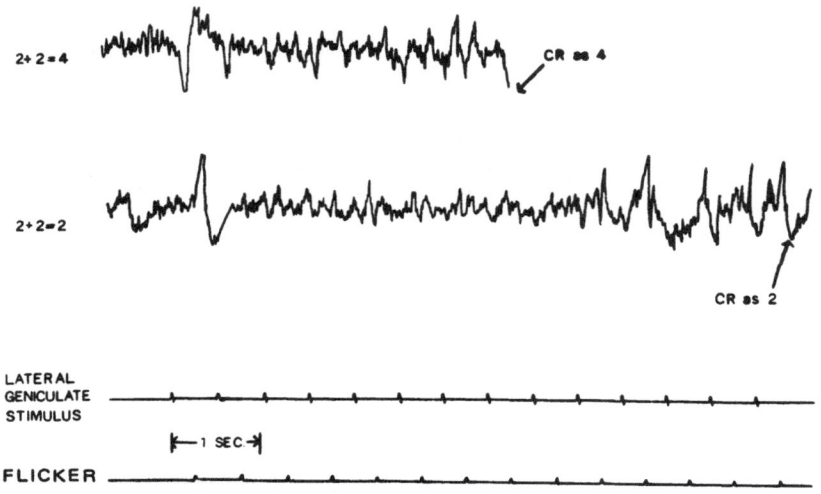

FIG. 11.4 Effects of out-of-phase 2-Hz stimulation (depicted in Fig. 11.2) delivered to different combinations of places in the visual system depicted in Fig. 11.3. Records from intralaminar thalamus when (top) animal summated lateral geniculate and flicker stimulus resulting in CR appropriate to 4 per sec stimulus; (bottom) intralaminar record when animal failed to summate, resulting in CR appropriate to a 2 per sec stimulus. (Unpublished observation from Kleinman & John, 1975.)

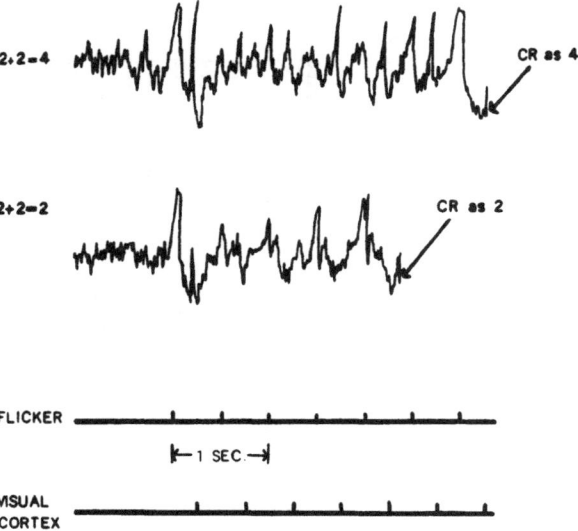

FIG. 11.5 Same as in Fig. 11.4 except brain stimulus was delivered to the visual cortex during flicker stimulation. (Unpublished observation from Kleinman & John, 1975.)

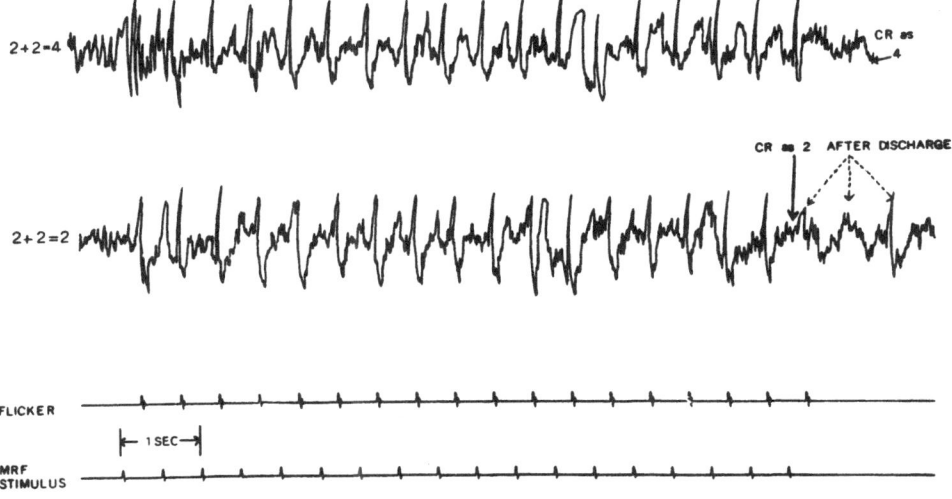

FIG. 11.6 Same as in Fig. 11.4 but brain stimulus delivered to reticular formation.

the brain stimulus was delivered to the lateral geniculate; in Fig. 11.5, to the visual cortex; and in Fig. 11.6, to the MRF. No matter what the site of the central stimulus, if the 2 per sec flicker stimulation interacted with the 2 per sec brain stimulation so as to elicit behavior indicating that the two signals had been informationally merged, 4 per sec rhythms were prominent in the intralaminar thalamic records. If the two signals were not effectively combined, the intralaminar activity was dominated by a 2 per sec rhythm. The anatomic basis for this observation is not obvious. Perhaps the effect of the lateral geniculate and visual cortex stimuli on the midline thalamus is via a corticofugal pathway, while the effect of the MRF is via an ascending pathway. This observation, thus far not

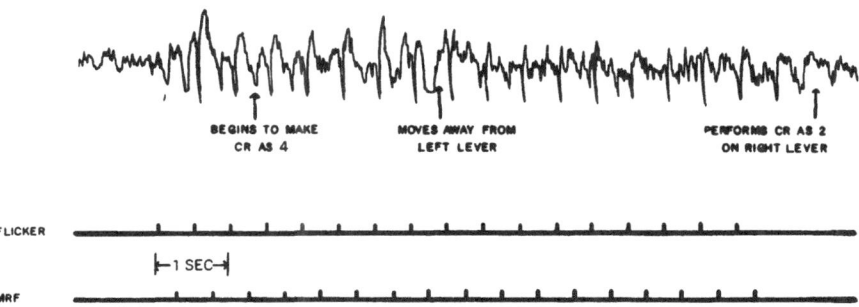

FIG. 11.7 Same conditions as in Fig. 11.6 illustrating the correlation between changes in EEG frequency and vacillation in animals behavior. (Unpublished observation from Kleinman & John, 1975.)

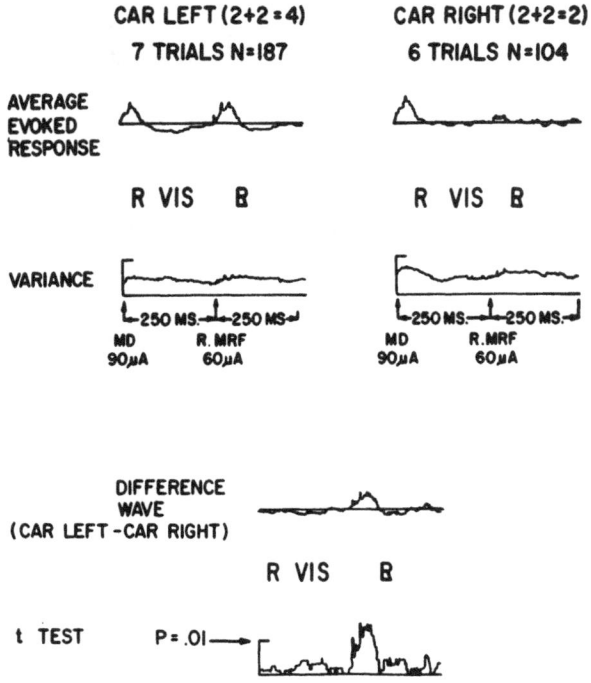

FIG. 11.8 Averaged evoked responses (top) recorded from the visual cortex to 2 per sec out-of-phase stimulation delivered to the medialis dorsalis and mesencephalic reticular formation. Evoked response waveshapes (left and right) are different depending on behavioral outcome. Difference wave and *t* tests between ERs on top, left, and right at bottom. See text for details. (Unpublished observation from Kleinman & John, 1975.)

paralleled by phenomena seen in recordings from other structures, suggests that the intralaminar and midline thalamus plays a central role in decision-making under the conditions of this experiment. This would be compatible with the known role of these structures in influencing the temporal fluctuations of excitability in widespread brain regions, especially since relative timing is the dimension differentiating the discriminanda used in this study.

Figure 11.7 shows that as the animal displays behavioral vacillation to compound flicker plus MRF stimulation, the intralaminar record shows a correlated shift from 4 per sec to 2 per sec rhythms. This correlation has been repeatedly observed in this cat.

Figure 11.8 shows that when compound stimulation of nucleus medialis dorsalis plus MRF at 2 per sec results in behavior appropriate to a 4 per sec signal, the ipsilateral visual cortex shows a clear average evoked potential to both brain stimuli, producing four distinct afferent volleys per sec into the visual cortex. However, when the compound brain stimulation produces behavior appropriate to a 2 per sec signal, only the medialis dorsalis stimulus elicits the cortical response. The evoked potential to the MRF stimulus is barely discernible. Thus, the cortex receives only two effective afferent volleys per sec in this case. The t-test computation at the bottom of Fig. 11.8 shows clearly that the differences in the average responses are statistically significant.

These observations suggest that when effective summation fails to occur, either the MRF stimulus fails to reach the visual cortex or the threshold of the visual cortex to that input has been raised. These results, taken together with those illustrated in Figs. 11.4–11.7, suggest that a cortical–intralaminar thalamus loop may play a central role in mediating the temporal discrimination required in this task. It is noteworthy that none of the other structures simultaneously recorded while the data in Fig. 11.8 were acquired showed a comparable phenomenon. No significant differences were obtained between the average responses recorded from auditory cortex, lateral geniculate, medial geniculate, centralis lateralis, dorsal hippocampus, hypothalamus, or contralateral visual cortex under these two conditions.

XI. CONCLUSIONS

The slow transfer from the initial compound CS to the flicker discrimination indicates that the mechanism established during training of discrimination between compound auditory–visual stimuli at two different repetition rates is not readily accessible to visual stimuli alone at the same two repetition rates, suggesting that the prepotent auditory component of the stimulus complex somehow inhibits establishment of learned relationships to the visual component. Nonetheless, sufficient savings result to indicate that stimuli which share common timing relationships gain access to common neuronal mediating mechanisms. Such mechanisms probably include nonsensory specific structures of the brain, a conclusion supported by the observation that electrophysiological responses to the CS in such situations are anatomically widespread, appear in the RF and the IMT, and acquire marked long latency components.

In contrast to the preponderantly negative results obtained in prior studies of SG to brain stimulation when high criterion levels and differentiated conditioned responses have been utilized, our results were strongly positive. Although high criterion levels (85% for many days) and differentiated approach–approach or avoidance–avoidance responses were used in these experiments, high levels of discriminated responses to RF stimuli were rapidly obtained in almost every animal. Significant response levels were also achieved to stimulation of other brain regions after RF training. These results are particularly impressive in comparison with the relatively slow transfer observed from the initial sensory CS to the flicker CS.

Stimulation of RF seems to provide uniquely effective access to the mechanism elaborated during sensory discrimination training. In most cases, the levels of discrimination performance observed during initial transfer of training to RF were higher than those observed during the subsequent transfer of training to other brain regions. The additional overtraining to the sensory discriminanda, and the general facilitation of central transfer which might result from prior experience with differential response to RF stimuli, were insufficient to overcome this apparently greater effectiveness of the RF.

One might question whether the RF stimuli were so effective because they actually simulated the perceptual effects of the peripheral cues or because they simulated the release of particular memories of the cues corresponding to the central stimulus repetition rate. Since the RF is a nonsensory specific system, it seems highly unlikely that RF stimulation would simulate a flicker or a click. In fact, we found that RF stimulation could successfully contradict either a concurrently presented flicker or click cue. We believe that the most plausible explanation for these findings is that the RF stimuli simulated the activation of particular memories. Significant levels of CR performance were obtained as soon as brain stimulation was delivered to any of the regions studied. In some cases, these initial responses were random with respect to direction, giving no evidence for differential transfer. A few cats showed a consistent side preference until stimulus currents were adjusted to proper levels, producing a spuriously low initial level of discrimination. However, for each site studied, at least one animal showed near perfect discrimination with little difficulty.[2]

This high level of SG may be attributed to the fact that the critical dimension distinguishing between the discriminanda used in this study was *timing*, a parameter which may be uniquely amenable to simulation by central stimuli. Although we concede that the ease of transfer from peripheral to central stimuli obtained in our studies may well be due to the fact that timing is a stimulus quality which lends itself to simulation more readily than the stimuli used in other studies of stimulus generalization, it does not follow that the relevance of the present studies to the primary question herein investigated is thereby nullified. Whether or not stimulus timing is readily analogized from modality to modality, it seems indisputably clear that the mechanism which encodes and stores information about stimulus–response contingencies in these experiments does not depend upon the activation of specific synaptic pathways. Conceivably, one might contend that these central stimuli are so massive that *all* synaptic pathways are driven at the same repetition rate, necessarily including that discrete subset upon which the discrimination supposedly depends. The relatively low current levels required for transfer, the small evoked responses observed in electrophysiological studies under these circumstances, and the ease with which RF stimuli but not LG stimuli contradict flicker or click stimuli

---

[2] As simultaneous central training at four different sites proceeded some animals seemed to show deterioration from their initial levels of performance. The reason for this is not understood.

(Kleinman & John, 1975) all argue against this rebuttal. The ease of transfer of training to other brain regions, observed in at least one and sometimes more animals for each structure studied, contradicting the general conclusion in the literature that stimulus generalization does not occur between brain structures which are not functionally interrelated, suggests that the different parts of the representational system become functionally integrated by virtue of their shared mediation of frequency discrimination required in this task.

The fact that the differential performance was unaltered when the frequency of central pulse trains was changed from 200 pulses per sec to 100 pulses per sec, when the individual pulse width changed from 200 to 100 nsec, or when the train duration was shortened from 30 to 10 to 2 or even 1 pulse establishes that the discrimination was independent of the "fine structure" of the brain stimulus. However, *these changes in stimulus fine structure undoubtedly were accompanied by changes in the subset of neurons locally excited by the electrical input, since radically different recovery cycles ensued.* This finding provides further support for the notion that the mediating mechanism is based upon temporal coherence in the stimulated neuronal ensembles rather than upon excitation in specific synaptic pathways which must fire for a particular learned response to be performed.

In comparison with prior reports in the literature, our findings seem unique. Most previous results indicated very limited access to the mechanisms mediating the memory of conditioned responses appropriate to stimuli of one sensory modality when other sensory modalities or particular brain regions are stimulated. We may have uncovered a unique common denominator, facilitating access to the system storing information about differentiated conditioned responses by virtue of the temporal parameter which defined the discriminanda used in our studies.

On one hand, the uniqueness of our results gives some cause for concern. We found ready interaction and rapid transfer of training between central sites, in contrast to most workers who used similarly high criteria and differential training. On the other hand, bearing in mind the reservation that temporal patterning is a special class of discriminanda, we seem to have found a stimulus parameter (timing) which provides ready access to brain mechanisms of information processing and storage. Our findings provide strong support for a statistical theory of information processing and are difficult to reconcile with theories which attribute learning to facilitation of a selected subgroup of synaptic pathways.

In view of these findings, it seems important to exploit the advantages offered by these methods for further elucidation of brain mechanisms of information storage and decision-making, while at the same time devising experiments to ascertain whether the brain mechanisms revealed by these studies are relevant to the coding, storage, and retrieval of other kinds of information.

12
Mental Experience

How it may be that ganglionic activity is transformed into thinking and how it is that thought is converted into the neuronal activity of conscious voluntary action we have no knowledge. Here is the fundamental question. Here physiology and psychology come face to face. We are far from this final understanding and life is short! [W. Penfield, 1954]

I. MAJOR AVENUES OF INVESTIGATION OF MENTAL EXPERIENCE

In the initial phases of the history of neuroscience, the paramount concern of workers in the field was with investigations of sensory, motor, and reflex functions. More recently, especially during the last quarter of a century, there has been steadily increasing attention to neural activity concerned with behavior. Research on the physiological basis of consciousness and mental experience has occupied a very small fraction of the attention of workers in neuroscience. If we examine the detailed content of the studies in this domain, there seem to be four lines of exploration which have had a major influence upon our thinking about the neural basis of subjective experience. The first avenue of investigation stemmed from two initial observations: (1) direct electrical stimulation of the reticular formation of the brain stem induced changes in the EEG seemingly identical with those observed in awakening from sleep or alerting to attention, and would cause behavioral arousal in a sleeping animal (Moruzzi & Magoun, 1949). This observation was followed by a series of observations that afferent paths subserving all sensory modalities make connections with the central reticular activating system (see Magoun, 1954). (2) The great functional importance of this system for the maintenance of consciousness was revealed by studies showing that lesions in its rostral portion caused a comatose state.

Animals with such lesions could not be aroused and displayed a chronically hypersynchronous EEG with no alteration by peripheral stimuli (Lindsley, Schreiner, Knowles, & Magoun, 1950). These findings provided the basis for interpretation of the earlier observation by Bremer (1936) that transection of the upper brain stem leaves the cerebral hemispheres in a permanent state of sleep. These observations led to a vast proliferation of studies of the reticular formation of the brain and its ascending and descending influences (Jasper *et al.,* 1958), revealing that a great amount of integrative activity takes place within as well as through the mediation of this system. However, the conclusion that the reticular formation was the site where conscious processes were localized or that it was essential for consciousness, a view to which many were and still are inclined, was invalidated by such studies as those of Adametz (1959) and Chow (1961), showing that if the reticular formation was damaged bilaterally with lesions far more extensive than those involved in the studies above *but inflicted in several stages,* no interference with consciousness or arousal was observed. Nonetheless, although the reticular formation may not be essential for consciousness, it seems to play an important role in the mediation of conscious processes in the intact brain.

The second major line of investigation related to study of natural alteration in consciousness as it takes place in passing from sleep to wakefulness and vice versa. Such studies, pioneered by the work of Kleitmann (1939), revealed that a complex alteration of functions occurred in sleep, some facilitatory and others inhibitory in nature. In detailed analyses of the functional organization mediating these complex changes, Hess showed the existence of a diencephalic region in which direct electrical stimulation caused a progressive decrease in activity leading to natural sleep (1949). These observations led to a still continuing series of studies on the mechanisms causing sleep, the levels or stages of sleep, and their neural concomitants. Of these, perhaps the one which has received greatest attention is the so-called REM (rapid eye movement) stage in which most dreams occur. Evidence (Dement, 1960) suggests that there is a specific need for this kind of desynchronized sleep, which some workers believe involves the consolidation of the day's experiences and recovery of the neural systems engaged in plastic activities (Moruzzi, 1966).

The third major line of investigation comes from observations of the effects of direct electrical stimulation of the exposed surface of the brain of conscious patients under local anesthesia during brain surgery and studies of the effects of removal of various brain regions. Analyzing the results of numerous such observations, Penfield divided the effects of stimulation into two classes: those which produced elementary sensations or alterations in immediate subjective experiences, and those which produced "flashback," psychical responses consisting of awareness of some previous experiences. Penfield considers these two types of phenomena to represent parts of a scanning mechanism which enables an individual to compare present experience with similar past experience. The

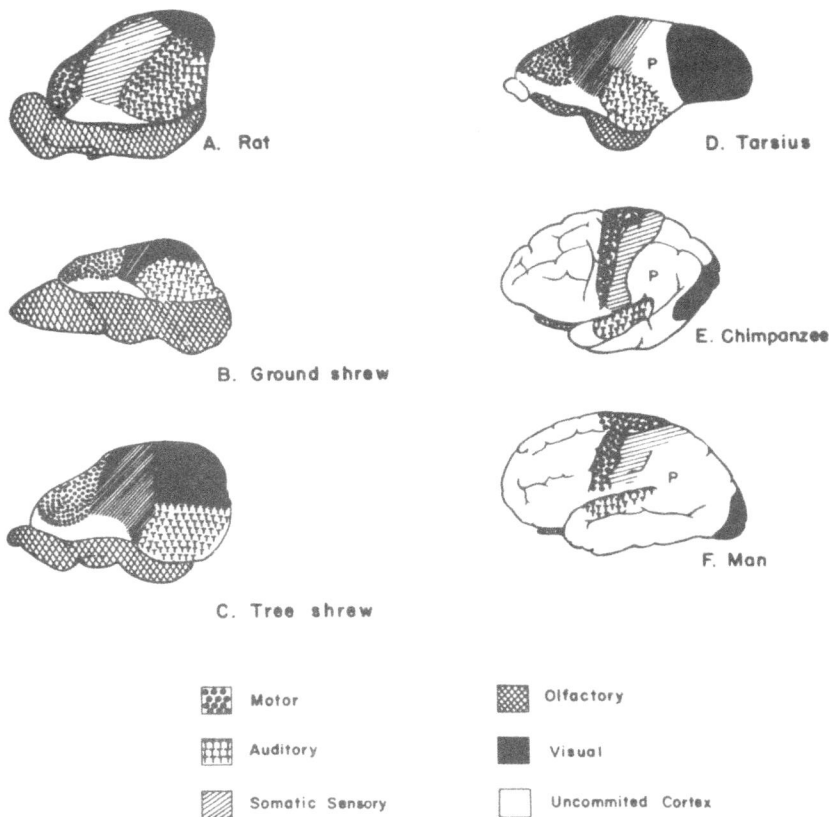

FIG. 12.1   Mammalian brains from rat to man prepared by Stanley Cobb to illustrate the proportional increase of uncommitted cortex (or undetermined cortex) as compared with sensory and motor cerebral cortex. (From Penfield, 1966.)

area of the cortex from which interpretative or flashback responses are obtained has been termed the "interpretive" cortex (Penfield, 1958) and corresponds well to what Stanley Cobb called the "uncommitted" cortex, that is, that portion of the cortex not concerned with sensory or motor processes. Figure 12.1 shows the phylogenetic increase of uncommitted cortex in mammalian brains, from rat to man, while Fig. 12.2 shows the extent of the interpretive cortex in man. Penfield concludes that the mechanism mediating these experiential responses to stimulation of the interpretive cortex is certainly *not* in the cortical convolution which is stimulated, but is at a distance from the point of stimulation because consciousness and memory persist no matter what region of cortex is removed. Largely because consciousness is lost when the midline thalamus or brain stem is damaged, Penfield postulated that the diencephalon and the cerebral

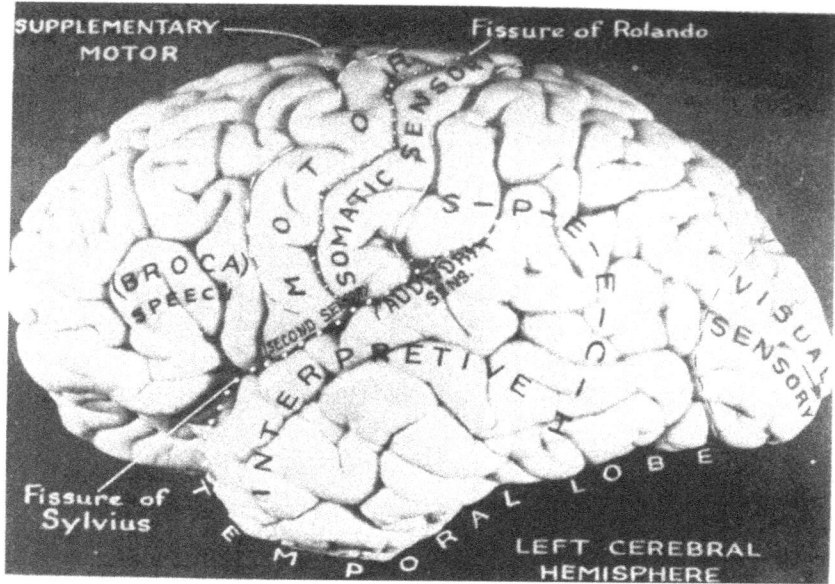

FIG. 12.2 Left cerebral hemisphere (dominant) showing areas devoted to speech, sensation, and voluntary movement. The area of cortex from which electrical stimulation produces experiential flashbacks and interpretive signals is labeled "interpretive." There was no overlap between interpretive and speech cortex as judged by stimulation. No experiential or interpretive responses were produced from the speech area. (From Penfield & Roberts, 1959.)

hemispheres functioned together as an integrated unit to which he gave the name "centrencephalic system." Penfield attributes consciousness to neuronal activity within this system (Penfield, 1954, 1966).

The fourth major line of investigation into the neuronal mechanisms mediating conscious experience consists of the observations conducted on the so-called "split-brain" preparation, particularly recent studies of the effects of surgical separation of the left and right hemispheres of cats, monkeys, and men. Some of these studies were discussed in Chapter 9. Such studies indicate that each of the surgically separated hemispheres must sense, perceive, learn, and remember quite independently of the other; that is, *brain bisection produces an individual with two separate minds, two independent spheres of consciousness.* (Sperry, 1966). It is interesting in this context to recall that Penfield observed that experiential responses relating the consequences of electrical stimulation to a reproduction of a definite past experience were *only* obtained from stimulation of the nondominant hemisphere (Penfield, 1966). As yet, there has been no analysis of the dependence of these two independent systems upon the common reticular system of the brain stem with which they both interact.

Examining the three major symposia concerned with mental activity and consciousness which have appeared in the last 20 years (Adrian, Bremer, & Jasper, 1954; Eccles, 1966; Hernández-Peón, 1963), we find that of the 58 papers therein contained, 13 focus upon analysis of the mechanisms of sleep, 12 upon the detailed function of sensory or motor mechanisms, 6 study anatomic circuitry, 6 report the effects of diverse lesions, 8 describe behavioral observations or psychiatric cases, 4 discuss the effects of brain stimulation, 2 discuss the functional role of various neurochemical substances, 1 is largely confined to description of the behavior of single cells, and 1 describes a variety of observations of animal behavior. Only 5 participants in these symposia, ostensibly focused upon the physiological basis of consciousness and mental experience, actually attempted to confront the question directly. Thus, although there has n substantial attention paid by neuroscientists to the analysis of the mechanisms which sustain consciousness or which alter the content of consciousness, remarkably little attempt has been made to deal with the question of *how* subjective experience is generated by the activity of neuronal masses.

A. Levels of Information

It is inordinately difficult to formulate answers to this question which seem adequate, evading the pitfalls of triviality on the one hand and of useless vague generality on the other. We propose the following definitions as first approximations which provide a basis for an experimental approach.

1. Sensations. Sensations are the spatiotemporal patterns of information arriving in the central nervous system because of excitation of exteroceptive and interoceptive organs. They are a product of the *irritability* of living matter and constitute *first-order* information. Such irritability is manifested throughout the phylogenetic scale, and is already present in protozoans. Sensations can elicit reflex responses adjusting the organism to its environment.

2. Perceptions. Perceptions are the interpretation of the meaning of sensations in the context of stored information about previous experiences. Perceptions constitute *second-order* information resulting from an interaction between sensations and memories.

Wundt and his contemporaries argued that the presence of consciousness was revealed when behavioral responses to stimuli ceased to be reflexive but displayed "purposiveness," by which they meant actions which were adaptive and resulted in the adjustment of the organism to its environment as a function of the experiential context of a stimulus rather than the action of the stimulus alone. For this reason, they considered identification of the lowest phylogenetic level showing learning as crucial for the decision as to the lowest level of organization capable of sustaining consciousness. In this regard, it is noteworthy that Corning, Dyal, and Willows, in their authoritative review of invertebrate learning (1973), reach the conclusion that, although the evidence for simple learning or associative conditioning remains highly controversial, there exists compelling evidence that protozoans display the ability to learn not to respond, that is, habituation, and some evidence for associative learning has been forthcoming. The capacity for complex learning clearly appears in the phylum Platyhelminthes, with the advent of a brain, defined sensory systems, and complex nerve bundles.

We choose to define perception, as well as sensation, provisionally as preconscious or *unfelt* categories of information processing. Sensations and perceptions are unimodal, referring to the detection and interpretation of stimuli within individual sensory modalities. These functions can be performed by machines which do not possess consciousness. We contend that under ordinary circumstances, fundamental sensations and much of perception, as defined, do not enter consciousness, although we can make ourselves aware of them by an analytic process.

3. *Consciousness.* Consciousness is a process in which information about multiple individual modalities of sensation and perception is combined into a unified multidimensional representation of the state of the system and its environment, and integrated with information about memories and the needs of the organism, generating emotional reactions and programs of behavior to adjust the organism to its environment. Consciousness is *third-level* information. Many stages of consciousness can exist in which these dimensions are present in variable amounts. The *content* of consciousness is the momentary constellation of these different types of information.

At the same time that consciousness is the product of an integration of preconscious sensations and perceptions structured in the light of previous experience and reflecting emotional state, drive level, and behavioral plans, feedback from consciousness to these more fundamental levels must take place. Memories are activated, attention is focused, perceptions influenced, emotions aroused, drive priorities altered, and plans of behavior revised as a result of this feedback, producing a continuous reorganization of basic processes due to the influence of higher level integrative and analytical functions.

4. *Subjective experience.* This derives from information about the content of consciousness. It is a process which reorganizes the sequential series of events into a single experiential *episode*, which merges sequential constellations of multisensory perceptions, memories, emotions, and actions into a unified and apparently continuous event, or "experience," that has a beginning and an end. Two critical transformations occur as a result of the process which generates this *fourth-order* information. First, although the information impinging upon the neuronal populations mediating each of the different dimensions of consciousness is represented by the same mechanism (spatiotemporal patterns of neural discharge) in every such population, the fourth-order information about each different dimension of consciousness is qualitatively distinct. Subjective experience consists of diverse colors, shapes, sounds, textures, smells, tastes, emotions, plans, movements, and *thoughts*, rather than a uniformly encoded description of

these disparate facets of experience. Somehow, qualitative diversity at this higher level of information is constructed out of representational uniformity at lower levels. At the same time, in spite of these qualitative distinctions between the different facets of consciousness and the capability to decompose experience into its constituent components, subjective experience merges these facets into an apparently simultaneous and continuous multidimensional unity.

As this unified subjective experience begins to take shape from a related series of episodes, memories relevant to this holistic event are activated, many of them in modalities not involved in the episodes taking place. Some of these memories are of rudimentary or fragmentary sensations, while some are of prior subjective experiences (see below).

5. *The self.* The second critical transformation, as subjective experience extends through time and an individual history is accumulated, is that memory of the sequence of episodes is constructed. This personal history, the accumulated memories of sets of fourth-order information, constitutes the basis for what we call the *self.* The concept of the self arises as a result of long-term memories constituting the record of an individual's subjective experiences. This individual historical record constitutes *fifth-order* information.

6. *Self-awareness.* If we consider subjective experiences as "higher order sensations," then "self-awareness" is analogous to the perception of those sensations. By this is meant the interpretation of subjective experience in terms of the previous history of life experiences of the individual. Self-awareness is the interpretation of present subjective experience in the context of the salient features, especially the more invariant features, of the pattern of previous subjective experiences. Self-awareness constitutes *sixth-order* information.

As the momentary content of consciousness is interpreted in the light of past experience, feedback to lower levels occurs which is probably more powerful than any described thus far. This feedback can be expected to activate trains of memories of other relevant life experiences, with a high probability that important occurrences (high drive level, high emotion events) will be followed by systematic or "rational" memory searches. The relatively global feedback resulting from integration of lower level information as it enters consciousness is modulated and made far more selective and better focused. Systematic evaluation of a flood of memories, identification of appropriate perceptions and rejection of more inappropriate perceptions which arose earlier in the experience, selection of the most appropriate emotional response, adjustment of drive levels to correspond to the exigencies and possibilities of the moment, and rational construction of the optimal program of behavior are among the consequences envisaged as resulting from this highest level of information. These processes are far more deliberate and analytical than those previously described.

A characteristic of self-awareness is the capacity for cognitive processes. By cognition or thought we mean the ability to have subjective experience vicariously by activating stored memories about perceptions and prior experiential episodes in a fashion which may be arbitrarily organized rather than occurring according to a previously established sequence. Because of this ability to manipulate, recombine, and reorganize the accumulated store of memories, the self is continuously in the process of modification and of analysis of its own experience.

A cognitive process is the representation of an *experience* in an abstract symbolic fashion, whether or not that experience actually occurred in that form in the personal history of the individual. The distinction between the memory of a rudimentary sensation postulated as essential for perception and the memory of a subjective experience is the amount and diversity of the stored information. The basic neurophysiological mechanisms may be quite the same, and even the anatomic loci may be shared. We see no compelling need to separate those mechanisms conceptually. Under some circumstances, particularly when cross-modal stimulation is utilized, generalization affords evidence for the presence of cognitive processes. In generalization, an organism interprets the meaning of a sensory stimulus as equivalent to some other sensation because of similarities in the abstract properties common to both stimuli. If the two stimuli are in different sensory modalities, it is clear that some nonsensory-specific abstraction has been performed. If the stimuli are in the same sensory modality, interpretation becomes more equivocal because of the possibility that similar receptors were activated.

Observational learning seems to constitute a more unequivocal type of evidence for the presence of cognitive processes. We and others (Grinberg-Zylberbaum *et al.*, 1974; John, Chesler, Bartlett, & Victor, 1968) have shown that naive animals can learn complex discriminative behaviors simply by observing the performance of trained animals. Since the observing animals do not directly experience the reinforcing stimuli, their acquisition of the discrimination must be attributed to the interpretation of what they observe by referring it to memories of previous experiential episodes which they did experience directly. Observational learning already requires consciousness, and probably requires self-awareness.

We have found it useful to postulate a series of different levels of information processing, each dependent upon all the levels below (feedforward) and each influenced by the levels above (feedback), in order to define sensation, perception, consciousness, the content of consciousness, subjective experience, the emergence of a self concept, and self-awareness. The proposed definitions treat each of these processes as fundamentally similar to all the others in that they are all representations of information, presumably in a common neuronal code. They are all different in that they constitute successively higher derivations extracted from the information representing the lower derivations. These ideas are illustrated in Fig. 12.3 which has been limited to two sensory modalities to simplify the diagram.

If we accept these formulations as working definitions, the task of experimental analysis of these processes may become easier. The processes representing

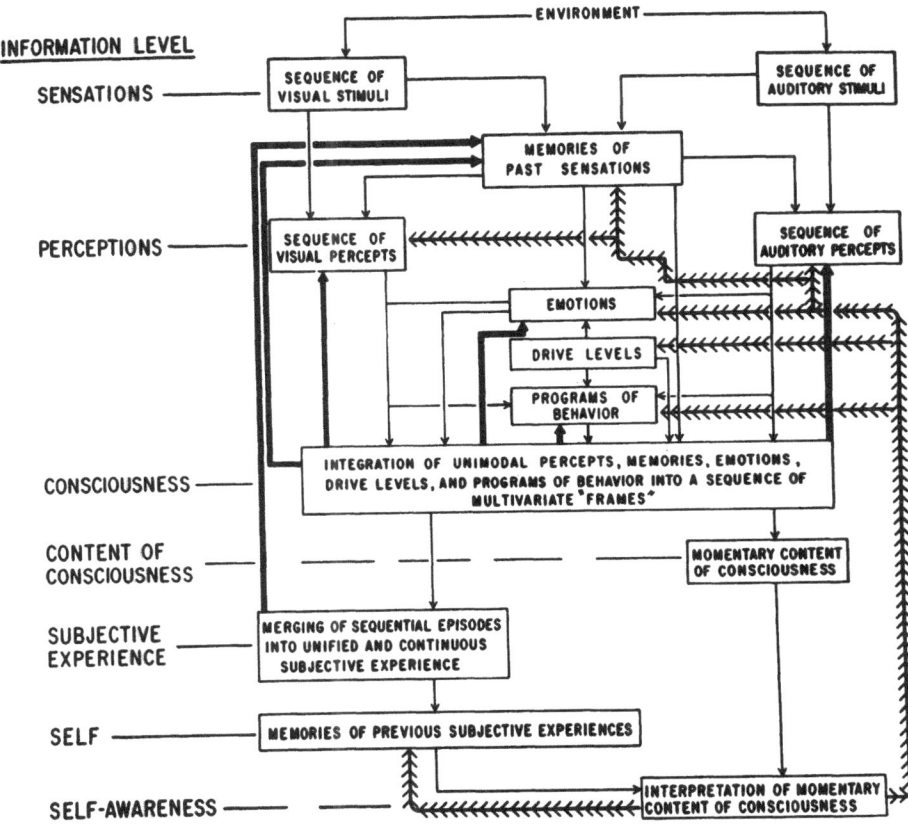

FIG. 12.3 Flow scheme for the levels of information involved in consciousness and self-awareness.

information at the lower levels must be analyzed first. As we gain insights into the representation of lower level information, it becomes possible to seek invariances across the representation of multiple items on the same level which share a common informational feature. Such invariances constitute the representation of information on a higher level. In this "bootstrap" fashion, it would appear possible to progress in a systematic development from initial studies of sensory mechanisms to eventual investigations into the neurophysiological basis of self-awareness.

B. "Experienced Integration" or Consciousness

Fessard (1954) is outstanding among the few who have undertaken a detailed examination of this question. Figure 12.4 illustrates his summary of the major conceptions about the integrative processes assumed to be essential to the

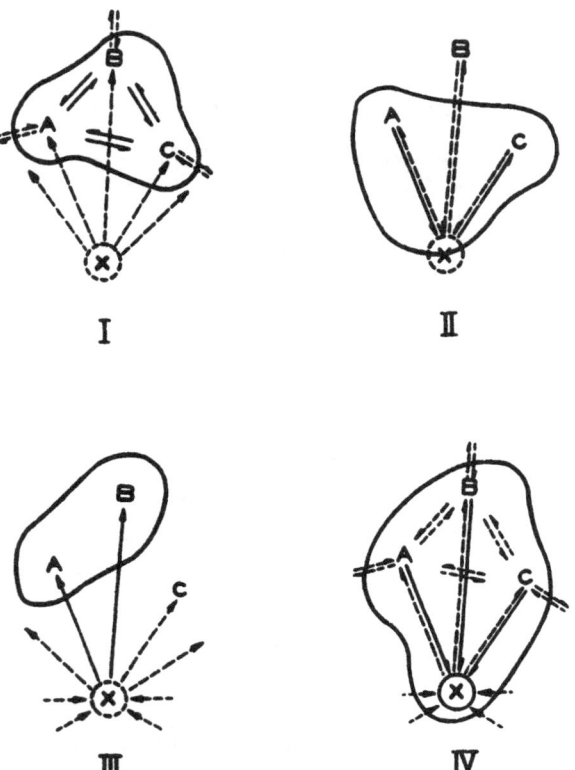

FIG. 12.4 Figuration, in a highly diagrammatic form, of different conceptions on the site of the nervous integrative processes that are assumed to be essential to the emergence of conscious experience. A, B, C: samples of cortical areas. X: the centrencephalic system. Connections that are supposed to be essential to EI at a certain instant are in solid lines, nonessential connections in stippled lines. Parts of the schematic brain that are assumed to participate "as a whole" in a certain instantaneous "experience" are enclosed within an irregular contour, supposedly of ever-changing aspect and location. I: Purely cortical integration. II: Cortical integration by means of pathways through X. III: Cortical integration under the unconscious control of subcortical centers. IV: Two conceptions in a single diagram: either integration of the whole cortico–subcortical system, or pure centrencephalic integration (X). (From Fessard, 1954.)

emergence of conscious experience. Note his presentation of the involvement of the centrencephalic system in integration processes (Fig. 12.4, IV).

Several points of Fessard's treatment of this problem, which the reader is urged to examine for himself, are particularly relevant to our discussion: He proposes that subjective experience, which he terms *experienced integration* or *EI*, is a property of an organized spatiotemporal pattern of neuronal activity and asks what might be the neuronal activity most likely to correspond to the existence of EI. What might be the integrative process which transforms the separately active neuronal ensembles in the brain into a unified experience? He identifies three divergent opinions: (1) the brain works as a whole and consciousness is a property of this total activity; (2) integrative experience is mediated separately by dynamic patterns within the undifferentiated mass of neurons comprising the bulk of the cerebral cortex; (3) there is a specific region involved in experienced integration, receiving inputs from other parts of the brain as communication lines.

C. Hypothesis of a Centralized Integrative System

Examining a variety of evidence, Fessard concludes that corticofugal projections toward a centralized integrative system, probably located in the reticular formation and in constant dynamic interaction with the cortex, probably offer the best structural diagram to represent the gross organization responsible for a conscious activity. This evaluation leads to a consideration of the intrinsic properties of reticular arrangements of neurons with short axons. Examination of these properties leads to two important conclusions: (1) in reticular systems, the dynamic properties of neurons working en masse should predominate over the consequences of activity of single cells; and (2) due to the shortness of most reticular axons, interactions of graduated somatodendritic potentials are more predominant in the control of activity in such systems than axonal spikes. Ephaptic and electrotonic transactions and field effects are discussed at some length.

Fessard next points out the role of the reticular system as a pacemaker controlling the electrical rhythms of the cortex, a consideration particularly relevant to the concepts of perceptual framing as discussed in Chapter 7, and proposes that this activity of the reticular formation plays an important role in the subjective sequence of experiences. Finally, and most importantly, Fessard addresses himself to the question of *the relationship between the neural phenomena essential to conscious experience and the forces that exist within organized matter and raises the question of whether experienced integration involves a transfer of order from the molecular domain to the population of neurons.* He suggests that *interneuronal electric fields could constitute a physical link between infraneuronal processes involving ionic distributions and patterns of excitation within ensembles of reticular elements,* and proposes that "*if consciousness be the most elaborate consequence of excitability, may it not be said that its deepest source likewise resides at molecular levels?*"

With this introduction, let us turn to our own analysis of the brain mechanisms mediating mental experience. The theoretical discussion of the neurophysiology of mental processes presented in this chapter presumes familiarity with the current concepts about learning and memory provided in the previous chapters and must be evaluated as speculations based largely upon such material.

II. REPRESENTATIONAL SYSTEMS

Neurons in the central nervous system fire for many reasons. A wide variety of different afferent inputs can cause discharge of the same neuron; the effects of a specific input on a given neuron are extremely variable, and fluctuations of homeostatic processes cause occasional "spontaneous" neuronal discharge in the absence of any discernible afferent input. Thus, little unique informational value or reliability attaches to the firing of an individual neuron in the brain. In contrast to the variability of response displayed by an individual neuron to a specific stimulus, responses averaged across neuronal populations appear to be very reliable. Information appears to be represented in the brain by the temporal pattern of nonrandomness, or *organization,* in the firing of ensembles of many neurons rather than by the activity of individual cells, which are important only insofar as they contribute to the statistical behavior of the population (see Chapter 10).

At any moment, a particular anatomic distribution of such nonrandom activity patterns exists in the brain. In certain regions, these patterns reflect the arrival of afferent input about sensory events. In other regions, nonrandom activity is due to interoceptors reporting internal state with respect to a wide number of bodily functions. Still other regions display activity patterns particularly related to levels of arousal (reticular formation), emotional affect or mood (limbic system), motivational levels (hypothalamus), and muscle tone, body position, and movements (pyramidal and extrapyramidal system and cerebellum).

From each of these regions of nonrandom activity, neural outflow propagates to numerous other regions where it exerts its influence to cause further nonrandom activity. This swirling constant interaction between patterns of organized neural discharge arising, interacting, and subsiding throughout the brain represents the instantaneous fluctuation of information in the system. The organized, nonrandom nature of the activity in an ensemble defines the activity as informational, the anatomic locus of the ensemble defines the modality of the information, and the details of the temporal pattern define the content of the information.

The information in the brain during a given time interval is defined as the anatomic system of temporally coordinated, nonrandom or organized neuronal activity occurring in all the different functional systems comprising the brain and the spatiotemporal patterns characterizing that set of organized processes. The similarity of electrical patterns observed in many different brain regions when a familiar event occurs reflects the functioning of this distributed anatomic network of organized neuronal patterns representing the different informationally significant facets of the momentary experience, which is called the *representational system* for that experience.

If the organized activity in a region and the transactions between that region and other regions processing information persist for a sufficiently long time to initiate the necessary chemical reactions, certain difficultly reversible changes occur in the participating neurons and *consolidation* occurs. Thereafter, that pattern of activity becomes easier to elicit. The focus of attention undoubtedly facilitates this process. A large body of evidence, reviewed herein and elsewhere (John, 1967a), establishes that when regions of the brain share a pattern of activity for a period of time, such temporal contiguity suffices to organize a new relationship between those regions. Subsequently, if any of these regions enters that mode of activity, the other regions are activated in the corresponding mode. Thus, *once consolidation has occurred in a representational system, activation of a portion of the system results in activation of the system as a whole.* As a result, *the same representational system can be activated via many different inputs.*

III. CONSCIOUSNESS AS A REPRESENTATIONAL SYSTEM

Consciousness about an experience is defined as information about the information in the system, that is, consciousness itself is a representational system. Consciousness, therefore, is a consequence of the occurrence of a set of temporal sequences of nonrandom activity in a set of interacting anatomic structures. The *form* of conscious experience, the modality of the facets of which it is composed, depends upon the anatomic location of the neuronal ensembles in which these statistical processes emerge. The *content* of conscious experience, the shape of these separate facets, depends upon the temporal pattern of the nonrandom activity in each of these regions.

Since the information in the brain at any moment consists of this set of anatomically distributed statistical processes, the information about that information, or consciousness, must be information about statistical processes in neuronal ensembles. Since no neuron in the brain can represent any information by the mere fact of its discharge but only contributes to the statistical behavior of the informational ensemble, it is necessary to conclude that conscious experience itself is the result of a statistical process; that is, information about information must be temporal patterns of organized activity in neuronal ensembles influenced by the occurrence of temporal patterns of organized activity in neuronal ensembles.

Whether or not the informational patterns which constitute the content of consciousness arise in some central integrative system receiving input from all other anatomic regions sustaining informational patterns or arise as a result of the interactions between informational patterns in these various regions is not clear. In either case, it is necessary to postulate that the content of consciousness, like any other kind of information, is defined by the statistical features of activity in an anatomically diffuse network of neurons.

The property of consciousness is the consequence of the emergence of organized patterns of activity in neuronal populations. This property is the result of a cooperative process between the neuronal elements of the system and is the direct result of neural activity. At the present stage of knowledge, we do not understand the precise physical nature of those consequences of this cooperative process which are responsible for this property. This poses no more serious a philosophical problem than a host of other cooperative processes in which the properties of the whole are greater than, and not predictable from, the sum of the properties of the parts. The separation of neurophysiological from "mental" phenomena, the so-called "mind–brain" dualism, merely represents a failure to recognize this fact. Action potentials in neurons, movements of sodium and potassium through membranes, alpha rhythms, evoked potentials, readout processes, and conscious experience all represent different aspects of neural activity. Although we may understand the physical chemistry of some better than others at this point in science, none of them merits being set apart as mysteriously different from the others.

IV. THE TRANSFORMATION FROM NEURONAL ACTIVITY TO SUBJECTIVE EXPERIENCE

Perhaps it will be useful at this point to summarize the major ideas which have thus far been presented and to identify the specific problem with which we are concerned.

1. We have marshalled a body of evidence that information in the nervous system is represented by the statistical behavior of neuronal ensembles rather than by the firing of any individual cell.

2. All of the constituent activities comprising the behavior of the ensemble are postsynaptic potentials and axonal spikes. Each of these unitary events is a transient electrical process, a shift in ionic concentrations and charge densities. None of the unitary events caused by a sensory stimulus, for example, can conceivably mediate the sensation. The subjective experience of a form in space cannot be derived from the firing of a neuron per se, especially if that neuron fires under many other circumstances when the form is absent. Neuronal ensembles possess *gestalt* properties, properties of the whole, which are not contained in their individual parts. Numerous examples exist of such properties on every level from atoms to societies. The liquid transparent properties of water are not inferable from the constituent atoms of hydrogen and oxygen, the properties of a semiconductor cannot be derived by considering the individual elements of a crystal lattice, the properties of a hologram cannot be inferred from the analysis of photosensitive elements in an emulsion, the functions of an electronic circuit are not inherent in any individual component, and the interactions of a group of people cannot be predicted by studying persons by themselves. Closed systems have certain types of system properties and open systems display quite different properties. On every level of analysis, we find that systems possess system properties not inherent in their individual constituents.

3. Internal experience, moods, emotions, are not properties of individual neurons but of the systems into which those neurons are combined. Like sensation or perception, internal experience is a gestalt process.

4. Such internal experience, like other gestalt processes, is mediated by the statistical behavior of neuronal ensembles. This statistical behavior is not a discrete event, an electrical charge, or a molecule or a synaptic transmission. In order to understand internal experience, it is necessary to identify some physical consequence of a statistical process which will plausibly mediate this integrative gestalt process. Integrative processes, however elusive their mechanisms, are assumed to arise from a material cause, a physical basis.

5. Subjective experience, awareness of sensations and of internal experiences, memories, thoughts, are merely another form of internal experience, information about information. These higher order representational systems are nonetheless statistical in nature and must have a comparable physical basis.

6. Thus, we argue that mental experience is the concomitant of neuronal activity, but arises from the cooperative behavior of ensembles of neurons. The ability of neuronal ensembles to generate subjective experience is neither to be derived from the examination of individual neurons nor is it a property possessed by the single cell. We reject the idea of mind–brain dualism and affirm that mental experience can be understood in terms of the processes sustained by neuronal ensembles.

7. Examining the physical properties of neuronal ensembles, it seems necessary to concede that the physical processes that generate subjective experience as the result of the cooperative behavior of neurons may relate to the ways in which energy is organized by neuronal masses, and may reflect general properties of matter. There seems no a priori reason to assume that the relevant physical processes can only be manifested by matter organized in a particular type of structure, that is, the mammalian (or human) brain. At this stage of our knowledge, it seems more appropriate to assume that these physical properties exist qualitatively as a consequence of the structure of matter and that any unique features of the human brain, with respect to subjective experience, arise from the quantitative characteristics of the system into which matter is organized in the brain. Thus, we refuse to concede mind–matter dualism unless and until evidence requiring that distinction is available.

With this outline of our argument, let us proceed to examine some further details.

V. PHYSICAL BASIS OF CONSCIOUSNESS

While the preceding passage insists upon the view that information about information, or consciousness, must be represented in the brain by the statistical properties of activity in interacting anatomically extensive neuronal ensembles with a content somehow related to the spatiotemporal patterning of these cooperative processes, it evades the question of just *how* organized cooperative processes in masses of neurons generate subjective experience. Mere denial of mind–brain dualism, the insistence that mental phenomena *must* somehow arise from cooperative neurophysiological processes, by itself amounts to little more than a statement of faith in the "religion" of science, the ultimate capability to submit problems to rational analysis. The evidence marshalled in Chapters 9 and 10 provides a substantial basis for the conviction that the representation of information of any sort, whether concrete information about a present event or abstractions about such information, must be statistical in nature. The evidence presented in Chapter 11 shows that predictions from such statistical formulations are supported by experiment. Direct stimulation of the brain by gross input of electrical current causes an animal to behave as if it were receiving a particular subjective experience.

Yet, what has all this told us about the basis of subjective experience? Unless we pursue this issue further, we shall merely have substituted a new "mind–statistical" dualism for the familiar mind–brain dualism. Confusion displaced to a different conceptual level is nonetheless confusion.

A. Mind–Matter Dualism

The critical issue here, in our opinion, is not our inability to abandon mind–brain dualism, but rather our failure to perceive that *we have implicitly accepted a dualism between the nature of mental experience and the nature of physical matter.* Philosophically, we have as little reason to presume a *fundamental* difference between the properties of the matter of which our brains are composed and other matter as we have for presuming that the mind exists independently from the brain. From the viewpoint of information theory, all of the discussions presented in this Volume about the representation of information in the brain are subsumed by the assertion that information in the brain consists of organized activity. The laws of thermodynamics guarantee that order is only a temporary fluctuation in the universal tendency to chaos. Information is negative entropy, whether that information be primary or derived.

Perhaps our philosophical quandary arises from the assumption that organized processes in human brains are *qualitatively* different from organized processes in other nervous systems or in even simpler forms of matter. Perhaps the difference is only *quantitative;* perhaps we are actually not as unique as we have assumed. Rather than assume that neuronal masses in brains are a unique form of matter which somehow engender the mysterious property of experienced integration, why not assume that matter possesses universal properties? If negative entropy, deviation from some random ground state, engenders the correlate of experienced integration, which Fessard called "consciousness" in man, perhaps there exists an integrative process which is a function of the state of energy in any form of matter. As the organization of the energy in that matter is altered, such alteration is reflected in this integrative property.

This conjecture, fanciful though it may at first appear, has at least the virtue of no special pleading for humankind. It is parsimonious. It suggests that all organized energy possesses a nonobvious attribute, an integrative process reflecting the overall state. The difference between a human being and an elementary particle becomes an enormous quantitative difference in the number of energy states which can be entered, but a qualitative *continuum* can be postulated. What is unique about the brains of human beings, from this viewpoint, is that they are exceptionally well suited to construct an almost infinite variety of energy states. Simpler organizations, whether lower animals, vegetables, molecules, or particles, are much more restricted in the number of different energy states which can be entered. The varieties of integrative processes of the hydrogen atom, from the human viewpoint, must be starkly limited to those defined by the hydrogen spectrum. Energy systems with so few states may be defined as *preconscious,* to place a threshold value on the complexity of quantitative states required for a qualitative property to emerge. We may postulate that the complexity of these integrative processes must excess some threshold level in order for *experienced integration* to be sustained. This level may only exist in systems with the incredibly high order of complexity of the centrencephalic system hypothesized by Penfield, or it may be sustained by more simple systems. The essential feature of this proposition is not the definition of the minimum system which can achieve the property of experienced integration, but the suggestion that this property emerges when the organization of energy reaches a certain complexity. Such definition, however, does *not* reintroduce dualism, but must be recognized as an arbitrary rather than an inherent distinction.

B. The "Hyperneuron"

The hypothesis just proposed eliminates what we will term mind–matter dualism as well as mind–brain dualism. It is, at least for the foreseeable future, totally untestable and therefore may be viewed by the reader as no more than a whimsical semantic exercise of no practical relevance. However, it generates a particular viewpoint when one reconsiders the problem of how consciousness of an experience arises from a representational system within a brain. A representational system, we have contended, consists of an anatomically dispersed system of neuronal sets engaged in temporally coordinated or simultaneous patterns of nonrandom behavior. As these temporal patterns of neuron discharge occur, the membranes of participating cells are depolarized and ionic shifts occur, with extrusion of potassium ions and ionic binding on extracellular mucopolysaccharide filaments. If we focus our attention not on the membranes of single neurons, but upon charge density distributions in the tissue matrix of neurons, glial cells, and mucopolysaccharide processes, we can envisage a complex, three-dimensional volume of isopotential contours, topologically comprised of portions of cellular membranes and extracellular binding sites and constantly changing over time. Let us call this volume of isopotential contours or convoluted surfaces a *hyperneuron.*

To every representational system corresponds a particular distribution of energy, a unique hyperneuron. The invariant features of a particular hyperneuron will be determined by the statistical processes in local ensembles and will be influenced insignificantly by the contribution of any individual neuron. Ensembles of neurons with regenerative circuitry will contribute a stable component to all or most hyperneurons, while ensembles with lower positive feedback will make more variable contributions. We can conceive of hyperneuron sequences with a composite of stable, invariant features and modulated, reactive features. As Fessard pointed out, the conditions for a hyperneuron to emerge may be uniquely satisfied by the anatomic featues of the reticular system of the brain stem, with its patterning modulated by inflow from regions of the limbic system and the cerebral cortex. Conceivably, this hyperstructure might extend into limbic and cortical regions. The work of Hess and Penfield, among others, has emphasized the capability of stimulation of these regions to alter the content of consciousness more profoundly than by simulating simple sensations. The invariant features of this hyperstructure constitute the "I" of subjective experience, while the variable features are the "here-now" of the subjective moment. The actual process by which subjective experience is generated would, in this view, be not at all mysterious. Subjective experience would be the inevitable concomitant of a particular organization of energy.

The most important feature of neuronal masses for the mediation of subjective experience may not be inherent in the neurons themselves, but may arise because they are uniquely well suited for the production of a wide variety of improbable distributions of energy. Were it possible to achieve comparable energy distributions without neurons, in other words to simulate a neuron-free hyperneuron, perhaps quite the same subjective experience would arise.

To a great extent the progress of science has settled issues of philosophy; thus, quantum mechanics precludes the completely deterministic universe of Laplace, Darwinian evolution establishes the time scale and genesis of the species, and so on. Perhaps the analogous transition, marking the coming of age of the neurosciences, is to settle the mind–body problem as follows: *body is the structure of the brain and mind is the momentary organization of energy flux in that structure.* Thus the anatomic and biochemical substrate of the complex spatio-temporal pattern of electrical and chemical activity, that is, our consciousness, may be considered from a unified point of view. At the very least, this formulation leads to a philosophical position as acceptable on purely logical grounds as the assumption that man possesses a unique "divine spark" which confers subjective experience upon him. If it fares better, it may serve as a working hypothesis leading to experimental exploration of the subjective consequences of altering the organization of energy in the central nervous system.

VI. REPRESENTATION OF AN EXPERIENCE

When a set of environmental events impinges upon an individual, constituting an experience, many different regions of the brain are contiguously active in organized patterns. Different brain regions represent the visual, auditory, and somatosensory aspects of the sensory barrage, the level of arousal, focus of attention, motivational valences, and drive states which coexist during that period and constitute the various facets of the subjective experience. Some of these facets have a sufficiently high signal-to-noise ratio to enter consciousness while others remain subliminal. Some of the resulting neural activity persists long enough for consolidation to occur, while the remainder decays leaving no permanent residue. That set of patterns which achieves consolidation constitutes the representational system for that experience. A hyperneuron has been established.

Activation of any local anatomic component of that representational system with the appropriate temporal pattern of activity in the neuronal ensemble can result in the activation of the representational system as a whole and the readout of the corresponding memory (John, 1967a). A particular memory can be activated in many different ways. A picture, a name, a smell, certain changes in internal state—all may elicit the same memory. Construction of a portion of a hyperneuron leads to activation of the total gestalt. Furthermore, the occurrence of endogenous patterns or readout can itself serve to activate representational systems or memories, and any such pattern, whether activated due to exogenous or endogenous processes, can itself serve to generate other patterns. Hyperneurons arise in most probable sequences.

VII. THE STREAM OF CONSCIOUSNESS

In this fashion, interaction of environmental input with internal state generates a sequence of successive and continuous activations of representational systems, a constant interaction between new experiences being consolidated and integrated into residues of previous experiences, producing a sequence of facets of conscious experience or a stream of consciousness. Many of the elements in this stream of consciousness will change from moment to moment, while certain elements will tend to be self-activating and will constitute a more or less constant stable framework within which a varying modulation occurs. Successive hyperneurons may share a portion of their energy states.

At this stage, we are ready to turn to a consideration of the influence of representational systems upon normal subjective experience and psychopathology.

13
Daily Subjective Experience and Psychopathology

In the previous chapter, we discussed in detail the development of representational systems and proposed a mechanism whereby the neuronal elements comprising portions of a representational system could be transformed into subjective experience. We presented the concept of a hyperneuron, a gestalt mechanism, and described how activation of a portion of a hyperneuron would lead to activation of the total gestalt. Finally, we proposed that occurrence of endogenous or readout processes could activate hyperneurons or gestalten, any of which could in turn serve to activate subsequent patterns. Overlap and feedback between these successive structures would define the most probable sequence of subjective processes.

Interaction of environmental events with preexisting hyperneuron structures activated by prior events or internal state would produce a *stream of consciousness*. Regenerative or self-activating portions of the constituent hyperneurons in the sequence would constitute a stable framework for this subjective sequence, undulated by changes in activity patterns resulting from refractoriness and afferent input.

I. BIAS OF CONSCIOUS EXPERIENCE

This stable framework, which is closely related to what might be called the "self-concept," not only provides subjective continuity across experiences but constitues the source of possible, indeed inevitable, biases upon the conscious perception of experience. A number of important factors which can contribute to such bias in different ways can be identified as the following.

A. Anatomic Factors

A voluminous literature shows that the anatomic structure of the brain and certain aspects of its metabolism including transmitter availability can be markedly altered by the systematic presence or absence of certain environmental stimuli, particularly early in life. A variety of regional patterns has been described. The evidence indicates that the effects are specific rather than general, with the particular pattern of anatomic growth or diminution reflecting the specific pattern of alteration in environmental features.

The genetic endowment of an individual thus seems to specify more or less stable species-specific features of brain anatomy and biochemistry, which are then appreciably modulated by the particular early experiences of that individual. Certain stimuli may cause increase in the actual number of brain cells in particular regions while decreasing the size of other regions. Thus, environmental deprivation or enrichment with respect to a particular kind of sensory experience may have opposite effects on different brain structures (Walsh *et al.*, 1972).

As a result of early experience, therefore, marked individual differences can arise in the actual anatomy and reactivity of various brain regions. The relative involvement of such regions in representational systems is thus seen to depend upon the early experience of the individual and its effect upon the anatomic structure of the brain.

B. Biochemical Factors

Regional availability of neurotransmitters arises as a result of the operation of complex homeostatic processes governing the rates of numerous biochemical reactions which are largely genetically specified. Fluctuations in local transmitter availability can occur due to a variety of metabolic factors, or even to the proliferation of a certain type of cell such as might arise due to early experience (Altman & Das, 1964). Changes in transmitter availability might either increase or decrease the probability that a given brain region would participate in the representational system established by a particular experience. These changes might be expected to drastically influence the ways in which that experience was perceived, especially with regard to its affective or motivational valences. Furthermore, changes in threshold related to shifts in transmitter availability once a representational system was established might result in altered probability that a particular kind of affect, for example, would consistently be activated, either adding its color to a wider set of experience than would otherwise be the case or systematically being absent or difficult to elicit.

C. Experiential Factors

An extensive body of evidence, reviewed elsewhere (John, 1967a), establishes that under certain conditions activation of an anatomic region once established persists for long periods of time. During such periods, this region acts as a "dominant focus" (Ukhtomski, 1945). A dominant focus, as the name so well implies, achieves a signal-to-noise ratio that is abnormally high and participates far more intensively in the representational systems constructed and activated, while it is dominant, than would otherwise be the case. Dominant foci can be established as the result of a learning experience. Thus, we can envisage particular kinds of experience which establish a high tonic level of excitability for a specific pattern of neural activity in an anatomic region. Such a sustained dominant pattern might be expected to contribute a consistent feature to the conscious perception of experience.

From this abbreviated summary, it can be seen that strong biases on conscious experience can arise from anatomic, biochemical, or experiential factors. The learned aspects of perception undoubtedly are to be attributed to such influences, as are many of the stable components of personality and self-concept, value systems, and aspirations.

II. "IMAGINARY" READOUT

The awareness of sensation represents that part of experience which is stimulus bound. This is perhaps the least interesting of the categories of mental phenomena which we must ultimately explain. We must also turn our attention to thinking, to imagination, even to fantasy, which represent departures from rigidly stimulus-bound experience and reflect some of the degrees of freedom of operation of the machinery of the brain. Stimulus-bound experience provides an initial vantage point from which to examine these other processes. Thought and imagination and fantasy consist of awareness of stimuli in their absence, of relating events into combinations different from those conventionally experienced, of distorting stimuli from the forms in which they normally impinge. Almost 150 years ago, intense controversy raged about the concept of "imageless thought." Psychologists debated the extent to which thought processes were a stream of memories and stimulus representations that constituted essential ingredients in the thought process. Psychology now offers experimental methods that make it possible to consider these problems from a biological viewpoint.

It seems reasonable to suggest that the mechanism normally mediating the release of the readout component is involved when memories are activated during the thought process, during imagination, and during fantasy. As noted in Chapter 10, similar phenomena have been reported in a number of human studies. In one such study (Sutton et al., 1967), subjects were presented with single or double clicks which could be either loud or soft. The subject was sometimes asked to guess whether the stimuli would be single or double. Figure 13.1 shows that when intensity was the relevant dimension, the evoked potential (EP) showed a marked deflection from the first click and little or no response to the second stimulus, even when the second click was present. Conversely, when

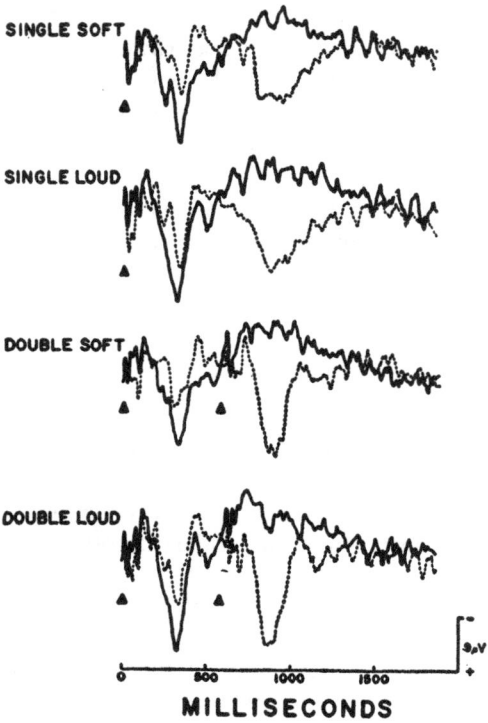

FIG. 13.1 Average response waveforms obtained to four types of clicks (as labeled) for one subject. (——) Waveforms obtained when the subject is guessing soft versus loud; (----) waveforms obtained when the subject is guessing single versus double; (Δ) points at which clicks were delivered. Labels at the left specify the physical characteristics of the stimuli. (From Sutton *et al.*, 1967.)

singleness versus doubleness was the relevant dimension, all EPs contained a clear second deflection at the latency when the second click occurred or ought to have occurred. In the case of the single clicks (two upper pairs of tracings), the clear responses seen in the latter part of the waveshape were caused by *the absence of an expected click*. Barlow, Morrell, and Morrell (1967) have similarly demonstrated the production of evoked potential by man when an expected event fails to occur, and numerous other examples were cited in Chapter 10.

Similar data are shown in Fig. 13.2 (Klinke *et al.*, 1968). The upper waveshape in this figure shows the average evoked response elicited by 80 repetitions of triplets of clicks. The second waveshape shows the average evoked response elicited by presentations of "triplets" from which the middle click had been

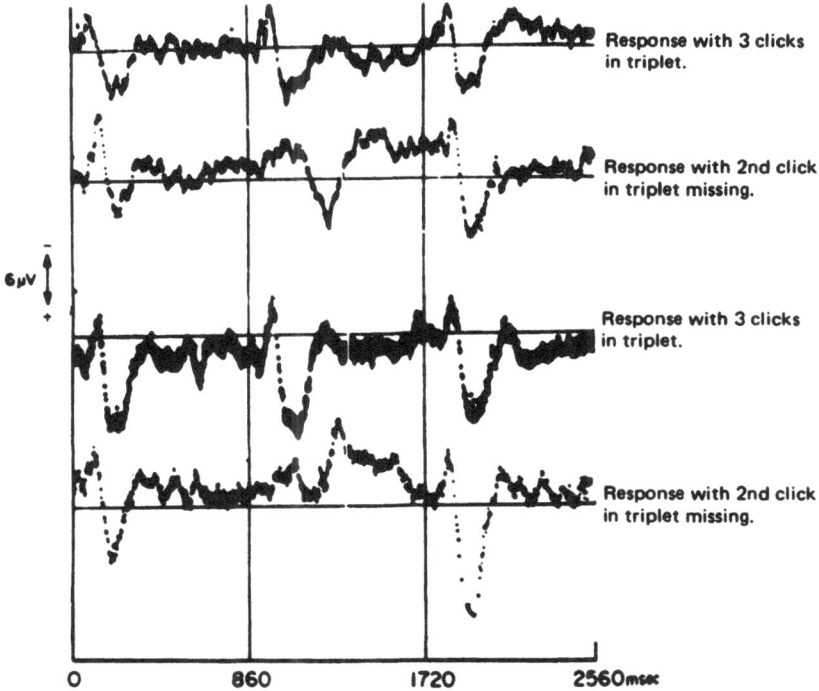

Response with 3 clicks
in triplet.

Response with 2nd click
in triplet missing.

Response with 3 clicks
in triplet.

Response with 2nd click
in triplet missing.

6 µV

−

+

0   860   1720   2560 msec

FIG. 13.2   Typical responses obtained from two subjects (curves A and B from RK, curves C and D from HF). A and C show the cerebral responses to three consecutive vibratory pulses at a presentation rate of 1 stimulus per 0.86 sec. The sweep starts with the onset of a stimulus. In B and D the second stimulus of each sweep is omitted. Each curve represents 80 summations. The voltage calibration is calculated for a single response. Upward deflection signifies negativity of vertex electrode. (From Klinke *et al.*, 1968.)

deleted. The response to the absent click is unequivocally clear. The third and fourth lines of the figure illustrate the repetition of this experiment with a different subject.

The examples given above indicate that a man can produce an evoked response at the time that a stimulus is expected. Work cited earlier by such investigators as John, Herrington, Sutton, and Clynes showed that stimuli having the same geometric form but differing with respect to physical size and stimulus energy elicited evoked responses of markedly similar waveshape. This similarity indicated that the brain was computing an invariant which reflected stimulus significance. The ability to perform this computation suggested that experience with a geometric form built a representational system capable of releasing the corresponding endogenous waveshape upon appropriate stimulation.

As noted in Chapter 10, seeking direct evidence that such was the case, Herrington and Schneidau (1968) recorded the responses of subjects to the

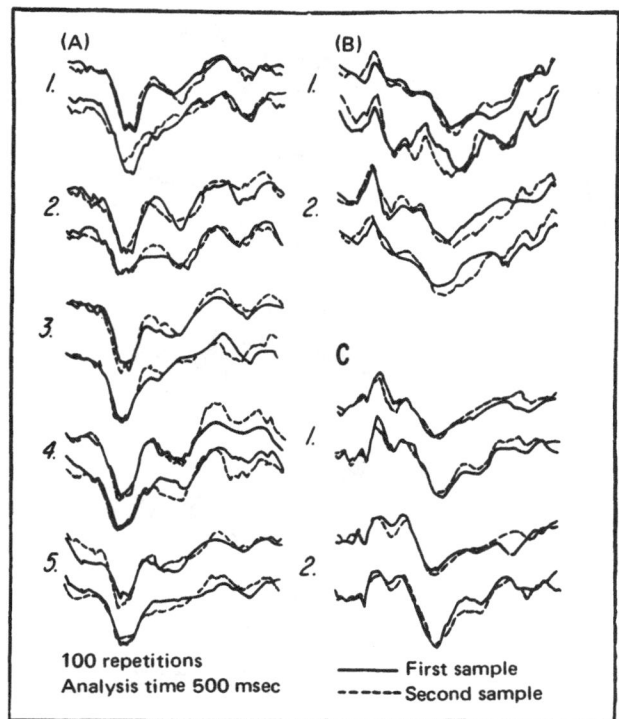

FIG. 13.3 (A) Top row of each block = ○, bottom = □. 2 + 4: seeing □ when looking at ○ and vice versa. (B) Top row of each block = □, bottom = ○. 2: seeing □ when looking at ○ and vice versa. (C) 1. top thinking of word SQUARE, bottom thinking of CIRCLE; 2. actually looking at words "SQUARE"(top) and "CIRCLE"(bottom).

presentation of squares and circles. When the evoked response typically elicited by a geometric form had been determined for a subject, the patterned stimuli were terminated. The subject was then asked to look at the screen and imagine that he saw a square or a circle appear before him whenever a blank flash was presented. The responses evoked by the blank flashes were averaged.

In some instances, as seen in Fig. 13.3, it was possible to tell which form the subject had imagined in his visual field because of the high similarity between the waveshape of the average evoked response elicited by actual presentation of the stimulus and that released from imagination during presentation of a blank flash.

Taken in conjunction with other data cited in Chapter 10, showing that the readout phenomenon discussed earlier has its counterpart in man, these findings show that during the process of thought or imagination the representational system literally recreates the electrical activity caused by past experience which is retrieved into consciousness.

III. "ABNORMAL" CONSCIOUSNESS

Some psychiatric disorders probably represent defects in the processes discussed above: defects in the way that past and present experience is perceived; defects in the way that thought is constructed from a sequence of released images of past, present, or future experience; defects in imagination in which the fluctuating content of consciousness deviates from whatever rules normally govern it, consistently leading to an aspect in the life of the individual that has assumed a larger amount of threat or a greater potential value than is usually the case for the content in question; defects in fantasy so that memories and stimulus-bound representations are combined in a fashion which consistently causes distress for the individual, eliciting anxiety, fear, or sadness.

The use of "defects" in this series of descriptions implies that it is meaningful to speak about "normal" thought, "normal" imagination, "normal" fantasy. The mechanisms of the brain must operate in basically the same fashion from individual to individual—memories are stored and released, and stimuli represented or distorted in ways which are determined largely by the characteristics of the machinery that has been developed in man to perform these functions. This implies that there is a range of probable performance of that machinery when constructing sequences of released memories and represented experiences. Certain kinds of sequences are more probable than others. The emotional consequences of certain kinds of sequences serve to make them more probable or less probable. The content of mental experience, although unknown and unpredictable in its fine detail, fluctuates within certain expected limits that are undoubtedly extremely broad.

A. Improper Brain Function

There must be a class of disordered thought process which reflects a disruption of that machinery such that the combinations in the stream of thought, the structure of imagination, the content of fantasy, are not only extremely improbable, but consistently so. Such disorders might be metabolic, or they might arise from traumatic injury to the brain. Certain parts of the machinery might either become hyperexcitable or function at a lower level than normally the case, biasing thought, imagination, or fantasy in a direction that reflected over- or underactivation of certain regions, or failing to exercise feedback mechanisms in which the content of mental experience was somewhat regulated by its emotional consequences.

B. Disordered Representational Systems

One can also envisage a second class of disorder arising not from malfunction of the machinery of the brain, but because the value systems, role models, goals, and memories accumulated by an individual, the blocks of imagery available to be released from storage to comprise the content of thought processes, were themselves abnormal, arising from situations so misperceived, so traumatic, so unpleasant, so emotion-laden and unusual as to be beyond the bounds of what we consider normal or realistic experience. An individual who has filled a warehouse with abnormal or invalid building blocks must build abnormal or unstable structures.

C. Societal Causes

In most societies, particularly industrialized societies, a predominant source of abnormal representation systems of this sort can be attributed to the systematic distortion of values, inculcation of unrealistic expectations, and the provision of undesirable role models arising from the deliberate manipulation of the individual by society. Artificial values which correspond to no legitimate goals of the individual, unrealistic expectations engendered by social myths, and the desire to emulate superficial role models misperceived as valid can contribute enormously to the contents of such a warehouse. It is beyond the purview of this chapter to explore in greater detail the innumerable ways in which an individual can be deformed by his family, peer group, teachers, or the larger society which molds all of these individuals. Suffice it to say that a substantial part of the abnormal content of these representational systems is inflicted upon the individual by society.

Thus we can discern three classes of potential disorder: abnormal mental processes arising from malfunction of the machinery of a brain processing normal material, abnormalities arising from proper functioning of the machinery of a brain processing abnormal material, and abnormalities arising from discordant relationships between the individual and society.

IV. ELECTROPHYSIOLOGICAL REFLECTIONS OF ABNORMAL PROCESSES

In most societies, particularly industrialized societies, a predominant source of the cat is processing information about a familiar event, many regions of the brain enter a mode of electrical activity which is shared by a large number of structures.

A comparable phenomenon seems to occur in man. Figure 13.4 presents results obtained by Livanov (1962, 1965). The upper left (a) shows the outline of a head on which a large number of electrodes have been placed. The correlation coefficient between each electrode and every other electrode has been calculated. Electrodes with a significant intercorrelation are marked in black with a

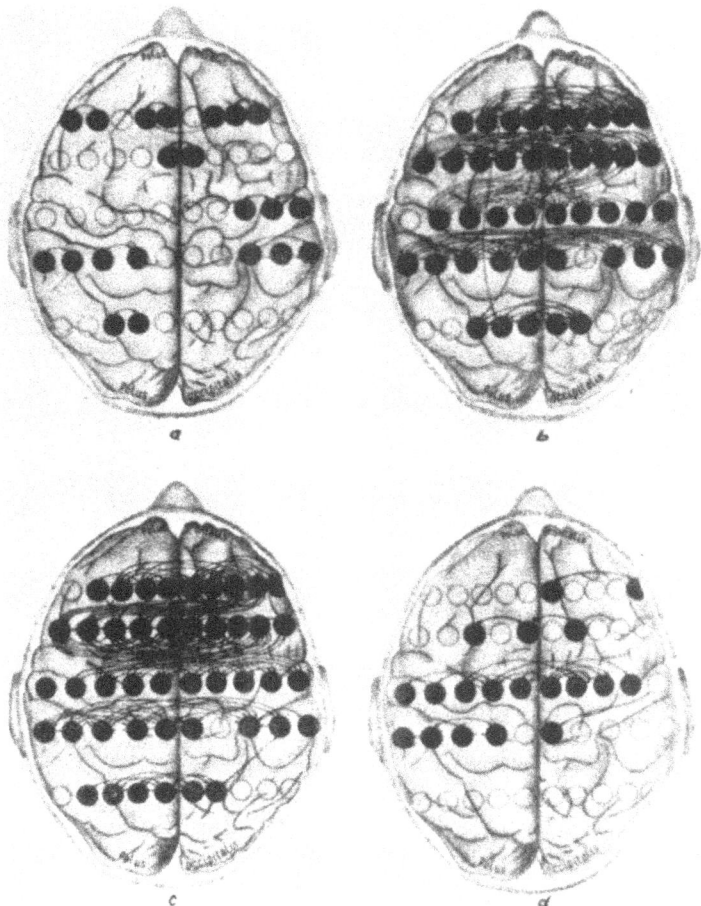

FIG. 13.4 Pair synchronization of biopotentials of human brain cortex during mental work. The circles on the brain cortex map indicate electrode placement. The arrows connect the areas of high degree of synchronization. (A) Before the arithmetic problem; (B) 15 sec after the problem was given; (C) 30 sec after the problem was given; (D) after solving of the problem. (From Livanov, 1965.)

line connecting them. These recordings were taken from a man sitting at rest. The upper right (b) shows the intercorrelations seen 15 sec after presentation of a task in mental arithmetic. Widespread synchronization of activity occurred. The activity of many regions of the cortex became closely coupled, as shown by the widespread distribution of dark circles. The lower left (c) shows that this state continues for 30 sec after the mental task was imposed. The lower right (d)

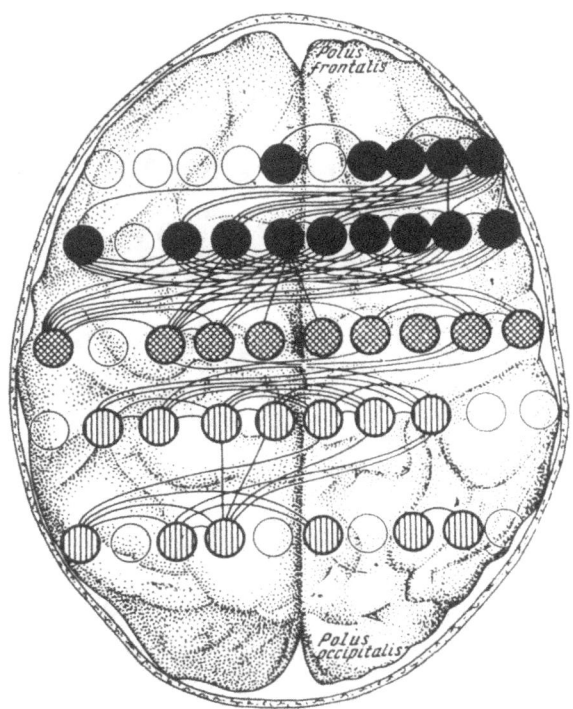

FIG. 13.5 High correlations between potentials in various parts of the cerebral cortex of a patient with obsessive neurosis at rest. (From Aslanov, 1970.)

illustrates the return to the original resting state after solution to the problem was correctly provided. These data show that in normal man during mental tasks a wide part of the brain enters a shared state with common modes of activity.

Figure 13.5 shows comparable data from a patient with "obsessive neurosis" (Aslanov, 1970). It is striking that the frontal regions display closely coordinated activity when the patient is at rest. Instead of the differentiated picture seen in Fig. 13.4a in which regions of the brain are relatively uncoupled, this individual seems to be engaged in intense mental activity while resting.

Figure 13.6b shows the distribution of correlation coefficients in a schizophrenic patient at rest; Fig. 13.6c the change in this distribution during the solving of a mental arithmetic problem (Gavrilova, 1970). In contrast with the normal subject (Fig. 13.5a), the frontal regions are coupled while this patient is at rest, and these regions are not integrated with the activity in the remainder of the cortex during the performance of a mental task.

These data suggest that the establishment of coherence is an essential feature of mental activity in a normal man, that the obsessive neurotic and the paranoid

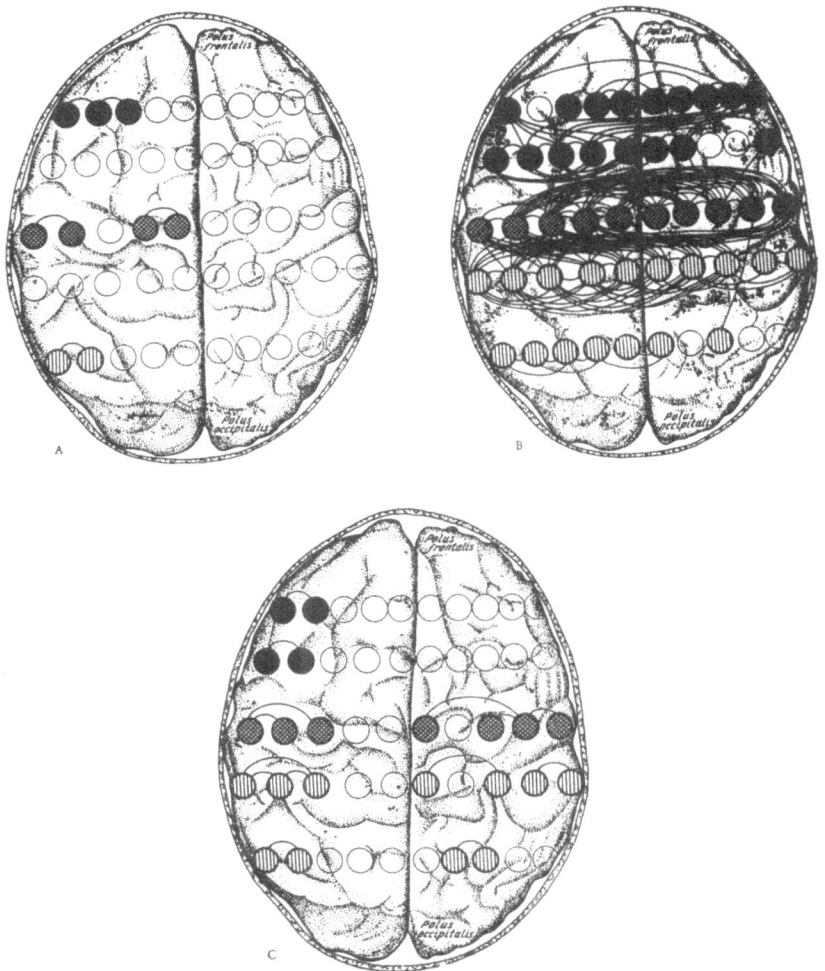

FIG. 13.6 High correlations between potentials in various points of the cortex of a healthy person at rest (A), of a patient with schizophrenia at rest (B), and during the solution of an arithmetical problem (C). Circles denote locations of recording electrodes. Points whose potentials give a high coefficient of correlation are joined by lines. Black circles denote points in the prefrontal area; crosshatched circles points in the motor area; vertically shaded circles points in posterior areas. (From Gavrilova, 1970.)

schizophrenic give indications of being involved in intense mental activity when we would expect them to be at rest, and that the schizophrenic does not integrate the activity of frontal regions of the cortex with more posterior regions under circumstances which normally accomplish this.

Such findings constitute electrophysiological evidence supporting the suggestions made earlier in this chapter that abnormal activity of certain anatomic regions might result in the sustained activation of particular representational systems. Other evidence supporting such speculations comes from consideration of the mode of action of antipsychotic pharmaceutical agents. In a fascinating study, Stevens (1973) has related the particular value of dopamine blocking agents in the treatment of schizophrenia, the localized distribution of dopamine axons in particular anatomic regions (neostriatum and limbic striatum), and the relationship of such structures to the limbic system. Analyzing these factors together with clinical EEG studies and other experimental data, Stevens suggests that preemption of consciousness by a characteristic fearfulness, a sense of unreality, and heightened sensory and sexual perception often seen in schizophrenia is due to a bias in certain structures, producing a tonically altered interaction of limbic and temporal lobe processes with activity in other brain regions processing sensory signals.

V. POSSIBLE TYPES OF PSYCHIATRIC DISORDER

In Section III, three possible classes of psychiatric disorder were described, one which arose from normal processing of experiences which could be considered abnormal, a second due to discordant or harmful relationships between the individual and society, and a third due to malfunction of the brain mechanisms necessary for processing, storing, and retrieving information and therefore essential to the processes of thinking, imagination, and fantasy. The first kind of abnormality provides little theoretical difficulty. Past traumatic experiences stored by representational systems constructed as described must inevitably arise in consciousness, influence the content of thought, be the subject of fantasy, and direct the imagination. In order to understand and deal with these disorders, it would seem essential to identify the traumatic experience and to make its effects explicit to the individual. Psychological and pharmacological techniques must be utilized to lessen or obliterate the effect that such experience has had on the representational systems of the brain.

The second kind of disorder must be considered a social disease. The disorder originates not within the individual but in the relationship between the individual and his family, peers, teachers, employers, or society at large. This disorder, manifested as distress from the inability to achieve goals set by others such as the difficulty in developing a life-style and relationships which are truly satisfactory, is the result of a systematic miseducation and deformation of some individuals by society. This class of disorder, increasingly prevalent among young people and members of minority groups, may well have no psychiatric solution. The only effective approach to this problem would seem to make the individual cognizant of the external origin of his distress, to aid him to systematically reevaluate his basic assumptions, but to recognize that meaningful amelioration of his distress can only be accomplished by essentially political, social, and economic changes in society, which is the fundamental cause of the problems he faces. Nonetheless, recognition of the contributing cause, explication of the deleterious effects on the individual, and identification of possible solutions constitute a significant service, which can be rendered in such cases. These problems, however, are not problems of neuroscience per se.

The third kind of psychiatric disorder, which has its origin in malfunctions of brain mechanisms, might have a number of causes. Changes might occur in the metabolic characteristics of the system, resulting in an alteration of the signal-to-noise ratio. The coherence levels necessary for information to be stored and retrieved, reflecting the ease of access and the probability of access to memories, might thus be either drastically raised or lowered, causing a change in the "confidence level" which the system requires to accept data. Suggestive supporting evidence can be found in the work of Callaway, Jones, and Layne (1965) on segmental set in schizophrenia.

Similarly, a representational system might be biased due to the pathological excess of activity in certain regions, perhaps arising from anatomic changes, metabolic disorders, or from prolonged, intense stimulation. The studies of the recovery cycle by Shagass (1968) suggest that in many psychiatric disorders the aftereffects of excitation are abnormal, with inhibitory processes not as effective as in the normal individual. Thus, a representational system might remain relatively close to the threshold level of excitation required for it to become activated as a whole because of the tonic bombardment of the system by the activity of some abnormal region. Evidence has been presented earlier that a representational system can be activated by activity in one of its parts. Changes in the levels of tonic activation of different recollections might occur such that the memories or actions represented by particular systems will play an abnormally frequent role in imagination, will compel fantasy, and will occupy thought more than one might consider normal. This possibility was illustrated by the pattern seen for the obsessive psychotic in the work of Aslanov, cited earlier. Conversely, defects in the activity of certain regions might cause failure to participate in the formation of representational systems. This possibility was illustrated by the reported failure of frontal regions of psychotic patients to become integrated with other parts of the brain during mental activity (Gavrilova, 1970). These deficits in activity might influence thought, imagination, and fantasy, because of the absence of certain affective or cognitive components that normally constitute a portion of such recollections, or because the feedback mechanisms that normally control the emergence of such content into consciousness are functioning inadequately.

At the present, these notions are in the realm of theoretical speculations. In order to evaluate such notions, it is essential to devise measures of neural integrative function to determine whether various brain regions participate appropriately in the establishment of representational systems and whether the relative involvement of various functional systems of the brain in the representation of experience is properly balanced. Measures of reactivity must be devised to establish whether different types of information have appropriate impact and contribute to the dynamic construction of representational systems in a fashion accurately reflecting their contribution to an experience. Such measures, and a new technology for their acquisition and analysis, are presented and discussed in Volume 2 of this work.

Pending such studies, these speculations are supported only by the argument that the same fundamental mechanisms must be responsible for the storage, retrieval, and representation of information in the brain, whether it be incorporated into realistically perceived sensations and experiences, normal thought images, plausible imagination, and enriched fantasy, or into misperceived or unhealthy experience, bizarre content of consciousness, distorted imagination, and distressing fantasies that dominate consciousness to an extent harmful to the individual. Whether useful insights into the nature of psychiatric disorder are provided by this theoretical viewpoint must await appropriate experimental investigation.

References

Abeles, M. Excitability of EEG "synchronizing" and "desynchronizing" neurones in the thalamus and brain-stem of the cat. III: Patterns of interaction between pulse pairs applied to the diffuse thalamic projection system. *Electroencephalography & Clinical Neurophysiology*. 1967, *23*, 35–40.

Abraham, F., Bryant, H., Mettler, M., Bergerson, B., Moore, F., Maderdrut, J., Gardiner, M., Walter, D., & Jennrich, R. Spectrum and discriminant analyses reveal remote rather than local sources for hypothalamic EEG: could waves affect unit activity? *Brain Research*, 1973, *49*, 349–366.

Adam, G., Adey, W. R., & Porter, R. W. Interoceptive conditional response in cortical neurones. *Nature*, 1966, *209*, 920–921.

Adametz, J. H. Rate of recovery of functioning in cats with rostral reticular lesions. *Journal of Neurosurgery*, 1959, *16*, 85–97.

Adey, W. R. Intrinsic organization of cerebral tissue in alerting, orienting, and discriminative responses. In G. C. Quarton, T. Melnechuk, & F. O. Schmitt (Eds.), *The neurosciences: A study program.* New York: Rockefeller University Press, 1967. Pp. 615–633.

Adey, W. R. The influence of impressed electrical fields at EEG frequencies. In N. Burch & H. L. Altshuler (Eds.), *Behavior and brain electrical activity.* New York: Plenum Press, 1975. Pp. 363–390.

Adey, W. R., Dunlop, C. W., & Hendrix, C. E. Hippocampal slow waves; distribution and phase relations in the course of approach learning. *Archives of Neurology* (Chicago), 1960, *3*, 74–90.

Adey, W. R., Elul, R., Walter, R. D., & Crandall, P. C. The cooperative behavior of neuronal populations during sleep and mental tasks. *Electroencephalography & Clinical Neurophysiology*, 1966, *23*, 88. (Abstract) (a)

Adey, W. R., Kado, R. T., & Didio, J. Impedance measurements in brain tissue of animal using microvolt signals. *Experimental Neurology*, 1962, *5*, 47–66.

Adey, W. R., Kado, R. T., Didio, J., & Schindler, W. J. Impedance changes in cerebral tissue accompanying a learned discriminative performance in the cat. *Experimental Neurology*, 1963, *7*, 259–281.

Adey, W. R., Kado, R. T., McIlwain, J. T., & Walter, D. O. The role of neuronal elements in regional cerebral impedance changes in alerting, orienting and discriminative responses: The role of neuronal elements in these phenomena. *Experimental Neurology*, 1966, *15*, 490–510. (b)

Adrian, E. D. *The basis of sensation: The action of the sense organs.* London: Christophers, 1928.

Adrian, E. D. Afferent discharges to the cerebral cortex from peripheral organs. *Journal of Physiology*, 1941, *100*, 159–191.

Adrian, E. D., Bremer, F., & Jasper, H. H. (Eds.). *Brain mechanisms and consciousness.* Springfield: Charles C Thomas, 1954.

Adrian, E. D., & Matthews, B. H. C. The Berger rhythm: Potential changes from the occipital lobes of man. *Brain*, 1934, *57*, 354–385.

Agranoff, B. W., & Klinger, P. D. Puromycin effect on memory fixation in the goldfish. *Science*, 1964, *146*, 952.

Aidley, A. J. *The physiology of excitable cells.* London: Cambridge University Press, 1971.

Akert, K., Koella, W. P., & Hess, R., Jr. Sleep produced by electrical stimulation of the thalamus. *American Journal of Physiology*, 1952, *168*, 260–267.

Akert, K., Pfenninger, K., Sandri, C., & Moore, H. Freeze-etching and cytochemistry of vesicles and membrane complexes in synapses of the CNS. In G. Pappas & D. Purpura (Eds.), *Structure and function of synapses.* New York: Raven Press, 1972.

Akimoto, H., Yamaguchi, N., Okabe, J., Nakagawa, T., Nakamura, I., Abe, K., Torri, H., & Masahashi, K. On sleep induced through electrical stimulation of dog thalamus. *Folia Psychiatry & Neurology* (Japan), 1956, *10*, 117–146.

Albert, D. J. The effect of spreading depression on the consolidation of learning. *Neuropsychologia*, 1966, *4*, 49–64. (a)

Albert, D. J. The effects of polarizing currents on the consolidation of learning. *Neuropsychologia*, 1966, *4*, 65–77. (b)

Allison, T., Goff, W. R., & Sterman, M. B. Cerebral somato-sensory responses evoked during sleep in the cat. *Electroencephalography & Clinical Neurophysiology*, 1966, *21*, 461–468.

Allport, D. A. Phenomenal simultaneity and perceptual moment hypothesis. *British Journal of Psychology*, 1968, *59*, 395–406.

Altman, J. Postnatal growth and differentiation of the mammalian brain, with implications for a morphological theory of memory. In G. C. Quarton, T. Melnechuk, & F. O. Schmitt (Eds.), *The neurosciences: A study program.* New York: Rockefeller University Press, 1967. Pp. 723–743.

Altman, J., Brunner, R. L., & Bayer, S. A. The hippocampus and behavioral maturation. *Biological Medicine*, 1951, *24*, 123–140. (a)

Altman, J., & Das, G. D. Autoradiographic examination of the effects of enriched environment on the rate of glial multiplication in the adult rat brain. *Nature*, 1964, *204*, 1161–1163.

Anand, B. K., & Brobeck, J. R. Hypothalamic control of food intake. *Yale Journal of Biological Medicine*, 1951, *24*, 123–140. (a)

Anand, B. K. & Brobeck, J. R. Localization of a feeding center in the hypothalamus of the rat. *Proceedings of the Society of Experimental Biological Medicine*, 1951, *77*, 323–324. (b)

Anand, B. K., & Dua, S. Blood sugar changes induced by electrical stimulation of the hypothalamus of the cat. *India Journal of Medical Research*, 1955, *43*, 123–127. (a)

Anand, B. K., & Dua, S. Feeding responses induced by electrical stimulation of the hypothalamus in cat. *India Journal of Medical Research*, 1955, *43*, 113–122. (b)

Anchel, H., & Lindsley, D. B. Differentiation of two reticulo-hypothalamic systems regulating hippocampal activity. *Electroencephalography & Clinical Neurophysiology*, 1972, *32*, 209–226.

Andersen, P., & Andersson, S. A. *Physiological basis of the alpha rhythm.* New York: Appleton-Century-Crofts, 1968.

Andersen, P., Andersson, S. A., & Lómo, T. Some factors involved in the thalamic control of spontaneous barbiturate spindles. *Journal of Physiology*, 1967, *192*, 257–281. (a)

Andersen, P., Andersson, S. A., & Lómo, T. The nature of thalamo-cortical relations during spontaneous barbiturate spindle activity. *Journal of Physiology*, 1967, *192*, 283–307. (b)

Andersen, P., & Eccles, J. C. Inhibitory phasing of neuronal discharge. *Nature* (London), 1962, *196*, 645–647.

Andersen, P., & Sears, T. A. The role of inhibition in the phasing of spontaneous thalamo-cortical discharge. *Journal of Physiology* (London), 1964, *173*, 459–480.

Andersson, S. A., Holmgren, E., & Manson, J. R. Synchronization and desynchronization in the thalamus of the unanesthetized decorticate cat. *Electroencephalography & Clinical Neurophysiology*, 1971, *31*, 335–345.

Andersson, S. A., & Manson, J. R. Rhythmic activity in the thalamus of the unanesthetized decorticate cat. *Electroencephalography & Clinical Neurophysiology*, 1971, *31*, 21–34.

Annau, Z., Heffner, R., & Snyder, H. S. The dose response relationship between d, and l, tranylcypromine and self-stimulation at three loci. *Society for the Neurosciences, 3rd Annual Meeting*, 1973, p. 364. (Abstract)

Aslanov, A. S. Correlation between cortical potentials in patients with obsessive neuroses. In V. S. Rusinov (Ed.), *Electrophysiology of the central nervous system*. New York: Plenum Press, 1970.

Asratyan, E. A. Changes in the functional state and pattern of electrical activity in cortical areas involved in the establishment of conditioned connection. *Proceedings of the 23rd International Congress of the Physical Sciences* (Tokyo), 1965, *4*, 629–636.

Bach, L. M. N., & Magoun, H. W. The vestibular nuclei as an excitatory mechanism for the cord. *Journal of Neurophysiology*, 1947, *5*, 331–337.

Baer, A. Uber Gleichzeitige Electrische Reizung Zweier Grosshirnstellen am Ungehemmten Hunde. *Archiv für die Gesamte Physiologie Pflügers*, 1905, *106*, 523–567.

Bálint, R. 'Seelenlaehmung der Schauens. *Monatschrift für Psychiatrie und Neurologie*, 1909, *25*. In A. R. Luria, *The working brain*. London: Penguin Press, 1973.

Bard, P. A diencephalic mechanism for the expression of rage with special reference to the sympathetic nervous system. *American Journal of Physiology*, 1928, *84*, 490–515.

Bard, P., & Mountcastle, V. B. Some forebrain mechanisms involved in the expression of rage with special reference to suppression of angry behavior. *Research Publication of Association of Research in Nervous and Mental Disease*, 1947, *27*, 362–404.

Barlow, H. B. Possible principles underlying the transformations of sensory messages. In W. A. Rosenblith (Ed.), *Sensory communication*. Cambridge, Massachusetts: MIT Press, 1961.

Barlow, H. B. Pattern recognition and responses of sensory neurons. *Annals of the New York Academy of Science*, 1969, *156*, 872.

Barlow, H. B. Single units and sensation: A neuron doctrine for perceptual psychology? *Perception*, 1972, *1*, 371–394.

Barlow, J. S. Auto-correlation and cross-correlation analysis in electroencephalography. *Institute of Electrical and Electronics Engineers, Transactions on Bio-Medical Engineering*, 1959, *6*, 179–183.

Barlow, J. S., Morrell, L., & Morrell, F. Some observations on evoked responses in relation to temporal conditioning to paired stimuli in man. *Proceedings of International Colloquium on Mechanisms of Orienting Reactions in Man*, 1967, (Bratislava-Smolenice, Czechoslovakia). (Published by Slovak Academy of Science, 1967.)

Barondes, S. H., & Cohen, H. D. Puromycin effect on successive phases of memory storage. *Science*, 1966, *151*, 594–595.

Barondes, S. H., & Cohen, H. D. Memory impairment after subcutaneous injection of acetoxycycloheximide. *Science*, 1968, *160*, 556–557.

Bartlett, F., & John, E. R. Means and variances of average-response waveforms. *Science,* 1970, *169,* 304–305.

Bartlett, F., & John, E. R. Equipotentiality quantified: The anatomical distribution of the engram. *Science,* 1973, *181,* 764–767.

Batini, C., Moruzzi, G., Palestini, M., Rossi, G. F., & Zanchetti, A. Effects of complete pontine transections on the sleep-wakefulness rhythm: the midpontine pretrigeminal preparation. *Archives of Italian Biology,* 1959, *97,* 1–12.

Bawin, S. M., Gavalas-Medici, R. J., & Adey, W. R. The effects of modulated very high frequency fields on specific brain rhythms in cats. *Brain Research,* 1973, *58,* 365–384.

Beck, E. C., Doty, R. W., & Kooi, K. A. Electrocortical reactions associated with conditioned flexion reflexes. *Electroencephalography & Clinical Neurophysiology,* 1958, *10,* 279–285.

Beck, E. C., Dustman, R. E., & Sakai, M. Electrophysiological correlates of selective attention. In C. R. Evans & T. B. Mulholland (Eds.), *Attention in neurophysiology.* London: Butterworth, 1969. Pp. 396–416.

Begleiter, H., Gross, M. M., & Kissin, B. Evoked cortical responses to affective visual stimuli. *Journal of Psychophysiology,* 1967, *3,* 336–343.

Begleiter, H., & Platz, A. Cortical evoked potentials to semantic stimuli. *Psychophysiology,* 1969, *6,* 91–100. (a)

Begleiter, H., & Platz, A. Modifications in evoked potentials by classical conditioning. *Science,* 1969, *166,* 769–771. (b)

Begleiter, H., Porjesz, B., Yerre, C., & Kissin, B. Evoked potential correlates of expected stimulus intensity. *Science,* 1973, *179,* 814–816.

Ben-Ari, Y., & La Salle, G. Plasticity at unitary level. II Modifications during sensory-sensory association procedures. *Electroencephalography & Clinical Neurophysiology,* 1972, *32,* 667–679.

Bennett, T. L. Hippocampal EEG correlates of behavior. *Electroencephalography & Clinical Neurophysiology,* 1970, *28,* 17–23.

Bennett, T. L., & Gottfried, J. Hippocampal theta activity and response inhibition. *Electroencephalography & Clinical Neurophysiology,* 1970, *29,* 196–200.

Berger, H. Uber das Elektrenkephalogramm des Menschen. *Archives Psychiatric Nervenkrankh,* 1929, *87,* 527–570.

Betz, V. A. Two centers in the cortical layer of the human brain. *Mosk. Vrachebn. Vestn.,* 1874, *24.*

Bianchi, L. The function of frontal lobes. *Brain,* 1895, *18.*

Bickford, R. G. Measurement and analysis of electrical activity (session summary). In N. R. Burch & H. L. Altshuler (Eds.), *Behavior and brain electrical activity.* New York: Plenum Press, 1975.

Bickford, R. G., Brimm, J., Berger, L., & Aung, M. Application of compressed spectral array in clinical EEG. In P. Kellaway & I. Petersen (Eds.), *Automation of clinical electroencephalography.* New York: Raven Press, 1973. Pp. 55–64.

Blakemore, C. Developmental factors in the formation of feature extracting neurons. In F. O. Schmitt & F. G. Worden (Eds.), *The neurosciences: Third study program.* Cambridge, Massachusetts: MIT Press, 1974. Pp. 103–113.

Blakemore, C., & Campbell, F. W. On the existence in the human visual system of neurons selectively sensitive to the orientation and size of retinal images. *Journal of Physiology* (London), 1969, *203,* 237–260.

Blakemore, C., & Cooper, G. Development of the brain depends on visual enrichment. *Nature,* 1970, *288,* 477–478.

Bloom, F. E., Nicholson, J., Ungerlei, J., & Ledley, R. S. Ultra-structural identification of catecholamine containing central synaptic terminals. *Journal of Histology & Cytology,* 1973, *21,* 333–340.

Blum, J. S., Chow, K. L., & Pribram, K. H. A behavioral analysis of the organization of the parieto-temporo-pre-occipital cortex. *Journal of Comparative Neurology*, 1950, *93*, 53–100.

Bogen, J. E., & Gazzaniga, M. S. Cerebral commissurotomy in man: Minor hemisphere dominance for certain visuo-spatial functions. *Journal of Neurosurgery*, 1965, *23*, 394–399.

Booth, D. A. Protein synthesis and memory. In J. A. Deutsch (Ed.), *The physiological basis of memory*. New York: Academic Press, 1973. Pp. 27–58.

Brady, J. V. Emotional behavior. In J. Field, H. W. Magoun, & V. E. Hall (Eds.), *Handbook of physiology*. Vol. 3. Washington, D.C.: American Physiological Society, 1960. Pp. 1529–1552.

Brady, J. V., & Nauta, W. J. H. Subcortical mechanisms in emotional behavior: Affective changes following septal forebrain lesions in the albino rat. *Journal of Comparative Physiology & Psychology*, 1953, *46*, 339–346.

Brain, Lord. Speech disorders (2nd ed.). London, and Washington, D.C.: Butterworth, 1965.

Brazier, M. A. B. Studies of evoked responses by flash in man and cat. In H. H. Jasper, L. D. Proctor, R. S. Knighton, W. C. Noshay, & R. T. Costello (Eds.), *Reticular formation of the brain*. Boston: Little, Brown & Co., 1958. P. 151.

Breen, R. A., & McGaugh, J. L. Facilitation of maze learning with post-trial injections of picrotoxin. *Journal of Comparative Physiology & Psychology*, 1961, *54*, 498–501.

Bremer, F. Nouvelles recherches sur le mécanisme du sommeil. *Comptes Rendus Société Biologie* (Paris), 1936, *122*, 460–464.

Bremer, F., & Stoupel, N. Facilitation et inhibition des potentiels evoques corticaux dans l'eveil cerebral. *Archives of International Physiology & Biochemistry*, 1959, *67*, 1–35.

Bridges, K. M. B. Emotional development in early infancy. *Child Development*, 1932, *3*, 324–341.

Brizzee, K. R., Vogt, J., & Kharetchko, X. Postnatal changes in glial neuron index with a comparison of methods of cell enumeration in the white rat. *Progress in Brain Research*, 1964, *4*, 136–149.

Brobeck, J. R., Tepperman, J., & Long, C. N. H. Experimental hypothalamic hyperphagia in the albino rat. *Yale Journal of Biological Medicine*, 1943, *15*, 831–853.

Broca, P. Remarques sur le siege de la faculte du language articule. *Bulletin of the Society Anthropology*, 1861, *6*.

Brodie, B. B., Spector, S., & Shore, P. A. Interaction of drugs with norepinephrine in the brain. *Pharmacology Review*, 1959, *11*, 548–564.

Broggi, G., & Margnelli, M. Dynamic properties of synaptic input from intralaminar nuclei to ventrolateral thalamus. *Brain Research*, 1971, *26*, 192–194.

Brooks, C. McC. The relative importance of changes in activity in the development of experimentally produced obesity in the rat. *American Journal of Physiology*, 1946, *147*, 708–716.

Brown, W. S., Marsh, J. T., & Smith, J. C. Contextual meaning effects on speech-evoked potentials. *Behavioral Biology*, 1973, *9*, 755–761.

Brown-Sequard, C. E. Existence de l'excitabilité motrice et de l'excitabilité inhibitoire dans les régions occipitales et sphenoidales de l'écorce cerebrale. *Comptes Rendus Memoires Société Biologie*, 1884, 8th Ser., Pt. 1, *36*, 301–303.

Buchsbaum, M., & Silverman, J. Stimulus intensity control and the cortical response. *Psychosomatic Medicine*, 1968, *30*, 12–22.

Buchwald, J. S., Halas, E. S., & Schramm, S. Progressive changes in efferent unit responses to repeated cutaneous stimulation in spinal cats. *Journal of Neurophysiology*, 1965, *28*, 200–215.

Buchwald, N. A., Romerosi, C., Hull, C. D., & Wakefield, C. Learned and unlearned responses to stimulation of the same subcortical site. *Experimental Neurology*, 1967, *17*, 451–465.

Bullock, T. H., & Quarton, G. C. Simple systems for the study of learning mechanisms. *Neurosciences Research Bulletin*, 1966, *4*, 105.

Bureš, J. Discussion. In D. P. Kimble (Ed.), *Anatomy of memory*. Palo Alto: Science Behavior Books, 1965. Pp. 49–50.

Bureš, J., & Burešova, O. Plasticity at the single neuron level. *Proceedings of the 23rd International Congress of Physiological Science* (Tokyo), 1965, *IV*, 359–364.

Bureš, J. and Burešova, O. Plastic changes of unit activity based on reinforcing properties of extracellular stimulation of single neurons. *Journal of Neurophysiology*, 1967, *30*, 98–113.

Bureš, J. and Burešova, O. Plasticity in single neurons and neural populations. In G. Horn & R. A. Hinde (Eds.), *Short-term changes in neural activity and behavior*, London and New York: Cambridge University Press, 1970.

Bureš, J., Burešova, O., & Weiss, T. Functional consequences of hippocampal spreading depression. *Physiologia Bohemoslovenica*, 1960, *9*, 219–227.

Burke, R. E. Composite nature of the monosynaptic excitatory postsynaptic potential. *Journal of Neurophysiology*, 1967, *30*, 1114–1137.

Burkhardt, D. A., & Riggs, L. A. Modification of the human visual evoked potential by monochromatic backgrounds. *Vision Research*, 1967, *7*, 453–459.

Burns, B. D. The production of afterbursts in isolated unanesthetized cerebral cortex. *Journal of Physiology* (London), 1954, *125*, 427–446.

Burns, B. D. *The mammalian cerebral cortex*. London: Arnold, 1958.

Burns, B. D. *The uncertain nervous system*. Baltimore: Williams & Wilkins, 1968.

Burns, B. D., Heron, W., & Grafstein, B. Response of cerebral cortex to diffuse monocular and binocular stimulation. *American Journal of Physiology*, 1960, *198*, 200–204.

Burns, B. D., & Pritchard, R. Contrast discrimination by neurones in the cat's visual cerebral cortex. *Journal of Physiology* (London), 1964, *175*, 445–463.

Burns, B. D., & Smith, G. K. Transmission of information in the unanesthetized cat's isolated forebrain. *Journal of Physiology* (London), 1962, *164*, 238–251.

Butler, R. A., Diamond, I. T., & Neff, W. D. Role of auditory cortex in discrimination of changes in frequency. *Journal of Neurophysiology*, 1957, *20*, 108–120.

Butters, N., & Rosvold, H. E. Effect of caudate and septal nuclei lesions on resistance to extinction and delayed alternation. *Journal of Comparative Physiology & Psychology*, 1968, *65*, 397–403.

Byrne, W. L. (Ed.). *Molecular approaches to learning and memory*. New York: Academic Press, 1970.

Callaway, E. Evoked potentials in psychopathology and psychiatric treatment. In N. R. Burch & H. L. Altshuler (Eds.), *Behavior and brain electrical activity*. New York: Plenum Press, 1975.

Callaway, E., & Halliday, R. A. Evoked potential variability: Effects of age, amplitude and methods of measurement. *Electroencephalography and Clinical Neurophysiology*, 1973, *34*, 125–133.

Callaway, E., Jones, R. T., & Layne, R. S. Evoked responses and segmental set in schizophrenia. *Archives General Psychiatry*, 1965, *12*, 83–89.

Calvet, J., Calvet, M. C., & Scherrer, J. Étude stratigraphique corticale de l'activité EEG spontanée. *Electroencephalography & Clinical Neurophysiology*, 1964, *17*, 109–125.

Cannon, W. B. *Bodily changes in pain, hunger, fear, and rage*. New York: Appleton-Century-Crofts, 1929. P. 20.

Cantril, H., & Livingston, W. K. The concept of transaction in psychology and neurology. *Journal of Individual Psychology*, 1963, *19*, 3–16.

Case, T. J. Alpha waves in relation to structures involved in vision. *Biological Symposium*, 1942, *7*, 107–116.

Cavonius, C. R. Evoked response of the human visual cortex: Spectral sensitivity. *Psychonomic Science*, 1965, *2*, 185–186.

Chang, H.-T. Dendritic potential of cortical neurons produced by direct electrical stimulation. *Journal of Neurophysiology*, 1951, *14*, 1–21.

Chang, H.-T. Cortical and spinal neurons. Cortical neurons with particular reference to the apical dendrites. *Cold Spring Harbor Symposium Quantitative Biology*, 1952, *17*, 189–202.

Chapman, W. P., Schroeder, H. R., Geyer, G., Brazier, M. A. B., Fager, C., Poppen, J., Solomon, H., & Yahovlev, P. Physiological evidence concerning importance of the amygdaloid nuclear region in the integration of circulatory function and emotion in man. *Science*, 1954, *120*, 949–950.

Chapouthier, G. Behavioral studies of the molecular basis of memory. In J. A. Deutsch (Ed.), *The physiological basis of memory*. New York: Academic Press, 1973.

Chow, K. L. Brain functions. *Annual Review of Psychology*, 1961, *12*, 281–310.

Chow, K. L., Lindsley, D. F., & Gollender, M. Modification of response patterns of lateral geniculate neurons after paired stimulation of contralateral and ipsilateral eyes. *Journal of Neurophysiology*, 1968, *31*, 729–739.

Chow, K. L., & Randall, W. Learning and retention in cats with lesions in reticular formation. *Psychonomic Science*, 1964, *1*, 259–260.

Chusid, J. G. *Correlative neuroanatomy and functional neurology* (15th ed.). Palo Alto, California: Lange Medical Publication, 1973.

Clark, G. M., Nathar, J. N., Kranz, H. G., & Maritz, J. S. A behavioral study on electrical stimulation of the cochlea and central auditory pathways of the cat. *Experimental Neurology*, 1972, *36*, 350–361.

Clynes, M. Brain space analysis of evoked potential components applied to chromaticity waves. *Proceedings 6th International Conference on Medical Electronics & Biological Engineering*, Tokyo, 1965.

Clynes, M., Kohn, M., & Gradijan, J. Computer recognition of the brain's visual perception through learning the brain's physiologic language. *Institute of Electrical and Electronics Engineers, International Conference Record*, Pt. *9*, 1967, pp. 125–142.

Cobb, W. A. The normal adult EEG. In D. H. Hill & G. Parr (Eds.), *Electroencephalography: A symposium on its various aspects*. New York: Macmillan, 1963. Pp. 232–239. (a)

Cole, K. S. Permeability and impermeability of cell membranes for ions. *Cold Spring Harbor Symposium Quantitative Biology*, 1940, *8*, 110–122.

Coombs, J. S., Curtis, D. R., & Eccles, J. C. The electrical constants of the motorneuron membrane. *Journal of Physiology* (London), 1959, *145*, 505–528.

Cooper, J. R., Bloom, F. E., & Roth, R. H. *The biochemical basis of neuropharmacology*. New York: Oxford University Press, 1974.

Cordeau, J. P., & Mahut, H. Some long-term effects of temporal lobe resections on auditory and visual discrimination in monkeys. *Brain*, 1964, *87*, 177–190.

Corning, W. C., Dyal, J. A., & Willows, A. O. D. *Inverterbrate learning* (Vol. I). New York: Plenum Press, 1973.

Costa, E., Grappetti, A., & Revuelta, A. Action of fen-fluramine on monoamine stores of rat tissues. *British Journal of Pharmacology*, 1971, *41*, 57–66.

Costa, E., & Revuelta, A. *Para*-chloroamphetamine and serotonin turnover in rat brain. *Neuropharmacology*, 1972, *11*, 291–302.

Coury, J. N. Neural correlates of food and water intake in the rat. *Science*, 1967, *156*, 1763–1765.

Cragg, B. G. The density of synapses and neurons in the motor cortex and visual areas of the cerebral cortex. *Journal of Anatomy*, 1967, *101*, 639–654.

Creelman, C. D. Human discrimination of auditory duration. *Journal Acoustical Society of America*, 1962, *34*, 582–593.

Creutzfeldt, O. D., Watanabe, S., & Lux, H. D. Relations between EEG phenomena and potentials of single cortical cells. I. Evoked responses after thalamic and epicortical stimulation. *Electroencephalography & Clinical Neurophysiology*, 1966, *20*, 1–18. (a)

Creutzfeldt, O. D., Watanabe, S., & Lux, H. D. Relations between EEG phenomena and potentials of single cortical cells. II. Spontaneous and convulsed activity. *Electroencephalography & Clinical Neurophysiology*, 1966, *20*, 19–37. (b)

Cushing, H. A note upon the faradic stimulation of the post-central gyrus in conscious patients. *Brain*, 1909, *32*, 44–53.

Dahlström, A., & Fuxe, K. Evidence for the existence of monoamine-containing neurons in the central nervous system. I. Demonstration of monoamines in the cell bodies of brain stem neurons. *Acta Physiologica Scandinavica*, 1965, *62* (suppl.), 232–251.

Davies, R. E., & Keynes, R. D. A coupled sodium-potassium pump. In A. Kleinzeller & A. Kotyk (Eds.), *Membrane transport and metabolism*. New York: Academic Press, 1961. Pp. 336–340.

Davis, A. E., & Wada, J. A. Hemispheric asymmetry: Frequency analysis of visual and auditory evoked responses to non-verbal stimuli. *Electroencephalography & Clinical Neurophysiology*, 1974, *37*, 1–9.

Davis, H., Hirsch, S. K., Shelnutt, J., & Bowers, C. Further validation of evoked response audiometry (ERA). *Journal of Speech & Hearing Research*, 1967, *10*, 717–732.

Dawson, R. G., & McGaugh, J. G. Drug facilitation of learning and memory. In J. A. Deutsch (Ed.), *The physiological basis of memory*. New York: Academic Press, 1973.

Dean, R. B. Theories of electrolyte equilibrium in muscle. *Biology Symposium*, 1941, *3*, 331–348.

Delgado, M. M. R. Study of some cerebral structures related to transmission and elaboration of noxious stimulation. *Journal of Neurophysiology*, 1955, *18*, 261–275.

Dement, W. The effect of dream deprivation. *Science*, 1960, *131*, 1705–1707.

Dempsey, E. W., & Morison, R. S. The production of rhythmically recurrent cortical potentials after localized thalamic stimulation. *American Journal of Physiology*, 1942, *135*, 293–300.

Denny-Brown, D. The frontal lobes and their functions. In A. Feiling (Ed.), *Modern trends in neurology*. New York: P. B. Hoeber, Inc., 1951.

Denny-Brown, D., & Chambers, R. A. The parietal lobe and behavior. *Research Publication, Association of Nervous & Mental Disorders*, 1958, *36*, 35–46.

Denny-Brown, D., Meyer, J., & Hornstein, S. The significance of perceptual rivalry resulting from parietal lesions. *Brain*, 1952, *75*, 433–438.

Desiraju, T., Broggi, G., Prelevic, S., Santini, M., & Purpura, D. P. Inhibitory synaptic pathways linking specific and nonspecific thalamic nuclei. *Brain Research*, 1969, *15*, 542–543.

Deutsch, J. A. Electroconvulsive shock and memory. In J. A. Deutsch, (Ed.), *The physiological basis of memory*. New York: Academic Press, 1973. (a)

Deutsch, J. A. (Ed.). *The physiological basis of memory*. New York: Academic Press, 1973. (b)

Dicheva, D., Atsev, E., & Popivanov, D. Investigations on mental work capacity by modern electrophysiological methods. *C. R. Academy Bulgarian Sciences*, 1971, *24*, 829–832.

Dichter, M., & Spencer, A. Penicillin-induced interictal discharges from the cat hippocampus, II. Mechanisms underlying origin and restriction. *Journal of Neurophysiology*, 1969, *32*, 663–687.

Dismukes, R. K., & Rake, A. V. Involvement of biogenic amines in memory formation. *Psychopharmacologia*, 1972, *23*, 17–25.

Disterhoft, J. F., & Olds, J. Differential development of conditioned unit changes in thalamus and cortex of cat. *Journal of Neurophysiology*, 1972, *35*, 665–679.

Donchin, E., & Cohen, L. Average evoked potentials and intramodality selective attention. *Electroencephalography & Clinical Neurophysiology*, 1967, *22*, 537–546.

Donovick, P. J. Effect of localized septal lesions on hippocampal EEG activity and behavior in rats. *Journal of Comparative Physiology & Psychology*, 1968, *66*, 569–578.

Dorry, G. W., & Zeaman, D. Teaching a simple reading vocabulary to retarded children: Effectiveness of fading and non-fading procedures. *American Journal of Mental Deficiency* 1975, *76*, 711–716.

Doty, R. W. Potentials evoked in cat cerebral cortex by diffuse and by punctiform photic stimuli. *Journal of Neurophysiology*, 1958, *21*, 437–464.

Doty, R. W. Conditioned reflexes elicited by electrical stimulation of the brain in macaques. *Journal of Neurophysiology*, 1965, *28*, 623–640.

Doty, R. W. Electrical stimulation of the brain in behavioral context. *Annual Review of Psychology*, 1969, *20*, 289–320.

Doty, R. W., & Kimura, D. S. Oscillatory potentials in the visual system of cats and monkeys. *Journal of Physiology*, 1963, *168*, 205–218.

Doty, R. W., & Rutledge, L. T. "Generalization" between cortically and peripherally applied stimuli eliciting conditioned reflexes. *Journal of Neurophysiology*, 1959, *22*, 428–435.

Douglas, R. J. The hippocampus and behavior. *Psychology Bulletin*, 1967, *67*, 416–422.

Drachman, D. A., & Ommaya, A. K. Memory and the hippocampal complex. *Archives of Neurology*, 1964, *10*, 411–425.

Dru, D., Walker, J. P., & Walker, J. B. Self-produced locomotion restores visual capacity after striate lesions. *Science*, 1975, *187*, 265–266.

Dumenko, V. N. The electrographic study of relationships between various cortical areas in dogs during the elaboration of a conditioned reflex stereotype. In I. N. Knipst (Ed.), *Contemporary problems of electrophysiology of the central nervous system*. Moscow: Academy of Science, 1967. Pp. 104–112.

Dumont, S., & Dell, P. Facilitation réticulaire des mécanismes visuels corticaux. *Electroencephalography & Clinical Neurophysiology*, 1960, *12*, 769–796.

Dustman, R. E., & Beck, E. C. Phase of alpha brain waves, reaction time and visually evoked potentials. *Electroencephalography & Clinical Neurophysiology*, 1965, *18*, 433–440.

Eccles, J. C. An electrical hypothesis of synaptic and neuromuscular transmission. *Annals of the New York Academy of Sciences*, 1946, *47*, 429–455.

Eccles, J. C. Spinal neurones: Synaptic connections in relation to chemical transmitters and pharmacological responses. In W. D. M. Paton (Ed.), *A symposium on pharmacological analysis of central nervous system action*. New York: Macmillan, 1962.

Eccles, J. C. *Brain and conscious experience*. New York: Springer-Verlag, 1966.

Economo, C. Von. Schlaftheorie. *Ergeb. Physiol. 28:* 312–339, 1929. Cited in S. Ochs, *Elements of neurophysiology*. New York: John Wiley & Sons, 1965.

Economo, C. Von. Die Encephalitis Lethargica, 1918. Deuticke, Vienna. Cited in S. Ochs, *Elements of Neurophysiology*. New York: John Wiley & Sons, 1965.

Efron, E. The effect of handedness on the perception of simultaneity and temporal order. *Brain*, 1963, *86*, 261–284. (a)

Efron, E. Temporal perception, aphasia and déja vu. *Brain*, 1963, *86*, 403–424. (b)

Efron, E. The duration of the present. *Annals of the New York Academy of Sciences*, 1967, *138*, 713–729.

Efron, R. The relationship between the duration of a stimulus and the duration of a perception. *Neuropsychologia*, 1970, *8*, 37–55. (a)

Efron, R. The minimum duration of a perception. *Neuropsychologia*, 1970, *8*, 57–63. (b)

Egger, M. D., & Flynn, J. P. Amygdaloid suppression of hypothalamically elicited attack behavior. *Science*, 1962, *136*, 43–44.

Eigen, M. Molecules, information and memory: From molecular to neural networks. In F. O. Schmitt & F. G. Worden (Eds.), *The neurosciences: Third study program*. Cambridge, Massachusetts: MIT Press, 1974. Pp. xix–xxvii.

Elazar, Z., & Adey, W. R. Electroencephalographic correlates of learning in subcortical and cortical structures. *Electroencephalography & Clinical Neurophysiology*, 1967, *23*, 306–319.

Ellen, P., & Wilson, A. S. Perseveration in the rat following hippocampal lesions. *Experimental Neurology*, 1963, *8*, 310–317.

Elul, R. Amplitude histograms of the EEG as an indicator of the cooperative behavior of neuron populations. *Electroencephalography & Clinical Neurophysiology*, 1967, *23*, 87. (a)

Elul, R. Statistical mechanisms in generation of the EEG. In L. Fogel & F. George (Eds.), *Progress in biomedical engineering*. Washington, D. C.: Spartan Books, 1967. (b)

Elul, R. Brain waves: Intracellular recording and statistical analysis help clarify their physiological significance. In R. Enslein (Ed.), *Data acquisition and processing in biology and medicine* (Vol. 5). New York: Pergamon Press, 1968.

Elul, R. Gaussian behavior of the EEG: Changes during performance of mental task. *Science*, 1969, *164*, 328.

Elul, R. The genesis of the EEG. *International Review of Neurobiology*, 1972, *15*, 228–272.

Elul, R., & Adey, W. R. The intracellular correlate of gross evoked responses. *23rd International Congress of Physiological Sciences*, 1965, *1034*. (Abstract)

Elul, R., & Adey, W. R. Nonlinear relationship of spike and waves in cortical neurons. *The Physiologist*, 1966, *8*, 98–104.

Evarts, E. V. Temporal patterns of discharge of pyramidal tract neurons during sleep and waking in the monkey. *Journal of Neurophysiology*, 1964, *27*, 152–171.

Evarts, E. V. Unit activity in sleep and wakefulness. In G. C. Quarton, T. Nelnechuk, & F. O. Schmitt (Eds.), *The neurosciences: A study program*. New York: Rockefeller University Press, 1967.

Feindel, W., & Penfield, W. Localization of discharge in temporal lobe automatism. *AMA Archives, Neurology & Psychiatry*, 1954, *72*, 605–630.

Feldman, M. H., & Purpura, D. P. Prolonged conductance increase in thalamic neurons during synchronizing inhibition. *Brain Research*, 1970, *24*, 329–332.

Ferrier, D. *The functions of the brain*. London: Smith, Elder, 1876.

Fessard, A. E. Mechanisms of nervous integration and conscious experience. In E. D. Adrian, F. Bremer, & H. Jasper (Eds.), *Brain mechanisms and consciousness*. Springfield, Illinois: Charles C Thomas, 1954. Pp. 200–236.

Fidone, S. J., & Preston, J. B. Inhibitory resetting of resting discharge of fusimotor neurons. *Journal of Neurophysiology*, 1971, *34*, 217–227.

Fields, C. Visual stimuli and evoked responses in the rat. *Science*, 1969, *165*, 1377–1379.

Filimonov, I. N. Localization of functions in the cerebral cortex and Pavlov's theory of higher nervous activity. *Klinicheskaya Meditsina*, 1951, *29*.

Filimonov, I. N. Architectonics and localization of functions in the cerebral cortex. In *Textbook of neurology* (Vol. 1). Moscow: Medgiz, 1957.

Flexner, J. B., & Flexner, L. B. Restoration of memory lost after treatment with puromycin. *Proceedings of National Academy of Sciences*, 1967, *57*, 1651–1654.

Flexner, J. B., & Flexner, L. B. Further observations on restoration of memory lost after treatment with puromycin. *Yale Journal of Biology and Medicine*, 1969, *42*, 235–240.

Flexner, J. B., Flexner, L. B., & Stellar, E. Memory in mice as affected by intracerebral puromycin. *Science*, 1963, *141*, 57–59.

Flood, J. F., Rosenszweig, M. R., Bennett, F. L. and Orme, A. E. Influence of training strength on amnesia induced by pretraining injections of cycloheximide. *Physiology and Behavior*, 9:589–593, 1972.

Flourens, M. J. P. *Recherches expérimentales sur les propriétés et les fonctions du système nerveux dans les animaux vertébrés.* Paris: Crevot, 1824.

Flourens, M. J. P. *Examen de phrénologie.* Paris: Hachette, 1842.

Flynn, J. P., & Wasman, M. Learning and cortically evoked movement during propagated hippocampal after discharges. *Science*, 1960, *131*, 1607.

Forester, O. Symptomatologie der Erkrankungen des Gehirns. Motorische Felder und Bahnen–Sensible corticale Felder. In O. Bumke & O. Foerster (Eds.), *Handbuch der Neurologie* (Vol. 6). Berlin: Springer-Verlag, 1936.

Fox, S. S., & Norman, R. J. A new measure of brain activity: Functional congruence; an index of neural homogeneity. *Proceedings of the 24th International Congress of Physiological Sciences,* 1968, *7*, 142.

Fox, S. S., & O'Brien, J. H. Duplication of evoked potential waveform by curve of probability of firing of a single cell. *Science*, 1965, *147*, 888–890.

Fraisse, P. *The psychology of time.* New York: Harper and Row, 1963.

Frankenhaeuser, M. *Estimation of time, an experimental study.* Amsterdam: Almqvist and Wiksell, 1959.

Freeman, W. J. Comparison of thresholds for behavioral and electrical responses to cortical electrical stimulation in cats. *Experimental Neurology*, 1962, *6*, 315–331.

Freeman, W. J. Waves, pulses and the theory of neural masses. In R. Rosen & F. M. Snell (Eds.), *Progress in theoretical biology.* New York: Academic Press, 1972. Pp. 88–166. (a)

Freeman, W. J. Linear analysis of the dynamics of neural masses. *American Review of Biophysics and Bioengineering*, 1972, *1*, 225–256. (b)

Freeman, W. J. Neural coding through mass action in the olfactory system. *Proceedings of Institute of Electrical & Electronic Engineering Conference on Biologically Motivated Automata Theory,* 1974.

Freeman, W. J. *Mass action in the nervous system.* New York: Academic Press, 1975.

Freeman, W., & Watts, J. W. *Psychosurgery, intelligence, emotion and social behavior following prefrontal lobotomy for mental disorders.* (2nd ed.) Springfield, Illinois: Charles C Thomas, 1949.

French, J. D., Hernández-Peón, R., & Livingston, R. B. Projections from cortex to cephalic brain stem (reticular formation) in monkey. *Journal of Neurophysiology*, 1955, *18*, 44–55.

Fried, P. A. Limbic system lesions in rats: Differential effects in an approach-avoidance task. *Journal of Comparative Physiology & Psychology*, 1971, *74*, 349–353.

Frigyesi, T. L., & Schwartz, R. Reciprocal innervation of ventrolateral and medical thalamic neurons by corticothalamic pathways. In T. L. Frigyesi, E. Rinvik, & M. D. Yahr (Eds.), *Corticothalamic projections and sensorimotor activities.* New York: Raven Press, 1972.

Fritsch, G., & Hitzig, E. Uber die elektrische Erregbarkeit des Grosshirns. *Archiv fur Anatomie, Physiologie und Wissenschaftliche Medicin*, 1870, *37*.

Fromm, G. H., & Bond, H. W. The relationship between neuron activity and cortical steady potentials. *Electroencephalography & Clinical Neurophysiology*, 1967, *22*, 159–166.

Frost, J. D. EEG intracellular unit potential relationships in isolated cerebral cortex. *Electroencephalography & Clinical Neurophysiology*, 1968, *24*, 434–443.

Frost, J. D., & Gol, A. Computer determination of relationships between EEG activity and single unit discharges in isolated cerebral cortex. *Experimental Neurology*, 1966, *14*, 506–519.

Frost, J. D., & Low, M. D. An electronic method for determining time-locked changes in neuronal firing rate. *Electroencephalography & Clinical Neurophysiology*, 1967, *23*, 176–178.

Fujita, V., & Sato, T. Intracellular records from hippocampal pyramidal cells in rabbit during theta rhythm activity. *Journal of Neurophysiology*, 1964, *27*, 1011–1025.

Fulton, J. F. *The frontal lobes*. Baltimore: Williams & Wilkins, 1948.

Fulton, J. F. *Frontal lobotomy and affective behavior*. New York: Norton, 1951.

Fulton, J. F., & Bailey, P. Tumors in the region of the third ventricle: Their diagnosis and relation to pathological sleep. *Journal of Nervous & Mental Disease*, 1929, *69*, 1–25, 145–164, 261–272.

Fuller, J. L., Rosvold, H. E., & Pribram, K. H. The effect on affective and cognitive behavior in the dog of lesions of the pyriform-amygdala-hippocampal complex. *Journal of Comparative Physiology & Psychology*, 1957, *50*, 89–95.

Fuster, J. M. Effects of stimulation of brain stems on tachistoscopic perception. *Science*, 1958, *127*, 150.

Galambos, R., & Sheatz, G. C. An electroencephalograph study of classical conditioning. *American Journal of Physiology*, 1962, *203*, 173–184.

Gall, F. J. *Sur les fonctions due cerveau et sur celles de chacune de ses parties*. Paris: Baillière, 1825.

Gamper, E. Structure and functional capacity of human mesencephalic monster (arhinencephalic with encephalocele); contributions to teratology and fiber system. *Zeitschrift Geselleschafte der Neurologie und Psychiatrie*, 1926, *102*, 154–235; referred to by Livingston (1967).

Gardner, E. *Fundamentals of neurology* (4th ed.). Philadelphia: Saunders, 1963.

Gastaut, H., Jus, A., Jus, C., Morrell, F., Storm Van Leeuwen, W., Dongier, S., Naquet, R., Regis, H., Roger, A., Bekkering, D., Kamp, A., & Werre, J. Etude topographique des reactions electroencephalographiques conditionness chez l'homme. *Electroencephalography & Clinical Neurophysiology*, 1957, *9*, 1.

Gavalas, R., Walter, D. O., Hamer, J., & Adey, W. R. Effect of low level, low frequency electric fields on EEG and behavior in *Macaca* nemestrina. *Brain Research*, 1970, *18*, 491–501.

Gavrilova, N. A. Spatial synchronization of cortical potentials in patients with disturbances of association. In V. S. Rusinov (Ed.), *Electrophysiology of the central nervous system*. New York: Plenum Press, 1970.

Gazzaniga, M. S. Psychological properties of the disconnected hemispheres in man. *Science*, 1965, *150*, 372.

Gazzaniga, M. S. *The bisected brain*. New York: Appleton-Century-Crofts, 1970.

Geller, A., & Jarvik, M. Unpublished data. Presented by Jarvik at neurosciences research program work session on consolidation of memory trace, MIT, November, 1967.

Gellhorn, E. *Autonomic imbalance and the hypothalamus*. Minneapolis: University of Minnesota Press, 1957.

Gellhorn, E. Prolegomena to a theory of emotions. In *Perspectives in biology and medicine* (Vol. 4). Chicago: Chicago University Press, 1961.

Gerard, R. W. Factors controlling brain potentials. *Cold Spring Harbor Symposium Quarterly, Biology*, 1936, *4*, 292–304.

Gerard, R. W. The fixation of experience. In J. F. Delafresnaye, A. Fessard, R. W. Gerard, & J. Konorski (Eds.), *CIOMS symposium on brain mechanisms and learning*. Oxford: Blackwell, 1961.

Gerard, R. W., & Libet, B. The control of normal and convulsive brain potentials. *American Journal of Psychiatry*, 1940, *96*, 1125.

Gerard, R. W., Marshall, W. H., & Saul, L. J. Electrical activity of the cat's brain. *Archives Neurology & Psychiatry*, 1936, *36*, 675–738.

Gerbrandt, L. K., Spinelli, D. N., & Pribram, K. H. The interaction of visual attention and temporal cortex stimulation on electrical activity evoked in striate cortex. *Electroencephalography & Clinical Neurophysiology*, 1970, *29*, 146–155.

Gerken, G. M. Behavioral measurement of electrical stimulation thresholds for medial geniculate nucleus. *Experimental Neurology*, 1971, *31*, 60–74.

Gerstein, G. L. Neuron firing patterns and the slow potentials. *Electroencephalography & Clinical Neurophysiology* (Suppl.), 1961, *20*, 68–71.

Glickman, S. E. Perseverative neural processes and consolidation of memory trace. *Psychological Bulletin*, 1961, *58*, 218–233.

Glivenko, E. V., Korol'kova, T. A., & Kuznetsova, G. D. Investigation of the spatial correlation between the cortical potentials of the rabbit during formation of a conditioned defensive reflex. *Fizicheskii Zhurnal SSSR Sechenova*, 1962, *48* (9), 1026.

Globus, A., & Scheibel, A. B. Pattern and field in cortical structure: The rabbit. *Journal of Comparative Neurology*, 1967, *131*, 155–172.

Gloor, P. Amygdala. In J. Field, H. W. Magoun, & V. E. Hall (Eds.), *Handbook of physiology* (Vol. 2). Baltimore: Williams & Wilkins, 1960.

Goddard, G. V. Functions of the amygdala. *Psychology Bulletin*, 1964, *62*, 89–109.

Goff, W. R., Sterman, M. B., & Allison, T. Cortical midline late response during sleep in the cat. *Brain Research*, 1966, *1*, 311–314.

Goldberg, J. M., Diamond, I. T., & Neff, W. D. Auditory discrimination after ablation of temporal and insular cortex in cat. *Federation Proceedings*, 1957, *16*, 47.

Goldberg, J. M., & Neff, W. D. Frequency discrimination after bilateral ablation of cortical auditory areas. *Journal of Neurophysiology*, 1964, *24*, 119–128.

Goldstein, K. *Language and language disorders*. New York: Grune & Stratton, 1948.

Goltz, F. Uber die Verrichtungen des Grosshirns. *Pflüger's Archiv Gesamte Physiologie*, 1884, *26*.

Gomulicki, B. R. The development and present status of the trace theory of memory. *British Journal of Psychology* (Monograph Suppl.), 1953, *29*, 1–94.

Goodman, L. S., & Gilman, A. *The pharmacological basis of therapeutics*. (3rd ed.) New York: Macmillan, 1965.

Goodwin, B. C. *Temporal organization in cells*. New York: Academic Press, 1963.

Gott, P. S., Rossiter, V. S., Galbraith, G. C., & Saul, R. E. Hemispheric evoked potentials to verbal and spatial stimuli in human commissurotomy patients. In *Cerebral dominance, BIS Conference Report #34*. Los Angeles: University of California, 1974.

Granit, R. Centrifugal and antidromic effects on the ganglion cells of the retina. *Journal of Neurophysiology*, 1955, *18*, 388.

Grastyán, E., & Angyan, L. Organization of motivation at thalamic level of the cat. *Physiology and Behavior*, 1967, *2*, 5–16.

Grastyán, E., & Karmós, G. The influence of hippocampal lesions on simple and delayed instrumental conditioned reflexes. *Physiological Hippocampe Colloquium International* (No. 107). Paris: C.N.R.S., 1962.

Grastyán, E., Lissak, L., Madarász, I., & Donhoffer, H. Hippocampal electrical activity during the development of conditioned reflexes. *Electroencephalography & Clinical Neurophysiology*, 1959, *11*, 409–430.

Green, J. D., Clemente, C. D., & DeGroot, J. Rhinencephalic lesions and behavior in cats. An analysis of the Kluver–Bucy syndrome with particular reference to normal and abnormal sexual behavior. *Journal of Comparative Neurology*, 1957, *108*, 505–545.

Green, J. S., & Arduini, A. Hippocampal electrical activity in arousal. *Journal of Neurophysiology*, 1954, *17*, 533–557.

Green, J. S., Maxwell, D. S., Schindler, W. J., & Stumpf, C. Rabbit EEG "theta" rhythm: Its anatomical source and relation to activity in single neurons. *Journal of Neurophysiology*, 1960, *23*, 403.

Grinberg-Zylberbaum, J., Carranza, M. B., Cepeda, G. V., Vale, T. C., & Steinberg, N. N. Caudate nucleus stimulation impairs the processes of perceptual integration. *Physiology & Behavior*, 1974, *12*, 913–918.

Grosser, G. S., & Harrison, J. M. Behavioral interaction between stimulated cortical points. *Journal of Comparative Physiology & Psychology*, 1960, *53*, 229–233.

Grossman, R. G., Whiteside, L., & Hampton, T. The time course of evoked depolarization of cortical glial cells. *Brain Research*, 1969, *14*, 401–415.

Grossman, S. P. Eating or drinking elicited by direct adrenergic of cholinergic stimulation of hypothalamus. *Science*, 1960, *132*, 301–302.

Grossman, S. P. Direct adrenergic and cholinergic stimulation of hypothalamic mechanisms. *American Journal of Physiology*, 1962, *202*, 872–882. (a)

Grossman, S. P. Effects of adrenergic and cholinergic blocking agents on hypothalamic mechanisms. *American Journal of Physiology*, 1962, *202*, 1230–1236. (b)

Grossman, S. P. Some neurochemical properties of the central regulation of thirst. In M. J. Wayner (Ed.), *Thirst, first international symposium on thirst in the regulation of body water*. New York: Pergamon Press, 1964.

Grossman, S. P. The VMH: A center for affective reactions, satiety, or both? *International Journal of Physiological Behavior*, 1966, *1*, 1–10.

Grossman, S. P., & Mountford, H. Learning and extinction during chemically induced disturbance of hippocampal functions. *American Journal of Physiology*, 1964, *207*, 1387–1393.

Guillery, R. W., & Stelzner, D. J. The differential effects of unilateral lid closure upon the monocular and binocular segments of the dorsal lateral geniculate in the cat. *Journal of Comparative Neurology*, 1970, *139*, 413–421.

Guyton, A. C. *Medical physiology* (4th ed.). Philadelphia: Saunders, 1971.

Hagbarth, K. E., & Kerr, D. I. B. Central influences on spinal afferent conduction. *Journal of Neurophysiology*, 1954, *17*, 295–307.

Haider, M., Spong, P., & Lindsley, D. B. Attention, vigilance and cortical evoked-potentials in humans. *Science*, 1964, *145*, 180–182.

Halle, M., & Stevens, K. Speech recognition: A model and a program for research. *IRE Transactions on Information Theory* 1962, *8*, 155–159.

Haller, A. *Elementa physiologiae corporis humani*. Lausanne, 1769.

Hamer, J. Effects of low level, low frequency electrical fields on human reaction time. *Communications in Behavior and Biology*, 1968, No. 2A.

Hanberry, J., & Jasper, H. H. Independence of diffuse thalamo-cortical projection system shown by specific nuclear destruction. *Journal of Neurophysiology*, 1953, *16*, 252–271.

Head, H. *Studies in neurology* (Vols. 1 & 2). London: Frowde, Hodder & Stoughton, 1920.

Head, H. *Aphasia and kindred disorders of speech*. London and New York: Cambridge University Press, 1926.

Heatherington, A. W., & Ranson, S. W. The spontaneous activity and food intake of rats with hypothalamic lesions. *American Journal of Physiology*, 1942, *136*, 609–617.

Hebb, D. O. *The organization of behavior*. New York: Wiley, 1949.

Henrickson, C. W., Kimble, R. J., & Kimble, D. P. Hippocampal lesions and the orienting response. *Journal of Comparative Physiology & Psychology*, 1969, *67*, 220–227.

Hernández-Peón, R. Central mechanisms controlling conduction along central sensory pathways. *Acta Neurologica Latinoamerica*, 1955, *1*, 256.

Hernández-Peón, R. Physiological basis of mental activity. *Electroencephalography & Clinical Neurophysiology* (Suppl.), 1963, *24*.

Hernández-Peón, R., Chávez-Ibarra, G., Morgane, P. J., & Timo-Iaria, C. Limbic cholinergic pathways involved in sleep and emotional behavior. *Experimental Neurology*, 1963, *8*, 93–111.

Hernández-Peón, R., & Chávez-Ibarra, G. Sleep induced by electrical or chemical stimulation of the forebrain. *Electroencephalography & Clinical Neurophysiology* (Suppl.), 1963, *24*, 188–198.

Hernández-Peón, R., Scherrer, R. H., & Jouvet, M. Modification of electrical activity in cochlear nucleus during "attention" in unanesthetized cats. *Science*, 1956, *123*, 331–332.

Hernández-Peón, R., Scherrer, R. H., & Velasco, M. Central influences on afferent conduction in the somatic and visual pathways. *Acta Neurologica Latinoamerica*, 1956, *2*, 8–22.

Herrington, R. N., & Schneidau, P. The effect of imagery of the visual evoked response. *Experientia*, 1968, *24*, 1136–1137.

Hess, W. R. *Diencephalon-autonomic and extra-pyramidal functions*. New York: Grune & Stratton, 1954.

Hess, W. R. Das Schlafsyndrom als Folge dienzephaler Reizung. *Helv. Physiol. Acta*, 1944, *2*, 305–344.

Hess, W. R. *Das Zwischenhirn: Syndrome, Lokalisationen, Functionen*. Basel: Schwabe, 1949.

Hess, R., Jr., Koella, W. P., & Akert, K. Cortical and subcortical recordings in natural and artificially induced sleep in cats. *Electroencephalography & Clinical Neurophysiology*, 1953, *5*, 75–90.

Hild, W., & Tasaki, I. Morphological and physiological properties of neurons and glial cells in tissue culture. *Journal of Neurophysiology*, 1962, *25*, 277–304.

Hilgard, E. R., & Marquis, D. G. *Conditioning and learning*. New York: Appleton, 1940.

Hirsch, H. V., & Spinelli, D. N. Modification of the distribution of receptive field orientation in cats by selective visual exposure during development. *Experimental Brain Research*, 1971, *12*, 509–527.

Hiscoe, H. B. Distribution of nodes and incisures in normal and regenerated nerve fibers. *Anatomical Record*, 1947, *99*, 447–475.

Hitzig, F. *Untersuchungen über des Gehirn*. Berlin: Unger, 1874.

Holmes, O., & Houchin, J. Units in the cerebral cortex of the anaesthetized rat and the correlations between their discharges. *Journal of Physiology* (London), 1966, *187*, 651–671.

Hori, Y., & Yoshii, N. Conditioned change in discharge pattern for single neurons of medial thalamic nuclei of cat. *Psychological Report*, 1965, *16*, 241.

Horn, G. Novelty, attention and habituation. In C. R. Evans & T. B. Mulholland (Eds.), *Attention in neurophysiology*. London: Butterworth, 1969.

Hubel, D. H. Single unit activity in striate cortex of unrestrained cats. *Journal of Physiology* (London), 1959, *147*, 226–238.

Hubel, D. H., & Wiesel, T. N. Receptive fields of single neurones in the cat's striate cortex. *Journal of Physiology*, 1959, *148*, 574–591.

Hubel, D. H., & Wiesel, T. N. Receptive fields, binocular interaction and functional architecture in the cat's visual cortex. *Journal of Physiology* (London), 1962, *160*, 106–154.

Hubel, D. H., & Wiesel, T. N. Receptive fields of cells in striate cortex of very young, visually inexperienced kittens. *Journal of Neurophysiology*, 1963, *26*, 994–1002.

Hubel, D. H., & Wiesel, T. N. Receptive fields and functional architecture in two non-striate visual areas (18 and 19) of the cat. *Journal of Neurophysiology*, 1965, *28*, 229–289.

Hudspeth, W., & Jones, G. B. Stability of neural interference patterns. In P. Greguss (Ed.), *Holography in medicine: Proceedings of the International Symposium on Holography in Biomedical Sciences*. IPC Science and Technology Press Ltd., 1975.

Humphrey, D. R. Re-analysis of the antidromic cortical response. II. On the contribution of cell discharge and PSP's to the evoked potentials. *Electroencephalography & Clinical Neurophysiology*, 1968, *25*, 421–442.

Hunt, H. F., & Diamond, I. T. Some effects of hippocampal lesions on conditioned avoidance behavior in the cat. *Proceedings of the 15th International Congress of Psychology*, Brussels, 1957.

Hunter, J., & Jasper, H. H. Effects of thalamic stimulation in unanesthetized animals. *Electroencephalography & Clinical Neurophysiology,* 1949, *1,* 305–324.

Hursh, J. B. Conduction velocity and diameter of nerve fibers. *American Journal of Physiology,* 1939, *127,* 131–139.

Hydén, H. RNA in brain cells. In G. C. Quarton, T. Melnechuk, & F. O. Schmitt (Eds.), *The neurosciences: A study program.* New York: Rockefeller University Press, 1967.

Igic, R., Stern, P., & Basagic, E. Changes in emotional behavior after application of cholinesterase inhibitor in septal and amygdala regions. *Neuropharmacology,* 1970, *9,* 73–80.

Irwin, S., & Benuazizi, A. Pentylenetetrazol enhances memory function. *Science,* 1966, *152,* 100–102.

Jackson, J. H. *Selected writings* (Vol. 2). New York: Basic Books, 1869.

Jacobsen, C. F. Studies of cerebral function in primates: I. The functions of the frontal association areas in monkeys. *Comprehensive Psychological Monographs,* 1936, *13,* 3–60.

Jacquet, Y. F., & Lajtha, A. Morphine action at central nervous system sites in rat: Analgesia or hyperalgesia depending on site and dose. *Science,* 1973, *182,* 490–492.

James, W. *The principles of psychology* (Vol. 1). New York: Dover, 1950. (Originally published, 1890.)

Jasper, H. H. Diffuse projection systems: The integrative action of the thalamic reticular system. *Electroencephalography & Clinical Neurophysiology,* 1949, *1,* 405–420.

Jasper, H. H., Proctor, L. D., Knighton, B. S., Noshay, W. C., & Costell, R. T. (Eds.). *Reticular formation of the brain.* Boston: Little, Brown & Co., 1958.

Jasper, H. H., Ricci, G., & Doane, B. Microelectrode analysis of cortical cell discharge during avoidance conditioning in the monkey. *Electroencephalography & Clinical Neurophysiology* (Suppl.), 1960, *13,* 137–155.

John, E. R. Higher nervous functions: Brain functions and learning. *Annual Review of Physiology,* 1961, *23,* 451.

John, E. R. Neural mechanisms of decision making. In W. S. Fields & W. Abbot (Eds.), *Information storage and neural control.* Springfield, Illinois: Charles C Thomas, 1963.

John, E. R. *Mechanisms of memory.* New York: Academic Press, 1967. (a)

John, E. R. Electrophysiological studies of conditioning. In G. C. Quarton, T. Melnechuk, & F. O. Schmitt (Eds.), *The neurosciences: A study program.* New York: Rockefeller University Press, 1967. (b)

John, E. R. Summary: Symposium on memory transfer American Association for the Advancement of Science. In W. L. Byrne (Ed.), *Molecular approaches to learning and memory.* New York: Academic Press, 1970. Pp. 335–342.

John, E. R. Brain mechanisms of memory. In J. McGaugh (Ed.), *Psychobiology.* New York: Academic Press, 1971.

John, E. R. Switchboard versus statistical theories of learning and memory. *Science,* 1972, *177,* 850–864.

John, E. R. Brain evoked potentials: Acquisition and analysis. In R. F. Thompson & M. M. Patterson (Eds.), *Bioelectric recording techniques, Part A: Cellular processes and brain potentials.* New York: Academic Press, 1973.

John, E. R. Neural correlates of readout from memory Abstracts, *Proceedings 25th International Congress Physiological Sciences,* 1974, New Delhi, India.

John, E. R., Bartlett, F., Shimokochi, M., & Kleinman, D. Neural readout from memory. *Journal of Neurophysiology,* 1973, *36,* 893–924.

John, E. R., Chesler, P., Bartlett, F., & Victor, I. Observation learning in cats. *Science,* 1968, *159,* 1489–1491.

John, E. R., Grastyán, E., Harmony, T., & Morrell, F. Unpublished observations at International Brain Research Organization conference on recent advances in brain research. Budapest, 1967.

John, E. R., Herrington, R. N., & Sutton, S. Effects of visual form on the evoked response. *Science*, 1967, *155*, 1439–1442.

John, E. R., & Killam, K. F. Electrophysiological correlates of avoidance conditioning in the cat. *Journal of Pharmacology & Experimental Therapeutics*, 1959, *125*, 252.

John, E. R., & Killam, K. F. Electrophysiological correlates of differential approach–avoidance conditioning in the cat. *Journal of Nervous & Mental Disease*, 1960, *131*, 183.

John, E. R., & Kleinman, D. Stimulus generalization between differentiated visual, auditory and central stimuli. *Journal of Neurophysiology*, 1975, *38*, 1015–1034.

John, E. R., Leiman, A., & Sachs, E. An exploration of the functional relationship between electroencephalographic potentials and differential inhibition. *Annals of the New York Academy of Sciences*, 1961, *92*, 1160–1182.

John, E. R., & Morgades, P. P. Neural correlates of conditioned responses studied with multiple chronically implanted moving microelectrodes. *Experimental Neurology*, 1969, *23*, 412–425. (a)

John, E. R., & Morgades, P. P. Patterns and anatomical distribution of evoked potentials and multiple unit activity elicited by conditioned stimuli in trained cats. *Communications in Behavioral Biology*, 1969, *3*, 181–207. (b)

John, E. R., & Morgades, P. P. A technique for the chronic implantation of multiple movable micro-electrodes. *Electroencephalography & Clinical Neurophysiology*, 1969, *27*, 205–208. (c)

John, E. R., Ruchkin, D. S., Leiman, A., Sachs, E., & Ahn, H. Electrophysiological studies of generalization using both peripheral and central conditioned stimuli. *23rd International Congress of the Physiological Sciences* (Tokyo), International Congress Series No. 87, 1965, 618–627.

John, E. R., Ruchkin, D. S., & Villegas, J. Signal analysis of evoked potentials recorded from cats during conditioning. *Science*, 1963, *141*, 421–429.

John, E. R., Ruchkin, D. S., & Villegas, J. Signal analysis and behavioral correlates of evoked potential configuration in cats. *Annals of the New York Academy of Sciences*, 1964, *112*, 362–420.

John, E. R., Shimokochi, M., & Bartlett, F. Neural readout from memory during generalization. *Science*, 1969, *164*, 1519–1521.

Johnson, R. W., & Hanna, G. R. The thalamocortical system as a neuronal machine: The interaction of ventrolateral nucleus with sensori motor cortex in the cat. *Brain Research*, 1970, *18*, 219–240.

Jouvet, M. Biogenic amines and the states of sleep. *Science*, 1969, *163*, 32–41.

Jowett, B. (Ed.). Theaetetus. In *The dialogues of Plato* (Vol. IV). London and New York: Oxford University Press, 1931, p. 254.

Jurko, M. F., Giurintano, L. P., Giurintano, S. L., & Andy, O. J. Spontaneous awake EEG patterns in three lines of primate evolution. *Behavioral Biology*, 1974, *10*, 377–384.

Kaada, B. R. Somato-motor, autonomical and electrocorticographic responses to electrical stimulation of "rhinencephalic" and other structures in primates, cat and dog. *Acta Physiologica Scandinavica*, 1951, *231*, 1–285. (Suppl. 83)

Kaada, B. B. Cingulate, posterior orbital, anterior insular and temporal polar cortex. In J. Field, H. W. Magoun, & V. E. Hall (Eds.), *Handbook of physiology: neurophysiology* (Vol. 2). Washington, D.C.: American Physiology Society, 1960, pp. 1345–1372.

Kaczmarek, L. K., & Adey, W. R. Efflux of Ca-452+ and H-3-gamma-aminobutyric acid from cat cerebral cortex. *Brain Research*, 1973, *63*, 331–336. (a)

Kaczmarek, L. K., & Adey, W. R. Calcium and gamma-aminobutyric acid fluxes in cat cerebral cortex. *Federation Proceedings*, 1973, *32*, 419. (b)

Kamikawa, K., McIlwain, J. T., & Adey, W. R. Response patterns of thalamic neurons during classical conditioning. *Electroencephalography & Clinical Neurophysiology*, 1964, *17*, 485–496.

Kandel, E. R., Frazier, W. T., & Wachtel, H. Organization of inhibition in abdominal ganglion of aplysia. I. Role of inhibition and disinhibition in transforming neural activity. *Journal of Neurophysiology*, 1969, *32*, 496–508.

Kandel, E. R., & Spencer, W. A. Electrophysiological properties of an archicortical neuron. *Annals of the New York Academy of Science*, 1961, *94* (2), 570–603.

Karahashi, Y., & Goldring, S. Intracellular potentials from "idle" cells in cerebral cortex of cat. *Electroencephalography & Clinical Neurophysiology*, 1966, *20*, 600–607.

Keidel, W. D., & Spreng, M. Audiometric aspects and multisensory power functions of electronically averaged evoked cortical responses in man. *Acta Otolaryngology*, Stockholm, 1965, *59*, 201–208.

Keidel, W. D., & Spreng, M. Recent status results and problems of objective audiometry in man. Parts I and II. *Journal Francaise o' Oto-rhinolaryngologie*, *19*, 1970.

Kelly, A. H., Beaton, L. E., & Magoun, H. W. A midbrain mechanism for faciovocal activity. *Journal of Neurophysiology*, 1946, *9*, 181–190.

Kelly, P. J., Dikmen, F. N., & Tarkington, J. A. Photically oriented conditioned reflexes elicited by electrical stimulation of the visual system in the cat. *Brain Research*, 1973, *51*, 293–305.

Kerr, D. I. B., & Hagbarth, K. E. An investigation of centrifugal olfactory fiber system. *Journal of Neurophysiology*, 1955, *18*, 362–374.

Kety, S. S. The central physiological and pharmacological effects of the biogenic amines and their correlations with behavior. In G. C. Quarton, T. Melnechuk, & F. O. Schmitt (Eds.), *The neurosciences: A study program*. New York: Rockefeller University Press, 1967.

Kety, S. S. The biogenic amines in the central nervous system: Their possible roles in arousal, emotion, and learning. In F. O. Schmitt, G. C. Quarton, T. Melnechuk, & G. Adelman (Eds.), *The neurosciences: Second study program*. New York: Rockefeller University Press, 1970.

Killam, K. F., & Hance, A. J. Analysis of electrographic correlates of conditional responses to positive reinforcement: I. Correlates of acquisition and performance. *Abstracts of the Proceedings of the 23rd International Congress of Physiological Science* (Tokyo), 1965.

Kimble, D. P. The effects of bilateral hippocampal lesions in rats. *Journal of Comparative Physiology & Psychology*, 1963, *56*, 273–283.

Kimble, D. P. Hippocampus and internal inhibition. *Psychology Bulletin*, 1968, *70*, 285–295.

Kimble, D. P., & Kimble, R. J. Hippocampectomy and response perseveration in the cat. *Journal of Comparative Physiology & Psychology*, 1965, *60*, 474–476.

King, F. A. Effects of septal and amygdaloid lesions on emotional behavior and conditioned avoidance responses in the rat. *Journal of Nervous & Mental Disease*, 1958, *126*, 57–63.

King, F. A., & Meyer, P. M. Effects of amygdaloid lesions upon septal hyperemotionality in the rat. *Science*, 1958, *128*, 655–656.

Kitai, S. T. Substitution of nitracranial electrical stimulation for photic stimulation during extinction procedure. *Nature*, 1965, *209*, 22.

Kitai, S. T. Generalization between photic and electrical stimulation to the visual system. *Journal of Comparative Physiology & Psychology*, 1966, *61*, 319–324.

Klee, M. R., & Hess, W. Strychnine effects on cell membrane properties. *Electroencephalography & Clinical Neurophysiology*, 1969, *27*, 683. (Abstract)

Klee, M. R., & Offenloch, K. Post-synaptic potentials and spike patterns during augmenting responses in cat's motor cortex. *Science*, 1964, *143*, 488–489.

Klee, M. R., Offenloch, K., & Tigges, J. Cross-correlation analysis of electroencephalographic potentials and slow membrane transients. *Science*, 1965, *147*, 519–521.

Kleinman, D., & John, E. R. Contradiction of auditory and visual information by brain stimulation. *Science*, 1975, *187*, 271–272.

Kleitmann, N. *Sleep and wakefurness.* Chicago: University of Chicago Press, 1939.

Kleist, K. *Gehimpathologie*, Barth, Leipzig, 1934.

Klinke, R., Fruhstorfer, H., & Finkenzeller, P. Evoked responses as a function of external and stored information. *Electroencephalography & Clinical Neurophysiology*, 1968, *26*, 216–219.

Kluver, H., & Bucy, P. C. Preliminary analysis of functions of the temporal lobes in monkeys. *Archives Neurological Psychiatry* (Chicago), 1939, *42*, 979–1000.

Knapp, S., & Mandell, A. J. Narcotic drugs: Effects on the serotonin biosynthetic systems of the brain. *Science*, 1972, *177*, 1209–1211.

Knight, J. M. Stimulus generalization in the hippocampus of the cat. Unpublished doctoral dissertation, University of Utah, 1964.

Knipst, I. N. (Ed.). Spatial synchronization of bioelectrical activity in the cortex and some subcortical structures in rabbit's brain during conditioning. In *Contemporary problems of electrophysiology of the central nervous system*. Moscow: Academy of Sciences, 1967. Pp. 160–167.

Knipst, I. N. Spatial synchronization of cortical and subcortical potentials in rabbits during formation of conditioned reflexes. In V. S. Rusinov (Ed.), *Electrophysiology of the central nervous system*. New York: Plenum Press, 1970.

Koikegami, H., Fuse, S., Hiroki, S., Kazami, T., & Kageyama, Y. On the inhibitory effect upon the growth of infant animals or on the obesity in adult cat induced by bilateral destruction of the amygdaloid nuclear region. *Folia Psychiatry Neurology* (Japan), 1958, *12*, 207–233.

Kopin, I. J. The adrenergic synapse. In G. C. Quarton, T. Melnechuk, & F. O. Schmitt (Eds.), *The neurosciences: A study program*. New York: Rockefeller University Press, 1967, pp. 427–432.

Korol'kova, T. A., & Shvets, T. B. Interrelation between distant synchronization and steady potential shifts in the cerebral cortex. In I. N. Knipst (Ed.), *Contemporary problems of electrophysiology of the central nervous system*. Moscow: Academy of Sciences, 1967. Pp. 160–167.

Krettek, J., & Price, J. L. An autoradiographic study of the thalamic and cortical projections of the amygdala in the rat. *Society for Neuroscience, 3rd Annual Meeting*, 1973. (Abstract)

Krieg, W. S. *Functional neuroanatomy*. New York: Blakiston Co., 1953.

Kristiansen, K., & Courtois, G. Rhythmic electric activity from isolated cerebral cortex. *Electroencephalography & Clinical Neurophysiology*, 1949, *1*, 265.

Krogh, A. The active and passive exchanges of inorganic ions through the surfaces of living cells and through living membranes generally. *Proceedings of the Royal Society*, 1946, *133*, 140–200.

Kruskal, J. B. Multidimensional scaling by optimizing goodness of fit to a nonmetric hypothesis. *Psychometrika*, 1964, *29*, 1–27.

Kuffler, S. S., Nicholls, J. G., & Orkand, R. E. Physiological properties of glial cells in the central nervous system of amphibia. *Journal of Neurophysiology*, 1966, *29*, 768–787.

Kuffler, S. W., & Potter, D. D. Glia in the leech central nervous system: Physiological properties and neuron-glia relationship. *Journal of Neurophysiology*, 1964, *27*, 290–320.

Kupferman, I., Castellucci, V., Pinsker, H., & Kandel, E. R. Neuronal correlates of habituation and dishabituation of the gill-withdrawal reflex in aplysia. *Science*, 1970, *167*, 1743–1745.

Kuhnt, U., & Creutzfeldt, O. D. Decreased post-synaptic inhibition in the visual cortex during flicker stimulation. *Electroencephalography & Clinical Neurophysiology*, 1971, *30*, 79–82.

Lajtha, A. Observations on protein catabolism in brain. In S. S. Kety & J. Elkes (Eds.), *Regional neurochemistry*. New York: Macmillan, 1961.

Lansing, R. W., Schwartz, E., & Lindsley, D. B. Reaction time and EEG activation under alerted and nonalerted conditions. *Journal of Experimental Psychology*, 1959, *58*, 1–7.

Lashley, K. S. Temporal variation in the function of the gyrus precentralis in primates. *American Journal of Physiology*, 1923, *65*, 585–602.

Lashley, K. S. Learning. I. Nervous mechanisms in learning. In C. Murchison (Ed.), *The foundations of experimental psychology*. Worcester, Massachusetts: Clark University Press, 1929.

Lashley, K. S. The problem of cerebral organization in vision. *Biology Symposium*, 1942, *7*, 301–332.

Lashley, K. S. In search of the engram. *Symposium of the Society of Experimental Biology*, 1950, *4*, 454–482.

Lashley, K. S. Cerebral organization and behavior. *Proceedings of the Association for Research on Nervous Mental Disease*, 1958, *38*, 1–18.

Laufer, M., & Verzeano, M. Periodic activity in the visual system of the cat. *Vision Research*, 1967, *7*, 215–229.

Lehninger, A. L. The neuronal membrane. *Proceedings of National Academy of Sciences*, 1968, *60*, 1055–1101.

Leiman, A. L. Electrophysiological studies of conditioned responses established to central electrical stimulation. Unpublished doctoral dissertation, University of Rochester, 1962.

Leiman, A. L., & Christian, C. N. Electrophysiological analyses of learning and memory. In J. A. Deutsch (Ed.), *The physiological basis of memory*. New York: Academic Press, 1973. Pp. 125–173.

Lettvin, J. Y., Maturana, H. R., McCulloch, W. S., & Pitts, W. What the frog's eye tells the frog's brain. *Proceedings of the Institute of Radio Engineers*, 1959, *47*, 1940–1951.

Lettvin, J. Y., Maturana, H. R., Pitts, H. R., & McCulloch, W. S. Two remarks on the visual system of the frog. In W. A. Rosenblith (Ed.), *Sensory communication*. New York: John Wiley & Sons, 1960.

Liberson, W. T., & Ellen, P. Conditioning of the driven brain wave rhythm in the cortex and the hippocampus of the rat. In J. Wortis (Ed.), *Recent advances in biological psychiatry* (Vol. 2). New York: Grune & Stratton, 1960.

Lickey, M. E. Seasonal modulation and non-twenty-four hour entrainment of a circadian rhythm in a single neuron. *Journal of Comparative Physiology & Psychology*, 1969, *68*, 9–17.

Lilly, J. C. Instantaneous relations between the activities of closely spaced zones on the cerebral cortex. *American Journal of Physiology*, 1954, *176*, 493–504.

Lilly, J. C., & Cherry, R. Surface movements of click responses from the acoustic cerebral cortex of the cat: The leading and the trailing edges of a response figure. *Journal of Neurophysiology*, 1954, *17*, 521–532.

Lilly, J. C., & Cherry, R. Surface movements of figures in the spontaneous activity of anesthetized cerebral cortex: The leading and the trailing edges. *Journal of Neurophysiology*, 1955, *18*, 18–32.

Lindsley, D. B. Emotions and the electroencephalogram. In M. L. Reymert (Ed.), *Feelings and emotions: The Mooseheart symposium*. New York: McGraw-Hill, 1950. Pp. 238–246.

Lindsley, D. B. The reticular system and perceptual discrimination. In H. H. Jasper, L. D. Proctor, R. S. Knight, W. S. Noshay, & R. J. Costello (Eds.), *Reticular formation of the brain*. Boston: Little, Brown & Co., 1958.

Lindsley, D. B. The reticular activating system and perceptual integration. In D. E. Sheer (Ed.), *Electrical stimulation of the brain*. Austin: University of Texas Press, 1961. Pp. 331–349.

Lindsley, D. B., Schreiner, L., Knowles, W. B., & Magoun, H. W. Behavioral and EEG changes following chronic brain stem lesions in the cat. *Electroencephalography & Clinical Neurophysiology*, 1950, *2*, 483–489.

Lindsley, D. B., Schreiner, L., & Magoun, H. W. An electromyographic study of spasticity. *Journal of Neurophysiology,* 1949, *12,* 197–205.

Lindsley, D. F., Carpenter, R. S., Killam, E. K., & Killam, K. F. EEG correlates of behavior in the cat. I. Pattern discrimination and its alternative by atropine and LSD-25. *Electroencephalography & Clinical Neurophysiology,* 1968, *24,* 497–513.

Lindsley, D. F., Chow, K. L., & Gollender, M. Dichoptic interactions of lateral geniculate neurons of cats to contralateral and ipsilateral eye stimulation. *Journal of Neurophysiology,* 1967, *30,* 628–644.

Livanov, M. N. Spatial analysis of the bioelectric activity of the brain. In *Proceedings of the 22nd International Congress of Physiological Science* (Leiden), 1962, 899–907.

Livanov, M. N. The significance of distant brain potential synchronization for realization of temporal connections. *Proceedings of the 23rd International Congress of Physiological Science* (Tokyo), 1965, *4,* 600–612.

Livanov, M. N., & Korol'kova, T. A. The influence of inadequate stimulation of the cortex with induction current on the bioelectric rhythm of the cortex and conditioned reflex activity. *Zh. Vyssh. Nerv. Deyat.,* 1951, *1,* 332–345.

Livanov, M. N., & Poliakov, K. L. The electrical reactions of the cerebral cortex of a rabbit during the formation of a conditioned defense reflex by means of rhythmic stimulation. *Izvestiya Akademiya Nauk. USSR Series Biology,* 1945, *3,* 286.

Livingston, R. B. Brain circuitry relating to complex behavior. In G. C. Quarton, T. Melnechuk, & F. O. Schmitt (Eds.), *The neurosciences: A study program.* New York: Rockefeller University Press, 1967. Pp. 499–515.

Longuet-Higgins, H. C. The nonlocal storage of temporal information. *Proceedings of the Royal Society* (B), 1968, *171,* 327–334.

Lorente de Nó, R. Analysis of the activity of the chains of internuncial neurons. *Journal of Neurophysiology,* 1938, *1,* 207–244.

Lubar, J. F. Effect of medial cortical lesions on the avoidance behavior of the cat. *Journal of Comparative Physiology & Psychology,* 1964, *58,* 38–46.

Luria, A. R. Disorders of simultaneous perception in a case of bilateral occipito-parietal brain injury. *Brain,* 1959, *82,* 437–449.

Luria, A. R. *Higher cortical functions in man.* New York: Basic Books, Consultants Bureau, 1966. (Original Russian text published by Moscow University Press, 1962.)

Luria, A. R. *The working brain.* Middlesex, England: Penguin Press, 1973.

Luria, A. R., & Tsvetkova, L. S. Towards the mechanisms of "dynamic aphasia." *Foundation of Language* (Amsterdam), 1968, *4.*

Macadar, A. W., & Lindsley, D. B. A search for the brainstem origin of two hypothalamic-hippocampal systems mediating hippocampal theta activity and desynchronization. *Society for Neurosciences, Third Annual Meeting,* 1973. (Abstract)

MacGregor, R. J., Prieto-Diaz, R., Miller, S. W., & Groves, P. M. Statistical properties of neurons in the rat reticular formation: Evidence for reverberating loops, widespread rhythmicities, and functional reorganization. *Brain Research,* 1973, *64,* 167–187.

Macht, M. B., & Bard, P. Studies on decerebrate cats, in the chronic state. *Federation Proceedings,* 1942, *1,* 55–56.

MacKay, D. M. *Information, mechanism and meaning.* London: MIT Press, 1969. (a)

MacKay, D. M. (Ed.). Evoked brain potentials as indicators of sensory information processing. *Neurosciences Research Program Bulletin,* 1969, *7,* 3. (b)

MacLean, P. D. Psychosomatic disease and the "visceral brain." Recent developments bearing on the Papez theory of emotion. *Psychosomatic Medicine,* 1949, *11,* 338–353.

MacLean, P. D. Studies on the limbic system ("visceral brain") and their bearing on psychosomatic problems. In E. Wittkower & R. Cleghorn (Eds.), *Recent developments in psychosomatic medicine.* London: Pitman, 1954. Pp. 101–125.

MacLean, P. D. Contrasting functions of limbic and neocortical systems of the brain and their relevance to psychophysiological aspects of medicine. *American Journal of Medicine,* 1958, *25,* 611–626.

MacLean, P. D. The limbic and visual cortex in phylogeny: Further insights from anatomic and microelectrode studies. In R. Hassler & H. Stephan (Eds.), *Evolution of the forebrain.* New York: Plenum Press, 1966.

MacLean, P. D. The paranoid streak in man. In A. Koestler & J. R. Smythies (Eds.), *Beyond reductionism.* London: Hutchinson and Co., 1969.

MacLean, P. D. The triune brain, emotion, and scientific bias. In F. O. Schmitt, G. C. Quarton, T. Melnechuk, & G. Adelman (Eds.), *The neurosciences: Second study program.* New York: Rockefeller University Press, 1970. Pp. 336–349.

Maekawa, K., & Purpura, D. P. Properties of spontaneous and evoked synaptic activities of thalamic ventrobasal neurons. *Journal of Neurophysiology,* 1967, *30,* 360–381.

Magni, F., Moruzzi, G., Rossi, G. F., & Zanchetti, A. EEG arousal following inactivation of the lower brain stem by selective injection of barbiturate into lower brain stem circulation. *Archives of Italian Biology,* 1959, *97,* 33–46.

Magoun, H. W. The ascending reticular system and wakefulness. In E. D. Adrian, F. Bremer, & H. H. Jasper (Eds.), *Brain mechanisms and consciousness.* Springfield, Illinois: Charles C Thomas, 1954, pp. 1–14.

Magoun, H. W., & Rhines, R. *Spasticity: The stretch reflex and extrapyramidal systems.* Springfield, Ill.: Charles C Thomas, 1948.

Majkowski, J. The electroencephalogram and electromyogram of motor conditioned reflexes after paralysis with curare. *Electroencephalography & Clinical Neurophysiology,* 1958, *10,* 503–514.

Majkowski, J. Electrophysiological studies of learning in split-brain cats. *Electroencephalography & Clinical Neurophysiology,* 1967, *23,* 521–531.

Marco, L. A., Brown, T. S., & Rouse, M. E. Unitary responses in ventrolateral thalamus upon intranuclear stimulation. *Journal of Neurophysiology,* 1967, *30,* 482–493.

Masland, R. L., Austin, G., & Grant, F. C. The electroencephalogram following occipital lobectomy. *Electroencephalography & Clinical Neurophysiology,* 1949, *1,* 273–282.

Mason, J. W. Plasma 17-hydroxycorticosteroid levels during electrical stimulation of the amygdaloid complex in conscious monkeys. *American Journal of Physiology,* 1959, *196,* 44–48.

Matousek, M., & Petersén, I. Frequency analysis of the EEG in normal children and adolescents. In P. Kellaway & I. Petersén (Eds.), *Automation of clinical electroencephalography.* New York: Raven Press, 1973. Pp. 75–102.

McCleary, R. A. Response specificity in the behavioral effects of limbic system lesions in the cat. *Journal of Comparative Physiology & Psychology,* 1961, *54,* 605–613.

McCleary, R. A., & Moore, R. Y. *Subcortical mechanisms of behavior.* New York: Basic Books, 1965.

McCulloch, W. F. Inter-areal interactions of the cerebral cortex. In P. C. Bucy (Ed.), *The precentral motor cortex.* Urbana: University of Illinois Press, 1943.

McCulloch, W. S. Why the mind is in the head. In L. A. Jeffress (Ed.), *Cerebral mechanisms in behavior.* New York: John Wiley & Sons, 1951. Pp. 42–111.

McGaugh, J. L. A multi-trace view of memory storage processes. Paper presented at International Symposium on recent advances in learning and retention. Rome, Italy, 1967.

McGaugh, J. L., & Petrinovich, L. Effects of drugs on learning and memory. *International Review of Neurobiology,* 1965, *8,* 139–191.

McGeoch, J. A., & Irion, A. L. *The psychology of human learning.* New York: Longmans, Green, 1952.

McGreer, P. L. The chemistry of mind. *American Scientist,* 1971, *59,* 227–229.

McNew, J. J., & Thompson, R. F. Role of the limbic system in active and passive avoidance conditioning in the rat. *Journal of Comparative Physiology & Psychology*, 1966, *61*, 173–180.

Mettler, F. A. *Neuroanatomy*. St. Louis: C. V. Mosby Co., 1948.

Meyer, W. J. The localization of extracellular substances in the brain with phosphotungstic acid. American Association of Anatomists, 82nd Session, Boston, 1969, 239. (Abstract)

Meynert, T. *Psychiatrie*. Vienna, 1884.

Michon, J. Tapping regularity as a measure of perceptual motor load. *Ergonomics*, 1966, *9*, 401–412.

Mill, J. S. *On liberty*. In M. Lerner (Ed.), *The essential works of John Stuart Mill*. New York: Bantom Books, 1961. (Originally published in 1859.)

Miller, G. A. The magical number seven, plus or minus two: Some limits on our capacity for processing information. *Psychology Review*, 1956, *63*, 81–97.

Miller, N. E. Experiments on motivation. Studies combining psychological, physiological, and pharmacological techniques. *Science*, 1957, *126*, 1271–1278.

Miller, N. E. Motivational effects of brain stimulation and drugs. *Federation Proceedings*, 1960, *19*, 846–854.

Milner, B. Interhemispheric differences in the localization of psychological processes in man. *British Medical Bulletin*, 1971, *27*, 272–277.

Milner, B., & Penfield, W. The effect of hippocampal lesions on recent memory. *Transactions of the American Neurology Association, 80th Annual Meeting*, 1955, 42–48.

Mishkin, M. Visual discrimination performance following partial ablations of the temporal lobe: II. Ventral surface vs. hippocampus. *Journal of Comparative Physiology & Psychology*, 1954, *47*, 187–193.

Mishkin, M., & Pribram, K. H. Visual discrimination performance following partial ablations of the temporal lobe: I. Ventral vs. lateral. *Journal of Comparative Physiology & Psychology*, 1954, *47*, 14.

Mittenthal, J. E., Kristan, W. B., & Tatton, W. G. Does the striate cortex begin reconstruction of the visual world? *Science*, 1972, *176*, 316–317.

Monakow, C., & Mourgue, R. *Introduction Biologique a l'etude du neurologie et de la psychopathologie*. Paris: Alcan, 1928.

Moore, G. P., Perkel, D. H., & Segundo, J. P. Statistical analysis and functional interpretation of neuronal spike data. *American Review of Physiology*, 1966, *28*, 493–522.

Moore, R. Y. Visual pathways and the central neural control of diurnal rhythms. In F. O. Schmitt & F. G. Worden (Eds.), *The neurosciences: Third study program*. Cambridge, Massachusetts: MIT Press, 1974. Pp. 537–542.

Moore, R. Y., Jones, B. E., & Halaris, A. E. Projection of noradrenergic neurons of locus coeruleus to telencephalon. *Society for Neuroscience*, 1973, p. 149. (Abstract)

Morgan, C. T. *Physiological psychology*. New York: McGraw-Hill, 1965.

Morison, R. S., & Dempsey, E. W. A study of thalamocortical relations. *American Journal of Physiology*, 1942, *135*, 281.

Morrell, F. Electrophysiological contributions to the neural basis of learning. *Physiological Review*, 1961, *41*, 443. (a)

Morrell, F. Effect of anodal polarization on the firing pattern of single cortical cells. *Annals of the New York Academy of Sciences*, 1961, *92* (3), 860–876. (b)

Morrell, F. Electrical signs of sensory coding. In G. C. Quarton, T. Melnechuk, & F. O. Schmitt (Eds.), *The neurosciences: A study program*. New York: Rockefeller University Press, 1967. Pp. 452–469.

Morrell, F., Engel, P., & Bouris, W. The effect of visual experience on the firing pattern of visual cortical neurons. *Electroencephalography & Clinical Neurophysiology*, 1967, *23*, 89.

Morrell, F., & Jasper, H. H. Electrographic studies of the formation of temporary connections in the brain. *Electroencephalography & Clinical Neurophysiology*, 1956, *8*, 201.

Morrell, L. EEG frequency and reaction time—a sequential analysis. *Neuropsychologica*, 1966, *4*, 41–48. (a)

Morrell, L. Some characteristics of simulus-provoked alpha activity. *Electroencephalography & Clinical Neurophysiology*, 1966, *21*, 552–561. (b)

Moruzzi, G. Synchronizing influences of the brain stem and the inhibitory mechanisms underlying the production of sleep by sensory stimulation. In H. H. Jasper & G. D. Smirnov (Eds.), The Moscow colloqium on electroencephalography of higher nervous activity. *Electroencephalography & Clinical Neurophysiology* (Suppl.), 1960, *13*, 231–256.

Moruzzi, G. Significance of sleep for brain mechanisms. In J. C. Eccles (Ed.), *Brain and conscious experience.* New York: Springer-Verlag, 1966.

Moruzzi, G., & Magoun, H. W. Brain stem reticular formation and activation of the EEG. *Electroencephalography & Clinical Neurophysiology*, 1949, *1*, 455–473.

Müller, G. E., & Pilzecker, A. Experimentalle beiträge zur Lehre vom Gedachtnis. *Zeitschrift für Psychologie* (Suppl.), 1900, *1*, 1–288.

Munk, H. *Über die Funktionen der Grosshirnrinde.* Berlin: Hirschwald, 1881.

Nakao, H., & Maki, T. Effect of electrical stimulation of the nucleus caudatus upon conditioned avoidance behavior in the cat. *Folia Psychiatry & Neurology* (Japan), 1958, *12*, 258–264.

Nauta, W. J. H. Hypothalamic regulation of sleep in rats; an experimental study. *Journal of Neurophysiology*, 1946, *9*, 285–316.

Nauta, W. J. H. Hippocampal projections and related neural pathways to the mid-brain in the cat. *Brain*, 1958, *81*, 319–340.

Nauta, W. J. H., & Karten, H. J. A general profile of the vertebrate brain, with sidelights on the ancestry of cerebral cortex. In F. O. Schmidt (Ed.), *The neurosciences, second study program.* New York: The Rockefeller University Press, 1970. Pp. 7–26.

Neider, P. C., & Neff, W. D. Auditory information from subcortical electrical stimulation in cats. *Science*, 1961, *133*, 1010–1011.

Nelson, P. G. Interaction between spinal motorneurons of the cat. *Journal of Neurophysiology*, 1966, *29*, 275–287.

Nelson, P. G., & Frank, K. Anomalous rectification in cat spinal motorneurons and effect on polarizing currents on excitatory postsynaptic potential. *Journal of Neurophysiology*, 1967, *30*, 1097–1113.

Nicholls, J. G., & Kuffler, S. W. Extracellular space as a pathway for exchange between blood and neurons in the central nervous system of the leech: Ionic composition of glial cells and neurons. *Journal of Neurophysiology*, 1964, *27*, 645–671.

Nielsen, J. M. *Agnosia, aphraxia, aphasia: Their value in cerebral localization.* (2nd ed.) New York: P. B. Hoeber, 1946.

Nielson, H. C. Evidence that electroconvulsive shock alters memory retrieval rather than memory consolidation. *Experimental Neurology*, 1968, *20*, 3–20.

Nielson, H. S., Knight, J. M., & Porter, P. B. Subcortical conditioning, generalization and transfer. *Journal of Comparative Physiology & Psychology*, 1962, *55*, 168–173.

Noback, C. R., & Demarest, R. J. *The human nervous system.* (2nd ed.) New York: McGraw-Hill, 1975.

Noda, H., & Adey, W. R. Firing of neuron pairs in cat association cortex during sleep and wakefulness. *Journal of Neurophysiology*, 1970, *33*, 672–683.

Noda, H., & Adey, W. R. Neuronal activity in association cortex of cat during sleep, wakefulness, and anaethesia. *Brain Research*, 1973, *54*, 243–259.

Noda, H., Manohar, S., & Adey, W. R. Correlated firing of hippocampal neuron pairs in sleep and wakefulness. *Experimental Neurology*, 1969, *24*, 232–247.

Norton, T., Frommer, G., & Galambos, R. Effects of partial lesions of optic tract on visual discriminations in cats. *Federation Proceedings*, 1966, *25*, 2168.

O'Brien, J. H., & Fox, S. S. Single-cell activity in cat motor cortex. I. Modifications during classical conditioning procedures. *Journal of Neurophysiology*, 1969, *32*, 267–284. (a)

O'Brien, J. H., & Fox, S. S. Single-cell activity in cat motor cortex. II. Functional characteristics of the cell related to conditioning changes. *Journal of Neurophysiology*, 1969, *32*, 285–296. (b)

Ochs, S. *Elements of neurophysiology.* New York: John Wiley & Sons, 1965.

Ojemann, R. G. Correlations between specific human brain lesions and memory changes. *Neurosciences Research Program Bulletin* (Suppl.), 1966, *4*, 1–70.

Olds, J. A preliminary mapping of electrical reinforcing effects in the rat brain. *Journal of Comparative Physiology & Psychology*, 1956, *49*, 281–285.

Olds, J., & Hirano, T. Conditioned responses of hippocampal and other neurons. *Electroencephalography & Clinical Neurophysiology*, 1969, *26*, 159–166.

Olds, J., & Olds, M. E. Interference and learning in paleocortical systems. In J. F. Delafresnaye, A. Fessard, R. W. Gerard, & J. Konorski (Eds.), *CIOMS symposium on brain mechanisms and learning.* Oxford: Blackwell, 1961.

Olds, J., Segal, M., Hirsh, R., Disterhoft, J. F., & Kornblith, C. L. Learning centers of rat brain mapped by measuring latencies of conditioned unit responses. *Journal of Neurophysiology*, 1972, *35*, 202–219.

Ornstein, R. *On the experience of time.* Baltimore: Penguin, 1969.

Osgood, C. E., & Miron, M. S. (Eds.). *Approaches to the study of aphasis.* Urbana: University of Illinois Press, 1963.

Overseth, O. E. Experiments in time reversal. *Scientific American*, 1969, *221*(4), 88–101.

Palestini, M., Pisano, M., Rosadini, G., & Rossi, G. F. Excitability cycle of the visual cortex during sleep and wakefulness. *Electroencephalography & Clinical Neurophysiology*, 1965, *19*, 276–283.

Palestini, M., Rossi, G. F., & Zanchetti, A. An electrophysiological analysis of pontine reticular regions showing different anatomical organization. *Archives of Italian Biology*, 1957, *95*, 97–109.

Palladin, A. V. (Ed.). *Problems of the biochemistry of the nervous system.* New York: Macmillan, 1964.

Papez, J. W. A proposed mechanism of emotion. *Archives of Neurology and Psychiatry*, (Chicago), 1937, *38*, 725–743.

Partridge, M. *Prefontal leucotomy: A study of 300 cases personally followed over 1½–3 years.* Oxford: Blackwell, 1950.

Pearlman, C. A., Sharpless, S. K., & Jarvik, M. Retrograde amnesia produced by anesthetic and convulsant agents. *Journal of Comparative Physiology & Psychology*, 1961, *54*, 109.

Penfield, W. Studies of the cerebral cortex of man. In E. D. Adrian, F. Bremer, & H. H. Jasper (Eds.), *Brain mechanisms and consciousness.* Springfield, Illinois: Charles C Thomas, 1954. Pp. 284–304.

Penfield, W. *The excitable cortex in conscious man.* Liverpool, England: Liverpool University Press, 1958.

Penfield, W. Speech, perception, and the uncommitted cortex. In J. C. Eccles (Ed.), *Brain and conscious experience.* New York: Springer-Verlag, 1966. Pp. 217–237.

Penfield, W. Consciousness, memory, and man's conditioned reflexes. In K. H. Pribram (Ed.), *On the biology of memory.* New York: Harcourt, Brace & World, Inc., 1969.

Penfield, W., & Jasper, H. H. *Epilepsy and the functional anatomy of the human brain.* Boston: Little, Brown & Co., 1954.

Penfield, W., & Milner, B. Memory deficit produced by bilateral lesions in the hippocampal zone. *AMA Archives Neurology & Psychiatry*, 1958, *79*, 475–497.

Penfield, W., & Perot, P. The brain's record of auditory and visual experience. A final summary and discussion. *Brain*, 1963, *86*, 595–609.

Penfield, W., & Roberts, L. *Speech and brain mechanisms*. Princeton, New Jersey: Princeton University Press, 1959.

Petrinovich, L., Bradford, D., & McGaugh, J. L. Drug facilitation of memory in rats. *Psychonomic Science*, 1965, *2*, 191.

Petsche, H., Stumpf, C., & Gogolak, G. The significance of the rabbit's septum as a relay station between the midbrain and the hippocampus. I. The control of hippocampus arousal activity by the septal cells. *Electroencephalography & Clinical Neurophysiology*, 1962, *14*, 202–211.

Pettigrew, J. D., Nikara, T., & Bishop, P. O. Responses to moving slits by single units in cat striate cortex. *Experimental Brain Research*, 1968, *6*, 373–390.

Phelps, R. W. The effect of spatial and temporal interactions on the responses of single units in the cat's visual cortex. *International Journal of Neuroscience*, 1973, *6*, 97–107.

Piaget, J. *The construction of reality in the child*. New York: Ballantine Books, 1971. (a)

Piaget, J. *Biology and knowledge*. Chicago: The University of Chicago Press, 1971. (b)

Picton, T. W., Hillyard, S. A., & Galambos, R. Cortical evoked responses to omitted stimuli. In M. N. Livanov (Ed.), *Major problems of brain electrophysiology*. Moscow: Union of Soviet Socialist Republics Academy of Sciences, 1973.

Pitts, W., & McCulloch, W. S. How we know universals. *Bulletin of Mathematics and Biophysics*, 1947, *9*, 127–147.

Pollen, D. A. On the generation of neocortical potentials. In H. H. Jasper, A. A. Ward, & A. Pope (Eds.), *Basic mechanisms of the epilepsies*. Boston: Little, Brown & Co., 1969.

Pollen, D. A., Lee, J. R., & Taylor, J. H. How does the striate cortex begin the reconstruction of the visual world? *Science*, 1971, *173*, 74–77.

Pollen, D. H., & Sie, P. G. Analysis of thalamic induced wave and spike by modifications in cortical excitability. *Electroencephalography & Clinical Neurophysiology*, 1964, *17*, 154–163.

Pollen, D. A., & Taylor, J. H. The striate cortex and the spatial analysis of visual space. In F. O. Schmitt & F. G. Worden (Eds.), *The neurosciences: Third study program*. Cambridge, Massachusetts: MIT Press, 1974, pp. 239–247.

Polyak, S. *The vertebrate visual system*. Chicago: University of Chicago Press, 1957. (translated by H. Kluver).

Porter, C. C., Totaro, J. A., & Stone, C. A. Effect of G-hydroxydopamine and some other compounds on the concentration of norepinephrine in the hearts of mice. *Journal of Pharmacology Exp. Ther.*, 1963, *140*, 308–316.

Pribram, K. H. The intrinsic systems of the forebrain. In J. Field & H. W. Magoun (Eds.), *Handbook of Physiology, Neurophysiology II*. Washington, D. C.: American Physiological Society, 1960, pp. 1323–1344.

Pribram, K. H. Four R's of remembering. In K. Pribram (Ed.), *On the biology of learning*. New York: Harcourt, Brace & Jovanovich, 1969.

Pribram, K. H. *Languages of the brain*. Englewood Cliffs, New Jersey: Prentice-Hall, 1971.

Pribram, K. H., & Bagshaw, M. Further analysis of the temporal lobe syndrome utilizing fronto-temporal ablations. *Journal of Comparative Neurology*, 1953, *99*, 347–357.

Pribram, K. H., & Kruger, L. Functions of the "olfactory brain." *Annals of the New York Academy of Sciences*, 1954, *58*, 109–138.

Pribram, K. H., & Mishkin, M. Analysis of the effects of frontal lesions in monkey: III. Object alternation. *Journal of Comparative Physiology & Psychology*, 1956, *49*, 41–45.

Pribram, K. H., Spinelli, D. N., & Kamback, M. C. Electrocortical correlates of stimulus response and reinforcement. *Science*, 1967, *157*, 94–95.

Pribram, K. H., & Tubbs, W. E. Short-term memory, parsing, and the primate frontal cortex. *Science*, 1967, *156*, 1765–1767.

Purpura, D. P. Nature of electrocortical potentials and synaptic organizations in cerebral and cerebellar cortex. In C. C. Pfeiffer & J. R. Smythies (Eds.), *International review of neurobiology* (Vol. 1). New York: Academic Press, 1959.

Purpura, D. P. Comparative physiology of dendrites. In G. C. Quarton, T. Melnechuk, & F. O. Schmitt (Eds.), *The neurosciences: A study program.* New York: Rockefeller University Press, 1967. Pp. 372–393.

Purpura, D. P. Interneuronal mechanisms in synchronization and desynchronization of thalamic activity. In M. A. B. Brazier (Ed.), *The interneuron.* UCLA Forum in Medical Sciences, No. 11, University of California Press, Berkeley and Los Angeles, 1969, 467–496.

Purpura, D. P. Operations and processes in thalamic and synaptically related neural subsystems. In F. O. Schmitt, G. C. Quarton, T. Melnechuk, & G. Adelman (Eds.), *The neurosciences: Second study program.* New York: Rockefeller University Press, 1970.

Purpura, D. P. Functional studies of thalamic internuclear interactions. *Brain and Behavior,* 1972, *6,* 203–234.

Purpura, D. P., & Cohen, B. Intracellelar recording from thalamic neurons during recruiting responses. *Journal of Neurophysiology,* 1962, *25,* 621–635.

Purpura, D. P., & Malliani, A. Spike-generation and propagation initiated in dendrites by transhippocampal polarization. *Brain Research,* 1966, *1,* 403–406.

Purpura, D. P., McMurty, J. G., Leonard, C. F., & Malliani, A. Evidence for dendritic origin of orthodromic spikes without depolarizing prepotentials in hippocampal neurons during and after seizure. *Journal of Neurophysiology,* 1966, *29,* 954–979.

Purpura, D. P., Scarff, T., & McMurtry, J. G. Intracellular study of internuclear inhibition in ventrolateral thalamic neurons. *Journal of Neurophysiology,* 1965, *28,* 487–496.

Purpura, D. P., & Shofer, R. J. Intracellular recording from thalamic neurons during reticulocortical activation. *Journal of Neurophysiology,* 1963, *26,* 494–505.

Purpura, D. P., & Shofer, R. J. Intracellular potentials during augmenting and recruiting responses. I. Effects of injections hyperpolarizing currents on evoked membrane potential changes. *Journal of Neurophysiology,* 1964, *27,* 117–132.

Purpura, D. P., & Shofer, R. J. Spike-generation in dendrites and synaptic inhibition in immature cerebral cortex. *Nature,* 1965, *206,* 833–834.

Purpura, D. P., Shofer, R. J., Housepian, E. M., & Noback, C. R. Comparative ontogenesis of structure-function relations in cerebral and cerebellar cortex. In D. P. Purpura & J. P. Schade (Eds.), *Progress in brain research. Growth and maturation of the brain* (Vol. 4). Amsterdam: Elsevier, 1964.

Pusakulich, R. L., & Nielson, H. C. Neural thresholds and state-dependent learning. *Experimental Neurology,* 1972, *34,* 33–44.

Quarton, G. C., Melnechuk, T., & Schmitt, F. O. (Eds.). *The neurosciences: A study program.* New York: Rockefeller University Press, 1967.

Rall, T. W., & Gilman, A. G. The role of cyclic AMP in the nervous system. *Neurosciences Research Progress Bulletin,* 1970, *8*(3), 221–323.

Ramon y Cajal, S. *Histologie du système nerveux de l'homme et des vertétés.* Paris: A. Malone, 1911, Vol. 2, 381–503.

Ramos, A., Schwartz, E., & John, E. R. Unit activity and evoked potentials during readout from memory. *26th International Congress of Physiological Science,* 1974, New Delhi, India.

Ramos, A., Schwartz, E., & John, E. R. Evoked potential-unit relationships in behaving cats. *Brain Research Bulletin,* 1976, *1,* 69–75. (a)

Ramos, A., Schwartz, E., & John, E. R. An examination of the participation of neurons in readout from memory. *Brain Research Bulletin,* 1976, *1,* 77–86. (b)

Ramos, A., Schwartz, E., & John, E. R. Stable and plastic unit discharge patterns during behavioral generalization. *Science,* 1976, *192,* 393–396. (c)

Randt, C. T., Quartermain, D., Goldstein, M., & Anagnoste, B. Norepinephrine biosynthesis inhibition: Effects on memory in mice. *Science*, 1971, *172*, 498–499.

Ranson, S. W. *The anatomy of the nervous system*. Philadelphia: Saunders, 1936.

Ranson, S. W. Somnolence caused by hypothalamic lesions in the monkey. *Archives Neurology and Psychiatry* (Chicago), 1939, *41*, 1–23.

Ranson, S. W., & Magoun, H. W. *The Hypothalamus. Ergebn. Physiol.*, 1939, *41*, 56–163.

Rapin, I., & Graziani, L. J. Auditory evoked responses in normal, brain damaged and deaf infants. *Neurology*, 1967, *17*, 881–894.

Rapin, I., Ruben, R. J., & Lyttle, M. Diagnosis of hearing loss in infants using auditory evoked responses. Paper presented at a meeting of the American Laryngology Rhinology and Otology Society (Eastern), 1970, Boston.

Ratliff, F. Inhibitory interaction and the detection and enhancement of contours. In W. A. Rosenblith (Ed.), *Sensory communication*. New York: John Wiley & Sons, 1960.

Regan, D. An effect of stimulus color on average steady-state potentials evoked in man. *Nature*, 1966, *210*, 1056–1057.

Regan, D. *Evoked potentials in psychology, sensory physiology and clinical medicine*. New York: Wiley-Interscience, 1972.

Reitan, R. M. Psychological deficits resulting from cerebral lesions in man. In J. M. Warren & K. Akert (Eds.), *The frontal granular cortex and behavior*. New York: McGraw-Hill, 1964.

Renshaw, B., Forbes, A., & Morison, B. R. Activity of isocortex and hippocampus: Electrical studies with micro-electrodes. *Journal of Neurophysiology*, 1940, *3*, 74–105.

Rheinberger, M., & Jasper, H. H. The electrical activity of the cerebral cortex in the unanesthetized cat. *American Journal of Physiology*, 1937, *119*, 186.

Riggs, L. A., & Sternheim, C. E. Human retinal and occipital potentials evoked by changes of the wavelength of the stimulating light. *Journal of the Optometry Society of America*, 1969, *59*, 635–640.

Riggs, L. A., & Whittle, P. Human occipital and retinal potentials evoked by subjectively faded visual stimuli. *Vision Research*, 1967, *7*, 441–451.

Roberts, W. W. Fear-like behavior elicited from dorsomedial thalamus of cat. *Journal of Comparative and Physiological Psychology*, 1962, *55*, 191–197.

Roberts, W. W., Dember, W. N., & Brodwick, M. Alternation and exploration in rats with hippocampal lesions. *Journal of Comparative and Physiological Psychology*, 1962, *55*, 695–700.

Robertson, A. D. Correlation between unit activity and slow potential changes in unanesthetized cerebral cortex of cat. *Nature*, 1965, *208*, 757–758.

Robinson, E. Effect of amygdalectomy on fear-motivated behavior in rats. *Journal of Comparative and Physiological Psychology*, 1963, *56*, 814–820.

Ruch, T. C., Patton, H. D., Woodbury, J. W., & Towe, A. L. *Neurophysiology*. Philadelphia: Saunders, 1965.

Ruchkin, D. S. Sorting of nonhomogeneous sets of evoked potentials. *Communications in Behavioral Biology*, 1971, *5*, 383–396.

Ruchkin, D. S., & John, E. R. Evoked potential correlates of generalization. *Science*, 1966, *153*, 209–211.

Rusinov, V. S. Electroencephalographic studies in conditioned reflex formation in man. In M. A. B. Brazier (Ed.), *The central nervous system and behavior*. New York: Macy Foundation, 1959.

Russell, B. *The autobiography of Bertrand Russell* (Vol. I). Toronto: McClelland & Stewart, Ltd., 1967.

Russell, W., & Nathan, P. Traumatic amnesia. *Brain*, 1946, *69*, 280–300.

Sakhuilina, G. T., & Merzhanova, G. K. Stable changes in the pattern of the recruiting response associated with a well-established conditioned reflex. *Electroencephalography & Clinical Neurophysiology*, 1966, *20*, 50–58.

Sanford, A. J. A periodic basis for perception and action. In W. P. Colquhoun (Ed.), *Biological rhythms and human performance*. New York: Academic Press, 1971. Pp. 179–209.

Sawa, M., Yukiharu, U., Masaya, A., & Toshio, H. Preliminary report on amygdaloidectomy in psychotic patients, with interpretation of oral–emotional manifestations in schizophrenics. *Folia Psychiatry & Neurology* (Japan), 1954, *7*, 309–316.

Sawyer, C. H. Rhinencephalic involvement in pituitary activation by intraventricular histamine in the rabbit under nembutal anesthesia. *American Journal of Physiology*, 1955, *180*, 37–46.

Scheibel, M. E., & Scheibel, A. B. Structural substrates for integrative patterns in the brain stem reticular core. In H. H. Jasper, L. D. Proctor, R. S. Knighton, W. S. Noshay, & R. T. Costello (Eds.), *The reticular formation of the brain*. Boston: Little, Brown & Co., 1958. Pp. 31–55.

Scheibel, M. E., & Scheibel, A. B. Some structural and function substrates of development in young cats. In W. A. Himwich & H. E. Himwich (Eds.), *Progress in brain research*. Vol. IX: *The developing brain*. Amsterdam: Elsevier Press, 1965.

Scheibel, M. E., & Scheibel, A. B. Patterns of organization in specific and nonspecific thalamic fields. In D. P. Purpura & M. D. Yahr (Eds.), *The thalamus*. New York: Columbia University Press, 1966. Pp. 13–46.

Scheibel, M. E., & Scheibel, A. B. Structural organization on nonspecific thalamic nuclei and their projection toward cortex. *Brain Research*, 1967, *6*, 60–94. (a)

Scheibel, M. E., & Scheibel, A. B. Anatomical basis of attention mechanisms in vertebrate brains. In G. C. Quarton, T. Melnechuk & F. O. Schmitt (Eds.), *The neurosciences: A study program*. New York: Rockefeller University Press, 1967. Pp. 577–602. (b)

Scheibel, M. E., & Scheibel, A. B. Elementary processes in selected thalamic and cortical subsystems: The structural substrates. In F. O. Schmitt, G. C. Quarton, T. Melnechuk, & G. Adelman (Eds.), *The neurosciences: Second study program*. New York: Rockefeller University Press, 1970. Pp. 443–457.

Schlag, J. D. Reactions and interactions to stimulation of the motor cortex of the cat. *Journal of Neurophysiology*, 1966, *29*, 44–71.

Schlag, J. D., & Chaillet, F. Thalamic mechanisms involved in cortical desynchronization and recruiting responses. *Electroencephalography & Clinical Neurophysiology*, 1963, *15*, 39–62.

Schlag, J. D., & Villablanca, J. A quantitative study of temporal and spatial response patterns in a thalamic cell population electrically stimulated. *Brain Research*, 1968, *8*, 255–270.

Schmitt, F. O., & Samson, F. E. Brain cell microenvironment. *Neurosciences Research Progress Bulletin*, 1969.

Schneider, A. M. Spreading depression: A behavioral analysis. In J. A. Deutsch (Ed.), *The physiological basis of memory*. New York: Academic Press, 1973, pp. 271–304.

Schreiner, L., & Kling, A. Behavioral changes following rhinencephalic injury in cat. *Journal of Neurophysiology*, 1953, *16*, 643–659.

Schreiner, L., & Kling, A. Effects of castration on hypersexual behavior induced by rhinencephalic injury in cat. *Archives Neurology Psychiat.*, 1954, *72*, 180–186.

Schreiner, L., & Kling, A. Rhinencephalon and behavior. *American Journal of Physiology*, 1956, *184*, 486–490.

Schuckman, H. Area and stimulus specificity in the transfer of responses conditioned to intracranial stimulation: A replication. *Psychological Reports*, 1966, *18*, 639–644.

Schuckman, H., & Battersby, W. S. Frequency specific mechanisms in learning. I. Occipital activity during sensory preconditioning stimulation. *Electroencephalography & Clinical Neurophysiology,* 1965, *18,* 44–55.

Schuckman, H., & Battersby, W. S. Frequency specific mechanisms in learning. II. Discriminatory conditioning induced by intracranial stimulation. *Journal of Neurophysiology,* 1966, *29,* 31–43.

Schuckman, H., Kluger, A., & Frumkes, T. E. Stimulus generalization within the geniculostriate system of the monkey. *Journal of Comparative Physiology & Psychology,* 1970, *73,* 494–500.

Schwindt, P. C., & Calvin, W. H. Equivalence of synaptic and injected current in determining the membrane potential trajectory during motor neuron rhythmic firing. *Brain Research,* 1973, *59,* 389–394.

Segal, M., & Olds, J. Behavior of units in hippocampal circuit of the rat during learning. *Journal of Neurophysiology,* 1972, *35,* 680–717.

Segundo, J. P., Roig, J. A., & Sommer-Smith, J. A. Conditioning of reticular formation stimulation effects. *Electroencephalography & Clinical Neurophysiology,* 1959, *11,* 471–484.

Seiden, L. S., & Peterson, D. D. Reversal of the reserpine-induced suppression of the conditioned avoidance response by L-dopa: Correlation of behavioral and biochemical differences in two strains of mice. *Journal of Pharmacology & Experimental Therapeutics,* 1968, *159,* 422–428.

Shagass, C. Averaged somatosensory evoked responses in various psychiatric disorders. In J. Wortis (Ed.), *Recent advances in biological psychiatry* (Vol. 10). New York: Plenum Press, 1968. Pp. 205–219.

Shallice, T. The detection of change and the perceptual moment hypothesis. *British Journal of Statistical Psychology,* 1964, *17,* 113–135.

Shepherd, G. *The synaptic organization of the brain.* New York: Oxford University Press, 1974.

Sherman, S. M. Visual field defects in monocularly and binocularly deprived cats. *Brain Research,* 1973, *49,* 24–45.

Sherman, S. M. Visual fields of cats with cortical and tectal lesions. *Science,* 1974, *185,* 355–357. (a)

Sherman, S. M. Monocularly deprived cats: Improvement of the deprived eye's vision by visual decortification. *Science,* 1974, *186,* 267–269. (b)

Sherrington, C. *The integrative action of the nervous system.* New Haven: Yale University Press, 1947.

Shipley, T., Jones, R. W., & Fry, A. Evoked visual potentials and human color vision. *Science,* 1965, *150,* 1162–1164.

Shipley, T., Jones, R. W., & Fry, A. Intensity and the evoked occipitogram in man. *Vision Research,* 1966, *6,* 657–667.

Skinner, J. E. Abolition of several forms of cortical synchronization during blockade in the inferior thalamic peduncle. *Electroencephalography & Clinical Neurophysiology,* 1971, *31,* 211–221.

Smith, D. E., King, M. B., & Hoebel, G. B. Lateral hypothalamic control of killing: Evidence for a cholinoceptive mechanism. *Science,* 1970, *167,* 900–901.

Smith, D. R., & Smith, G. K. A statistical analysis of the continual activity of single cortical neurons in the cat unanesthetized isolated forebrain. *Biophysical Journal,* 1965, *5,* 47–74.

Smith, G. K., & Smith, D. R. Spike activity in cerebral cortex. *Nature,* 1964, *202,* 253–255.

Smith, T. R., Wuerker, R. B., & Frank, K. Membrane impedance changes during synaptic transmission in cat spinal motorneurons. *Journal of Neurophysiology,* 1967, *30,* 1072–1193.

Smith, W. K. Non-olfactory functions of the pyriform–amygdaloid–hippocampal complex. *Federation Proceedings*, 1950, *9*, 118.

Sokolov, E. N. Neuronal models and the orienting reflex. In M. A. B. Brazier (Ed.), *CNS and behavior* (Vol. III). New York: Macy Foundation, 1960. Pp. 187–276.

Sokolov, E. N. Higher nervous functions: The orienting reflex. *Annual Review of Physiology*, 1963, *25*, 545–580.

Spencer, W. A., & Brookhart, J. M. Electrical patterns of augmenting and recruiting waves in depths of sensorimotor cortex of cat. *Journal of Neurophysiology*, 1961, *24*, 26–49.

Sperry, R. W. Some general aspects of interhemispheric integration. In V. B. Mountcastle (Ed.), *Interhemispheric relations and cerebral dominance*. Baltimore: Johns Hopkins Press, 1962.

Sperry, R. W. The great cerebral commissure. *Scientific American*, 1964, *210*, 42–52.

Sperry, R. W. Brain bisection and mechanisms of consciousness. In J. C. Eccles (Ed.), *Brain and conscious experience*. New York: Springer-Verlag, 1966.

Sperry, R. W. Hemispheric deconnection and unity in conscious awareness. *American Journal of Psychology*, 1969, *23*, 723–733.

Spiegel, E. A., Wycis, H. T., Freed, H., & Archinik, C. The central mechanisms of emotions. *American Journal of Psychiatry*, 1951, *108*, 426–432.

Spinelli, D. N. Receptive field organization of ganglion cells in the cat's retina. *Experimental Neurology*, 1967, *19*, 291–315.

Spinelli, D. N., & Barrett, T. Visual receptive field organization of single units in the cat's visual cortex. *Experimental Neurology*, 1969, *24*, 76–98.

Spinelli, D. N., & Pribram, K. H. Changes in visual recovery functions produced by temporal lobe stimulation in monkeys. *Electroencephalography & Clinical Neurophysiology*, 1966, *20*, 44–49.

Spinelli, D. N., & Pribram, K. H. Changes in visual recovery function and unit activity produced by frontal cortex stimulation. *Electroencephalography & Clinical Neurophysiology*, 1967, *22*, 143–149.

Spinelli, D. N., Pribram, K. H., & Weingarten, M. Visual receptive field modification induced by non-visual stimuli. *Federation Proceedings*, 1966, *25*, 2173.

Spinelli, D. N., & Weingarten, M. Afferent and efferent activity in single units of the cat's optic nerve. *Experimental Neurology*, 1966, *15*, 347–362.

Spong, P., Haider, M., & Lindsley, D. B. Selective attentiveness and cortical evoked responses to visual and auditory stimuli. *Science*, 1965, *148*, 395–397.

Sprague, J. M. Interaction of cortex and superior colliculus in mediation of visually guided behavior in the cat. *Science*, 1966, *153*, 1544–1547. (a)

Sprague, J. M. Visual, acoustic, and somaesthetic deficits in the cat after cortical and midbrain lesions. In D. P. Purpura & M. Yahr (Eds.), *The thalamus*. New York: Columbia University Press, 1966. (b)

Stefanis, C. Relations of the spindle waves and the evoked cortical waves to the intracellular potentials in pyramidal motor neurons. *Electroencephalography & Clinical Neurophysiology*, 1963, *15*, 1054.

Stein, D. G., Rosen, J. J., & Butters, N. (Eds.). *Plasticity and recovery of function in the central nervous system*. New York: Academic Press, 1974.

Stein, L. Anticholinergic drugs and the central control of thirst. *Science*, 1964, *139*, 46–48.

Stein, L. Chemistry of reward and punishment. In D. H. Efron (Ed.), *Psychopharmacology, a review of progress: 1957–1967*. Washington, D. C.: Government Printing Office, 1968. Pp. 105–123.

Stein, L., & Seifter, J. Possible mode of antidepressive action of imipramine. *Science*, 1961, *134*, 286–287.

Stein, L., & Wise, D. Possible etiology of schizophrenia: Progressive damage to the noradrenergic reward system by 6-hydroxydopamine. *Science*, 1971, *171*, 1032–1036.

Stevens, J. An anatomy of schizophrenia. *Archives General Psychiatry*, 1973, *29*, 177–189.

Stewart, D. L., & Resen, A. H. Adult versus infant brain damage: Behavioral and electrophysiological effects of striatectomy in adult and neonatal rabbits. In G. Newton & A. H. Riesen (Eds.), *Advances in psychobiology* (Vol. I). New York: Academic Press, 1972.

Strumwasser, F. The demonstration and manipulation of a circadian rhythm in a single neuron. In J. Aschoff (Ed.), *Circadian clocks.* Amsterdam: North-Holland Publ., 1965. Pp. 442–462.

Strumwasser, F. Neurophysiological aspects of rhythms. In G. C. Quarton, T. Melnechuk, & F. O. Schmitt (Eds.), *The neurosciences: A study program.* New York: Rockefeller University Press, 1967. Pp. 516–528.

Strumwasser, F. Membrane and intracellular mechanisms governing endogenous activity in neurons. In F. D. Carlson (Ed.), *Physiological and biochemical aspects of nervous integration.* Englewood Cliffs, New Jersey: Prentice-Hall, 1968. Pp. 329–341.

Stumpf, C. The fast component in the electrical activity of rabbit's hippocampus. *Electroencephalography & Clinical Neurophysiology*, 1965, *18*, 447–487.

Stutz, R. M. Stimulus generalization within the limbic system. *Journal of Comparative Physiology & Psychology*, 1968, *65*, 79–82.

Sutton, S., Braren, M., Zubin, J., & John, E. R. Evoked potential correlates of stimulus uncertainty. *Science*, 1965, *150*, 1187–1188.

Sutton, S., Tueting, P., Zubin, J., & John, E. R. Information delivery and the sensory evoked potential. *Science*, 1967, *155*, 1436–1439.

Tasaki, I. *Nervous transmission.* Springfield, Ill.: Charles C Thomas, 1953.

Teitelbaum, H. A comparison of effects of orbitofrontal and hippocampal lesions upon discrimination learning and reversal in the cat. *Experimental Neurology*, 1964, *9*, 452–462.

Teitelbaum, P. Motivational correlates of hypothalamic activity. *Proceedings of the 22nd International Physiological Congress*, Vol. I, Pt. 2., Leyden, Holland, 1962.

Teitelbaum, P., & Epstein, A. N. The lateral hypothalamic syndrome. *Psychology Review*, 1962, *69*, 74–90.

Teitelbaum, P., & Stellar, E. Recovery from the failure to eat, produced by hypothalamic lesions. *Science*, 1954, *120*, 894–895.

Tepas, D. I., Guiteras, V. L., & Klingaman, R. Variability of the human average evoked response to visual stimulation: A warning. *Electroencephalography & Clinical Neurophysiology*, 1974, *36*, 533–537.

Terzian, H., & Ore, G. D. Syndrome of Klüver and Bucy produced in man by bilateral removal of the temporal lobes. *Neurology*, 1955, *5*, 373–380.

Terzuolo, C. A., & Bullock, T. H. Measurement of imposed voltage gradient adequate to modulate neuronal firing. *Proceedings of the National Academy of Science*, 1956, *42*, 687–694.

Teuber, H. L. Some alterations in behavior after cerebral lesions in man. In *Evolution of nervous control.* Washington, D.C.: American Association for the Advancement of Science, 1959, pp. 157–194.

Teuber, H. L. The riddle of frontal lobe function in man. In J. M. Warren & K. Akert (Eds.), *The frontal granular cortex and behavior.* New York: McGraw-Hill, 1964.

Thatcher, R. W. A demonstration of anatomical and temporal changes in the organization of the brain which occur as a function of CS duration. Unpublished doctoral dissertation, University of Waterloo, 1970.

Thatcher, R. W. A quantitative electrophysiological analysis of human memory. (abstract) Paper presented at symposium on brain mechanisms and memory processes: an overview. Eastern Psychological Association, Philadelphia, 1974. (a)

Thatcher, R. W. Evoked potential correlates of human short-term memory. *Fourth Annual Neuroscience Convention*, 1974, p. 450. (Abstract) (b)

Thatcher, R. W. Electrophysiological correlates of information retrieval in rats. In: *Holography in Medicine: Proceedings of the international symposium on Holography in Biomedical Sciences*, P. Greguss (ed.), London, England: IPC Science and Technology Press Lt., 1975.

Thatcher, R. W. Evoked potential correlates of semantic information processing. In S. Harnard (Ed.), *Lateralization in the nervous system*. New York: Academic Press, 1976. (a)

Thatcher, R. W. Electrophysiological correlates of animal and human memory, In: *The Neurobiology of Aging*, R. Terry & S. Gershon (Eds.), New York: Raven Press, 1976. (b)

Thatcher, R. W. The neural representation of experience and time. In G. G. Haydu (Ed.), *Experience forms: Their cultural and individual place and function*. New York: Mouton Press, in press.

Thatcher, R. W., & Cadell, T. E. A demonstration of time-dependent processes associated with recall. *Proceedings of the 77th Annual Convention, APA*, 1969, pp. 231–232.

Thatcher, R. W., & John, E. R. Information and mathematical quantification of brain state. In N. R. Burch & H. L. Altshuler (Eds.), *Behavior and brain electrical activity*. New York: Plenum Press, 1975.

Thatcher, R. W., & Purpura, D. P. Maturational status of inhibitory and excitatory synaptic activities of thalamic neurons in neonatal kitten. *Brain Research*, 1972, *44*, 661–665.

Thatcher, R. W., & Purpura, D. P. Postnatal development of thalamic synaptic events underlying evoked recruiting responses and electrocortical activation. *Brain Research*, 1973, *60*, 21–34.

Thomas, G. J., & Slotnick, B. Effects of lesions in the cingulum on maze learning and avoidance conditioning in the rat. *Journal of Comparative Physiology & Psychology*, 1962, *55*, 1085–1091.

Thomas, G. J., & Slotnick, B. Impairment of avoidance responding by lesions in cingulate cortex in rats depends on food drive. *Journal of Comparative Physiology & Psychology*, 1963, *56*, 959–964.

Thompson, A. F., & Walker, A. E. Behavioral alterations following lesions of the medial surface. *AMA Archives of Neurological Psychiatry*, 1951, *65*, 251–252.

Thompson, R. F., Patterson, M. M., & Tyler, T. J. Neurophysiology of learning. *Annual Review of Psychology*, 1972, *23*, 73–113.

Thompson, R. F., Smith, H. E., & Bliss, D. Auditory, somatic, sensory, and visual response interactions and interrelations in association and primary cortical fields of the cat. *Journal of Neurophysiology*, 1963, *26*, 365–378.

Thompson, R. F., & Spencer, W. A. Habituation: A model phenomenon for the study of neuronal substrates of behavior. *Psychology Review*, 1966, *73*, 16–43.

Torii, S. Two types of pattern of hippocampal electrical activity induced by stimulation of hypothalamus and surrounding parts of the rabbits brain. *Japan. Journal of Physiology*, 1961, *11*, 147–157.

Travis, R. P. Jr., Hooten, T. F., & Sparks, D. L. Single unit activity related to behavior motivated by food reward. *Physiology & Behavior*, 1968, *3*, 309–318.

Travis, R. P. Jr., & Sparks, D. L. Changes in unit activity during stimuli associated with food and shock reinforcement. *Physiology & Behavior*, 1967, *2*, 171–177.

Travis, R. P., Jr., & Sparks, D. L. Unitary responses and discrimination learning in the squirrel monkey. *Physiology & Behavior*, 1968, *3*, 187–196.

Travis, R. P., Jr., Sparks, D. L., & Hooten, T. F. Single unit response related to sequences of food motivated behavior. *Brain Research*, 1968, *1*, 455–458.

Truex, R. C., & Carpenter, M. B. *Human neuroanatomy.* Baltimore: Williams & Wilkins, 1964.

Truex, R. C., & Carpenter, M. B. *Human neuroanatomy,* (6th ed.). 1973.

Tucker, T., & Kling, A. Differential effects of early vs. late brain damage on visual duration discrimination in cats. *Federation Proceedings,* 1966, *25,* 106.

Ukhtomski, A. A. Essays on the physiology of the nervous system. In *Collected works* (Vol. 4). Leningrad, 1945.

Ungar, G. Molecular approaches to neural coding. *International Journal of Neurology,* 1972, *3,* 193–200.

Ungar, G., Galvan, L., & Clark, R. H. Chemical transfer of learned fear. *Nature,* 1968, *217,* 1259–1261.

Ungerstedt, U. Stereotaxic mapping of the mono-amine pathways in the rat brain. *Acta Physiologica Scandinavia,* Suppl. 367, 1971.

Urbaitis, J. C., & Hinsey, J. C. Ablations of cortical and collicular areas in cats: Effects on a visual discrimination. *Federation Proceedings,* 1966, *25,* 1167.

Ursin, H. The temporal lobe substrate of fear and anger. *Acta psychiatrica et neurologica scandinavica,* 1960, *35,* 378–396.

Ursin, H., & Kaada, B. R. Functional localization within the amygdaloid complex in the cat. *Electroencephalography & Clinical Neurophysiology,* 1960, *12,* 1–20.

Ursin, H., Linck, P., & McCleary, R. A. Spatial differentiation of avoidance deficit following septal and cingulate lesions. *Journal of Comparative Physiology & Psychology,* 1969, *68,* 74–79.

Vanderwolf, C. H. Effects of experimental diencephalic damage on food hoarding and shock avoidance behavior in the rate. *Physiology & Behavior,* 1967, *2,* 399–402.

Vanderwolf, C. H. Effects of medial thalamic damage on initiation of movement and learning. *Psychonomic Science,* 1969, *17,* 23–25.

Vanderwolf, C. H. Limbic-diencephalic mechanisms of voluntary movement. *Psychology Review,* 1971, *78,* 83–113.

Van Heerden, P. J. The basic principles of artificial intelligence. In P. J. Van Heerden (Ed.), *The foundation of empirical knowledge.* Wassenaar, The Netherlands: N. V. Vitgeveriz Wistik (Royal van Gorcum, Ltd.), 1968.

Vasilevs, N. N. Relationship between background impulse activity of cortical neurons and electrocorticogram phases. *Bulletin of Experimental Biology and Medicine* (USSR), 1965, *59,* 597.

Vaughan, H. G., Jr., & Hull, R. C. Functional relation between stimulus intensity and photically evoked cerebral responses in man. *Nature,* 1965, *206,* 720–722.

Vaughan, H. G. Jr., & Silverstein, L. Metacontrast and evoked potentials: A reappraisal. *Science,* 1968, *160,* 207–208.

Verzeano, M. Sequential activity of cerebral neurons. *Archives of International Physiology and Biochemistry,* 1955, *63,* 458–476.

Verzeano, M. Las funciones del sistema nervioso; correlaciones entre estructura bioguimica y electrofisiologica. *Acta Neurology Latinoamerica,* 1963, *9,* 297–307.

Verzeano, M. Evoked responses and network dynamics. In R. E. Whalen, R. F. Thompson, M. Verzeano, & N. M. Weinberger (Eds.), *The neural control of behavior.* New York: Academic Press, 1970. Pp. 27–54.

Verzeano, M. Pacemakers, synchronization, and epilepsy. In H. Petsche & M. A. B. Brazier (Eds.), *Synchronization of EEG activities in epilepsies.* New York: Springer-Verlag, 1972. Pp. 154–188.

Verzeano, M., Laufer, J., Spear, P., & McDonald, S. L'activitie des reseaux neuroniques dans le thalamus du singe. In A. Monnier (Ed.) *Actualites Neurophysiologigues* (Vol. 6). Paris: Masson, 1965, pp. 223–252.

Verzeano, M., & Negishi, K. Neuronal activity in cortical and thalamic networks. *Journal of General Physiology* (Suppl.), 1960, *43*, 177.

Verzeano, M., & Negishi, K. Neuronal activity in wakefulness and in sleep. In G. E. W. Wolstenholme & M. O'Connor (Eds.), *The nature of sleep.* Longon: Churchill, 1961. Pp. 108–126.

Voronin, L. G., & Sokolov, E. N. Cortical mechanisms of the orientating reflex and its relation to the conditioned reflex. In H. H. Jasper & G. D. Smirnov (Eds.), The Moscow Colloquium on Electroencephalography of Higher Nervous Activity. *Electroencephalography & Clinical Neurophysiology* (Suppl. 13), 1960.

Walker, A. E., Thompson, A. F., & McQueen, J. D. Behavior and the temporal rhinencephalon in the monkey. *Bulletin, Johns Hopkins Hospital,* 1953, *93,* 65–93.

Walsh, R. N., Cummins, R. A., Budtz-Olsen, O. E., & Torok, A. Effects of environmental enrichment and deprivation on rat frontal cortex. *International Journal of Neuroscience,* 1972, *4,* 239–242.

Walter, W. G., Cooper, R., McCallum, C., & Cohen, J. The origin and significance of the contingent negative variation or "expectancy wave." *Electroencephalography & Clinical Neurophysiology,* 1965, *18,* 720.

Waziri, R., Kandel, E. R., & Frazier, W. T. Organization of inhibition in abdominal ganglion of aplysia. II. Post-tetanic potentiation, heterosynaptic depression, and increments in frequency of inhibitory postsynaptic potentials. *Journal of Neurophysiology,* 1969, *32,* 509–519.

Webster, D. B., & Voneida, T. J. Learning deficits following hippocampal lesions in split-brain cats. *Experimental Neurology,* 1964, *10,* 170–182.

Wedensky, N. E. Zeitschrift der russ. Gesellschaft für Volkshygiene, 1897. Cited in N. E. Wedensky, Die Erregung, Hemmung, und Narkose. *Archives Gesamte Physiology Pflügers,* 1903, *100,* 1–144.

Weinberg, H., Grey-Walter, W., & Crow, H. H. Intracerebral events in humans related to real and imaginary stimuli. *Electroencephalography & Clinical Neurophysiology,* 1970, *29,* 1–9.

Weinberg, H., Grey-Walter, W. G., Cooper, R., & Aldridge, V. J. Emitted cerebral events. *Electroencephalography & Clinical Neurophysiology,* 1974, *36,* 449–456.

Weingarten, M., & Spinelli, D. N. Retinal receptive field changes produced by auditory and somatic stimulation. *Experimental Neurology,* 1966, *15,* 363–376.

Weiss, P. 1 + 1 ≠ 2 (when one plus one does not equal two). In G. C. Quarton, T. Melnechuk, & F. O. Schmitt (Eds.), *The neurosciences: A study program.* New York: Rockefeller University Press, 1967.

Wernicke, C. Der aphasische Symptomenkomplex. Breslau: Cohn & Weigart, 1874.

Westbrook, W. H., & McGaugh, J. L. Drug facilitation of latent learning. *Psychopharmacologia,* 1964, *5,* 440–446.

Whalen, R. E., Thompson, R. F., Verzeano, M., & Weinberger, N. M. (Eds.). *The neural control of behavior.* New York: Academic Press, 1970.

Wheatley, M. D. The hypothalamus and affective behavior in cats: A study of the effects of experimental lesions, with anatomic correlations. *Archives of Neurology and Psychiatry,* Chicago, 1944, *52,* 296–316.

White, C. T. Temporal numerosity and the psychological unit in duration. *Psychological Monographs,* 1963, *12,* 77.

White, C. T. Evoked cortical responses and pattern stimuli. *American Psychology,* 1969, *24,* 212–214.

Wicke, J. D., Donchin, E., & Lindsley, D. B. Visual evoked potentials as a function of flash luminance and duration. *Science,* 1964, *146,* 83–85.

Wickelgren, B. G., & Sterling, P. Influence of visual cortex on receptive fields in the superior colliculus of the cat. *Journal of Neurophysiology*, 1969, *32*, 16–23.

Wiesel, T. N., & Hubel, D. H. Effects of visual deprivation on morphology and physiology of cells in the cat's lateral genicolate body. *Journal of Neurophysiology*, 1963, *2B*, 978–993. (a)

Wiesel, T. N., & Hubel, D. H. Single-cell responses in striate cortex of kittens deprived of vision in one eye. *Journal of Neurophysiology*, 1963, *2B*, 1003–1017. (b)

Wiesel, T. N., & Hubel, D. H. Comparison of the effects of unilateral and bilateral eye closure on cortical unit responses in kittens. *Journal of Neurophysiology*, 1965, *28*, 1029–1040.

Wilson, M., & Wilson, W. A., Jr. Intersensory facilitation of learning sets in normal and brain operated monkeys. *Journal of Comparative Physiology & Psychology*, 1962, *55*, 931–934.

Winans, S. S., & Meikle, T. H. Visual pattern discrimination after removal of the striate visual cortex in cats. *Federation Proceedings*, 1966, *25*, 2167.

Wise, C. D., & Stein, L. Facilitation of brain self-stimulation by central administration of norepinephrine. *Science*, 1969, *163*, 299–301.

Wisniewski, H. M., Coblentz, J. M., & Terry, R. D. Pick's disease: A clinical and ultrastructure study. *Archives of Neurology*, 1972, *26*, 97–108.

Woody, C. D., & Engel, J. Changes in unit activity and thresholds to electrical microstimulation at coronal-pericruciate cortex of cat with classical conditioning of different facial movements. *Journal of Neurophysiology*, 1972, *35*, 230–242.

Woody, C. D., Vassilevsky, N. N., & Engel, J. Conditioned eye blink-unit: Activity at coronal-precruciate of cat. *Journal of Neurophysiology*, 1970, *33*, 838.

Woody, C. D., & Yarowsky, P. J. Conditioned eye blink using electrical stimulation of coronal-precruciate cortex as conditioned stimulus. *Journal of Neurophysiology*, 1972, *35*, 242–252.

Worden, F. G., & Livingston, R. B. Brain stem reticular formation. In D. E. Sheer (Ed.), *Electrical stimulation of the brain*. Austin Texas: University of Texas Press, 1961. Pp. 263–276.

Yamada, T., & Greer, M. A. The effect of bilateral ablations of the amygdala on endocrine function in the rat. *Endocrinology*, 1960, *66*, 565–574.

Yasukochi, G. Emotional responses elicited by electrical stimulation of the hypothalamus in cat. *Folia Psychiatry & Neurology* (Japan), 1960, *14*, 260–267.

Yoshii, N. Electroencephalographic study on experimental neurosis, a conditioned partly awake state. *Proceedings of the 22nd International Congress of Physiological Sciences*, Leiden, 1962.

Yoshii, N., & Hockaday, W. J. Conditioning of frequency-characteristic repetitive electroencephalographic response with intermittent photic stimulation. *Electroencephalography & Clinical Neurophysiology*, 1958, *10*, 487–502.

Yoshii, N., & Ogura, H. Studies on the unit discharge of brain stem reticular formation in the cat. I. Changes of reticular unit discharges following conditioning procedure. *Medical Journal of Osaka University*, 1960, *11*, 1.

Yoshii, N., Pruvot, P., & Gastaut, H. Electroencephalographic activity of the mesencephalic reticular formation during conditioning in the cat. *Electroencephalography & Clinical Neurophysiology*, 1957, *9*, 595.

Zanchetti, A. Subcortical and cortical mechanisms in arousal and emotional behavior. In G. C. Quarton, T. Melnechuk & F. O. Schmitt (Eds.), *The neurosciences: A study program*. New York: Rockefeller University Press, 1967. Pp. 602–615.

Zucker, I., & McCleary, R. A. Perseveration in septal cats. *Psychonomic Science*, 1964, *1*, 387–388.

Author Index

Numbers in *italics* refer to the pages on which the complete references are listed.

Subject Index